W9-CIP-501

INTRODUCTION TO
EMBRYONIC DEVELOPMENT

Steven B. Oppenheimer

California State University, Northridge

Allyn and Bacon, Inc.   Boston   London   Sydney   Toronto

INTRODUCTION TO
EMBRYONIC DEVELOPMENT

Library of Congress Cataloging in Publication Data

Oppenheimer, Steven B 1944-
    Introduction to embryonic development.

    Bibliography: p.
    Includes index.
    1. Embryology. I. Title. [DNLM: 1. Embryology.
QS604.3 062i]
QL955.066   596'.03'3   79-26339
ISBN 0-205-06899-5
ISBN (International) 0-205-07348-4

10 9 8 7 6 5 4 3 2   85 84 83 82 81

Production editor: Mary Beth Finch
Cover designer: Linda Knowles
Preparation buyer: Linda Card
Series editor: Joseph E. Burns

    Cover photo:
Freeze fracture of the periphery of an unfertilized sea urchin egg showing cortical granules.
Transmission electron micrograph, courtesy of Edward Pollock.

CONTENTS

CHAPTER 7 · ORGANOGENESIS *150*

CHAPTER 8 · EMBRYONIC INDUCTION *234*

PREFACE

WHEN I UNDERTOOK to write this book, I did so with the hope that sharing my own enthusiasm for the subject matter would excite the curiosity and interest of students approaching the study of embryology and developmental biology. To this end, wherever possible I have introduced topics by asking questions: What turns on the complex series of metabolic reactions that occur in eggs immediately after fertilization? What is the nature of the molecules that control the process of sperm-egg recognition? What are the mechanisms of morphogenesis that shape the embryo? What factors turn on specific genes during differentiation? What is the fascinating relationship between cancer cells and embryonic cells? It is clear that all of the answers to these sorts of questions are not yet at hand. The student, I hope, will be left with more than knowledge of some answers, and will gain an understanding of how the experimental method leads to the answers.

In this text, I have attempted to provide the reader with a logical approach to the subject. The book begins at the beginning—with gametogenesis—and moves through fertilization, cleavage, and each subsequent step in embryonic development. Basic embryology is covered in the first half of the text. With this fundamental background behind the student, the second half of the book examines molecular mechanisms involved in controlling morphogenesis and differentiation. The text ends with a discussion of the intriguing relationship between embryonic cells and tumor cells. The role of the cell surface in controlling developmental events is one major theme stressed throughout the text. The large number of micrographs and drawings should help to convey a concrete sense of work in various aspects of the field. The key terms that appear in the margins, and the glossary at the end of the book, should assist the student in understanding and reviewing the subject matter. The readings suggested at the ends of the chapters have, for the most part, been chosen to help the undergraduate gain a better understanding of the topics discussed.

Acknowledgements. When I began to write this embryology-developmental biology text, I did so with the encouragement of many individuals, including my wife, Carolyn, and my colleagues. With the confidence I received from others and the knowledge that I, myself, have had some of the best teachers in the country who have given me insights into teaching the subject, I decided to undertake the venture. At this time, I wish to thank Heinrich Ursprung, Malcolm Steinberg, Michael Edidin, Saul Roseman, Stephen Roth, and Robert DeHaan, who served as the nucleus of individuals at Johns Hopkins who provided me with the foundations needed to write this text, and helped mold my approach to the subject.

Special thanks are given to Gary Folven, Managing Editor, who skillfully steered this book through the critical final stages, and to Joseph Burns, Editor, who also did a great deal in the final phases involved in the production of this book. I also wish to thank Frank Ruggirello, Editor, who provided initial stimulation to me, and Harvey Pantzis who helped get the book started. I thank Mary Beth Finch, Production Editor, for an extremely meticulous job, and the entire staff at Allyn and Bacon for their excellent assistance.

I would like to thank all my colleagues who have kindly provided photos and permission to use illustrations of their work for this text. I am particularly indebted to Peter Armstrong, Patricia Calarco, David Epel, Garth Nicolson, Steven Rosen, and Victor Vacquier for providing collections of superb micrographs; their time spent in the darkroom to enhance my text is truly appreciated. I would also like to express my gratitude to George Morris of Scientific Illustrators, who did an especially excellent job of drawing numerous figures contained in this text.

I would like to make special mention and express my gratitude to the many fine reviewers who have helped shape and guide me in the writing of this book: Patricia G. Calarco, University of California, San Francisco; David Epel, Hopkins Marine Station, Stanford University; John Morris, Oregon State University; Garth Nicolson, University of California, Irvine; Charles W. Porter, San Jose State Unviersity; Ralph S. Quatrano, Oregon State University; J. R. Shaver, Michigan State University; Fred Wilt, University of California, Berkeley. I have taken much of the advice of these reviewers. Any errors, however, are my own. Finally, I wish to thank my wife Carolyn for excellent typing assistance and a superb sense of humor.

INTRODUCTION

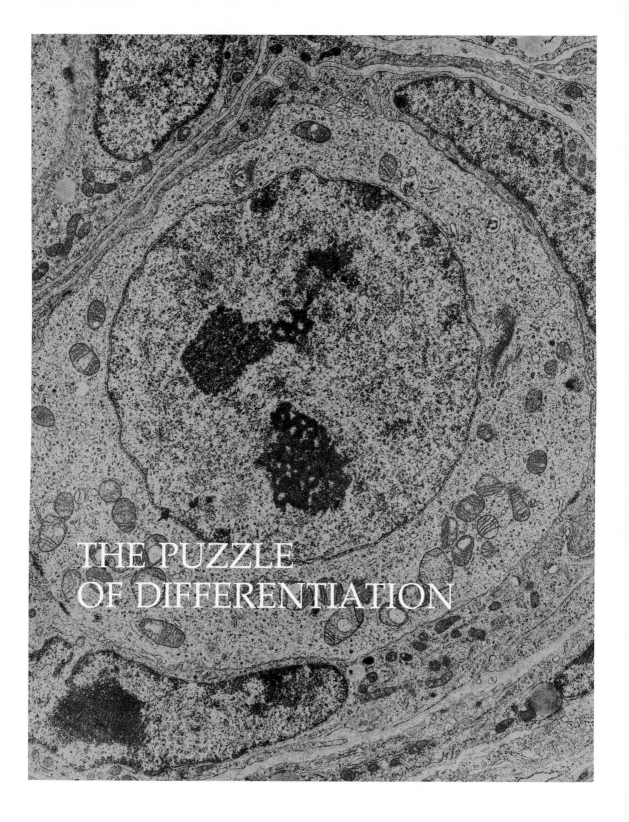

THE PUZZLE
OF DIFFERENTIATION

YOU AND I BEGAN LIFE as fertilized eggs (zygotes). Fertilized eggs each contain a set of chromosomes, half from the mother and half from the father. As the fertilized egg begins to divide (cleave) the chromosomes duplicate themselves and all daughter cells usually receive identical sets of chromosomes. The chromosomes contain genes—DNA units that code for the synthesis of messenger RNA. The messenger RNA, in turn, directly codes for the synthesis of specific proteins. We as students of modern biology realize that if all daughter cells usually receive identical sets of chromosomes during early embryonic development, they also receive identical sets of genes. How, then, can cells with the same genes become different? We have muscle, nerve, fat, blood, bone, all different tissues. The "puzzle of differentiation" is one of the most intriguing questions facing embryologists and developmental biologists today.

The question of how cells differentiate, or become different, has, in part, been solved. Much, however, is still to be learned. Throughout this text we will consider the problem of differentiation in different systems and at different levels. First, we will look at the problem superficially to gain some insight into some of the factors that allow cells with identical genes to form all of the tissues necessary for us to function as living organisms.

Cells become different because only certain genes become activated in certain cells. For example, although all of our cells contain the gene that codes for the protein hemoglobin, only red blood cells contain hemoglobin. Thus the hemoglobin genes must have become activated in the red blood cells to allow this cell to produce hemoglobin messenger RNA and, in turn, hemoglobin. In nerve cells or muscle cells, however, although they possess the hemoglobin gene, this gene is inactive and hemoglobin is not synthesized.

What are the factors that activate different genes in different cells? How do these factors work? The first question can, in part, be answered. The answers to the second question, however, are not well understood. The factors that activate specific genes include interaction of the genes with: (1) molecules contained in specific regions of cytoplasm, (2) molecules provided by interaction with neighboring cells, and (3) environmental factors. Let us briefly examine these three factors here. We will return to more in-depth discussions on the nature of differentiation throughout the text.

Interaction of Genes with Molecules in Specific Regions of Cytoplasm

If all daughter cells in developing embryos usually receive the same genes during division of the fertilized egg, don't they also receive the same cytoplasm during the division process? The answer to this question is no, not necessarily. Daughter cells can receive different types of cytoplasm and therefore can receive different molecules that may interact with their genes. One very clear example of how cells in early embryos can receive different cytoplasm is illustrated in Figure I–1. In eggs of organisms such as *Dentalium*, one can distinguish three layers of cytoplasm: a clear layer, a granular layer, and a second clear layer. Before the first division of the fertilized *Dentalium* egg one of the clear cytoplasmic regions is extruded from the zygote (fertilized egg). This extrusion is called a polar lobe. The first division (cleavage) then occurs and the polar **polar lobe** lobe containing all of one of the clear regions of cytoplasm is drawn back into only one of the daughter cells (CD cell). Thus, only one of the daughter cells contains this specific cytoplasm. This type of cytoplasmic segregation continues to occur during *Dentalium* development, resulting in the parceling out of specific cytoplasm to specific cells in the embryo. So we can see that although all daughter cells may contain the same

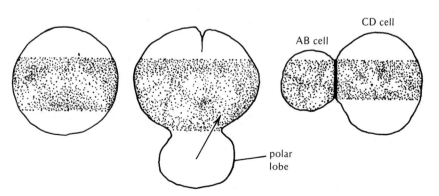

Figure I–1. Cleavage of the mollusc *Dentalium*. After E. B. Wilson, *J. Exp. Zool.* 1:1-72 (1904).

genes, they do not contain the same cytoplasm. The special cytoplasm contained in some of the daughter cells of the embryo appears to play an important role in turning on specific genes in these cells. Only the cells with the special cytoplasm are able to form normal embryos if separated from the other cells. The special cytoplasm of a cell may therefore help that cell become different from other cells by activating specific genes. Evidence to support this statement will be presented in Chapter 8.

We have looked at only one example of cytoplasmic specialization. Later on we will examine other examples, such as the gray crescent of amphibian embryos. Gray crescent cytoplasm is special material localized in only one region of the fertilized egg, needed by specific amphibian embryo cells to differentiate into specific tissue that plays a key role in controlling normal development of the embryo. We will examine these specializations in detail in Chapter 8.

Interaction of Genes with Molecules Provided by Interaction with Neighboring Cells

We saw how special cytoplasm is parceled out to only certain cells in the embryo, causing differences in cells that contain similar genes. Let us now briefly consider an example of a situation in which contact with other cells can cause differentiation to occur in the responding cells. Many examples of this type of cell-cell interaction will be described throughout the text. Here we will consider one of these examples, the amphibian embryo. We just mentioned that the gray crescent in the amphibian embryo appears to contain special cytoplasm. This gray crescent material becomes localized in a specific region of the amphibian embryo that begins to migrate from the surface to the inside of the embryo. This region (the prospective notochord) finally underlies the prospective nervous system of the embryo. Thus the special cytoplasm contained in the original gray crescent is now found in cells that underlie the area that will become the nervous system (Figure I–2). Unless contact is made between the prospective notochord and prospective nervous system, the nervous system does *not* develop. If a sheet of mica is placed between the prospective notochord and prospective nervous system, preventing contact of the two regions, nervous system differen-

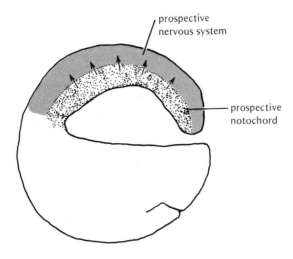
prospective
nervous system

prospective
notochord

Figure I–2. Contact of two regions in amphibian embryo. Prospective notochord and prospective nervous system (only areas shown) must contact each other for nervous system differentiation to occur.

tiation (formation of the brain, spinal cord, and accessory structures) does not take place. In Chapter 8, we will examine the interaction between these regions in greater detail. We can conclude that in order for the genes that cause nervous system differentiation to become activated in the prospective nervous system cells, interaction with prospective notochord cells must occur. This is an example of differentiation caused by interaction with other cells. The nature of this interaction at a molecular level, however, is not yet well understood.

Environmental factors such as light and temperature can also influence differentiation. Flowering in plants, for example, is controlled by length of light exposure. The development of flowers, in turn, controls the whole reproductive process in plants. Insect development is clearly influenced by light and temperature. As will be seen in Chapter 8, nervous system differentiation can also be influenced by environmental factors such as acidity and salt concentration.

We have introduced embryology by briefly examining cell differentiation caused by rather poorly understood components in special cytoplasm, neighboring cells, and the environment. We will return to this theme in more depth throughout this text to help understand some of the events that occur during the development of different parts of the organism. Toward the end of the text we will return to study differentiation at the molecular level, after the component processes of development have been explored.

In Chapter 1, we will start our study of embryology at the beginning—with the gametes, the sperm and egg, that unite to form the new organism. A study of the fertilization process will help us to understand the complex series of events that must occur before life begins as a dividing embryo. We will continue the study of embryology by examining the factors involved in determining how fertilized eggs cleave. An examination of the events that transform the early embryo, a simple ball of cells, into the complex being that begins to resemble the adult, will follow. Throughout the text, emphasis will be placed upon attempts at understanding the mechanisms involved in controlling the component processes involved in embryonic development. For example, how do cells rearrange themselves in embryos to get their final destinations? Evidence will be presented that suggests that cellular rearrangements in embryos are, in part, caused by adhesive recognition between the cells. Cells stop migrating because they specifically adhere tenaciously to certain other cells. We will examine what is known about the molecular nature of such cell-cell interactions in embryos.

In this text, we will deal with many intriguing questions. How does an egg cell grow to 100,000 times its original volume in preparation for nourishing the new embryo? How does a sperm cell become transformed from a little sphere into a sleek swimming gamete that is able to touch home to fertilize the egg? What turns on the complex series of metabolic reactions that occur in eggs immediately after fertilization? How can a tiny sperm that fuses with only 0.0002 percent of the egg surface trigger the multitude of changes that occur in the new zygote? What is the nature of the molecules involved in the process of sperm-egg recognition? What controls cleavage? How does a conglomeration of cells become transformed into a layered embryo that begins to resemble a "real" organism? What structures are derived from each germ layer and what is the mechanism of their formation? What are the forces that shape the embryo—the mechanisms of morphogenesis? What is the nature of the molecules that appear to "turn on" specific genes during differentiation? How do amphibian limbs regenerate and might this information eventually help us develop means of promoting regeneration of limbs in man? How do our eyes, heart, kidneys, and limbs develop? How does German measles virus cause blindness in human babies? How does one go about studying differentiation? How have modern techniques in molecular, cell, and developmental biology been used to examine the nature of differentiation? How does differentiation at the protein level come about? How are higher orders of structure such as microtubules, microfilaments, flagella, and ribosomes assembled in differentiating cells? What is the nature of the intriguing relationship between cancer cells and embryonic cells? Under what conditions will certain tumor cells become transformed into normal differentiating tissue?

This text will not answer all of these questions completely. We will, however, deal in depth with these questions and many more in what is hoped to be an organized, clear and enthusiastic manner. Some of you may become excited about this field and eventually participate in uncovering the unsolved problems that still remain in developmental biology.

Readings

Clement, A. C., Cell Determination and Organogenesis in Molluscan Development: A Reappraisal Based on Deletion Experiments in *Ilyanassa. Amer. Zool.* 16:447–453 (1976).

Ebert, J. D., and I. M. Sussex, *Interacting Systems in Development*, 2nd ed. Holt, Rinehart & Winston, New York (1970).

Freeman, G., The Role of Cleavage in the Localization of Developmental Potential in the Ctenophore *Mnemiopsis Ieidyi. Develop. Biol.*, 49:143–177 (1976).

Spemann, H., and H. Mangold, Induction of Embryonic Primordia by Implantation of Organizers from a Different Species (1924). Reprinted in *Foundations of Experimental Enbryology*, B. H. Willier and J. M. Oppenheimer, eds., Prentice Hall, Englewood Cliffs, N.J., p. 144 (1964).

Wessells, N. K., *Tissue Interactions and Development*. W. A. Benjamin, Menlo Park, California (1977).

Wilson, E. B., Experimental Studies on Germinal Localization. I. The Germ Regions in the Egg of *Dentalium*. II. Experiments on the Cleavage-Mosaic in *Patella* and *Dentalium. J. Exp. Zool.*, 1:1–72 (1904).

CHAPTER 1

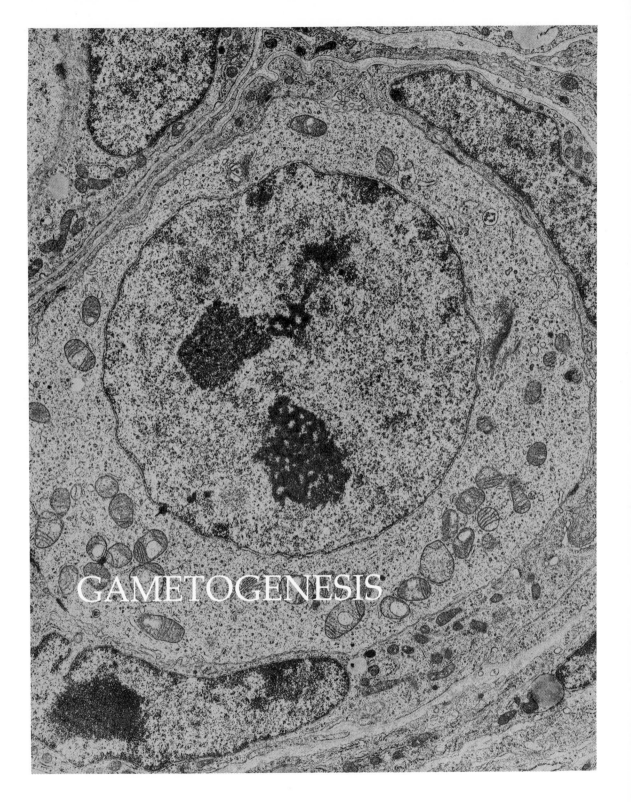

GAMETOGENESIS

HOW DOES AN EGG CELL GROW to 100,000 times its original volume in preparation for nourishing the new embryo? How does a sperm cell become transformed from a little sphere into a sleek swimming gamete that is able to touch home to fertilize the egg? How does the chromosome number in sex cells become reduced to half so that it can be restored to the normal number during fertilization by contributions of genetic information from both parents? These are some of the central themes of the story of gametogenesis.

In this chapter we will be concerned with the formation of the gametes, the sperm and egg cells. These cells will unite with each other (in the next chapter on fertilization) to mark the start of the developing embryo. It is appropriate to begin our study of embryology with formation of the gametes because the story of embryonic development begins with eggs and sperm. In addition to examining the basic components of gametogenesis, we will consider some thoughts concerning methods of interfering with the normal events that occur in gamete formation. This work is of interest because of the growing population problems in the world. At this time there is no one completely safe and effective method of preventing fertilization. By studying gametogenesis and examining the various phases of the processes we will attempt to define vulnerable phases that may be sensitive to specific drugs. Thus, we will interweave basic concepts of gamete formation with utilization of this knowledge in practical ways. I hope that the young developmental biologist will begin to realize that the study of embryology can lead to far more than just an understanding of the science itself. No science is isolated from the modern world. Developmental biology, especially, is leading to exciting developments in medicine that can only result in bettering the state of mankind on earth. Let us begin our study of gametogenesis by looking at spermatogenesis, the formation of sperm cells. After this excursion to the testis, we will examine oogenesis—the development of eggs.

Spermatogenesis

Summary of Events

Let us outline the events involved in spermatogenesis so that we have a framework for discussing some of the more recent and interesting aspects of the process. The area where sperm are produced, the seminiferous tubules, consists of two important types of cells: the germ cells, which are in various stages of meiosis as described below, and the Sertoli cells, which support and nourish the developing sperm cells. The Sertoli cells probably provide the developing sperm cells with specific factors that cause growth, division, and differentiation.

Sertoli cells

Primordial germ cells do not originate in the gonad but migrate or are carried by the blood to the gonad from other regions in the body. Most evidence suggests that the sperm and egg cells are derived from the primordial germ cells that have taken up residence in the gonads. In Chapter 7 we will examine the origin of primordial germ cells in more detail.

primordial germ cells

The primordial germ cells in the sperm-forming or seminiferous tubules of the testis appear to give rise to the spermatogonia, the cells that eventually develop into the sperm. In many vertebrates the spermatogonia are located in the outer region of the seminiferous tubules. As these cells develop and mature they often move towards the inner region where the lumen or canal of the tubule is located (Figure 1–1).

seminiferous tubules

spermatogonia

Some spermatogonia continue to divide and are thus the source of new sex cells. Others begin to grow. These growing cells are now called primary spermatocytes. The DNA of the chromosomes is duplicated.

primary spermatocytes

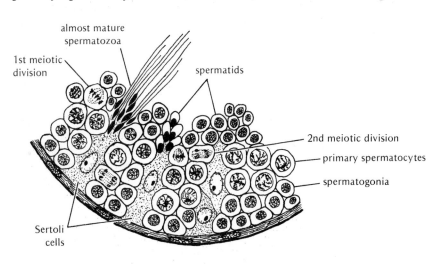

almost mature spermatozoa

1st meiotic division

spermatids

2nd meiotic division

primary spermatocytes

spermatogonia

Sertoli cells

Figure 1–1. Mammalian seminiferous tubule, showing cells undergoing meiosis. From B. I. Balinsky, *Introduction to Embryology*, 4th edition, © 1975 by W. B. Saunders Co., p. 19.

The spermatogonia start with the same amount of chromosomal DNA as normal body cells, so that once this DNA is duplicated in the primary spermatocytes, they possess double the amount of chromosomal DNA as other body cells.

The primary spermatocytes now undergo two division cycles. By the end of these divisions, the final cell progeny will have only half the number of chromosomes found in the normal body cells. Cells with half **haploid** the body cell chromosome number are termed haploid, in contrast to the term diploid that refers to the normal number of chromosomes found in **diploid** the spermatogonia or body cells. Thus, the net result of spermatogenesis is to form cells with half the number of chromosomes found in normal body cells.

Meiosis. Let us briefly list the major steps involved in the two divisions, **meiosis I and II** termed meiosis I and II, that result in the formation of haploid sperm cells. Later in the chapter, when we discuss oogenesis, we will return to some of the interesting aspects of meiosis. We should stress that sex cells must be haploid because at fertilization, combination of sperm and egg restore the normal diploid chromosome number by providing the chromosome contributions of the mother and father. In this way genetic information from both parents contributes to the development of the new and unique offspring.

The primary spermatocytes begin the first meiotic division by enter- **prophase** ing the prophase, the stage in which the chromosomes become visible as double, thread-like structures (leptotene phase). During the prophase **homologous pairs** the chomosomes arrange themselves in homologous pairs (zygotene phase). That is, each chromosome pairs up with another of identical size and shape. Since the DNA of each chromosome has also been dupli- **chromatids** cated, each pair of chromosomes contains four chromatids, because each pair of chromosomes is really a pair of double chromosomes. This com- **tetrad** plex is called a tetrad. Each member of a pair of homologous chromosomes is derived from a different parent. Some genetic material is exchanged between the homologous chromosomes in the tetrad by a **crossing over** process called crossing over (pachytene phase) (Figure 1–2). In this way genetic information from the mother and father becomes mixed on single chromosomes. Crossing over increases the genetic variation in the offspring which, in turn, promotes the development of more fit individuals through natural selection. After crossing over has occurred the homologous chromosomes begin to separate from one another again (diplotene phase). The phases of the meiotic prophase can be remembered by noting what the words mean:

leptotene   Leptotene: *lepto*, thin; *tene*, thread (chromosomes visible as thin threads).

Zygotene: *zygo,* pair; *tene,* thread (homologous chromosomes pair).

Pachytene: *pachy,* thick; *tene,* thread (crossovers occur).

Diplotene: *diplo,* double; *tene,* thread (homologous chromosomes are clearly seen as they spread apart).

zygotene

pachytene

diplotene

As the prophase comes to an end the spindle is formed, the nuclear membrane disintegrates and the nucleoli disperse. The next step in meiosis I is the metaphase in which the homologous pairs of chromosomes line up and become attached to the spindle fibers. Anaphase of meiosis I follows. In this stage the homologous chromosomes separate. The chromatids of each homologous chromosome do *not* separate. In the final stage, telophase, the cell separates into two. Each daughter cell has only half the number of chromosomes as the primary spermatocyte. These cells are called secondary spermatocytes.

metaphase

anaphase

telophase

secondary spermatocyte

In the second meiotic prophase, the chromosomes become visible again, the nuclear membrane (if present) dissolves, and new spindle fibers form again. The chromosomes line up at the equator and attach to the spindle fibers during metaphase, and at anaphase, the chromatids of each chromosome separate and move toward the poles of the cell. The spindle disappears and the nuclear membranes reform around each set

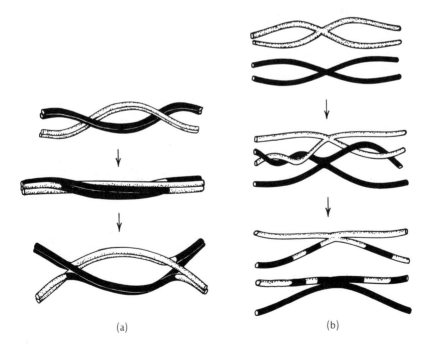

Figure 1–2. Crossing over of two chromosomes. (a) After T. H. Morgan, *The Physical Basis of Heredity,* J. B. Lipincott, 1919. (b) Simplified drawing showing sections transferred.

(a)   (b)

Figure 1–3. Summary of spermatogenesis.

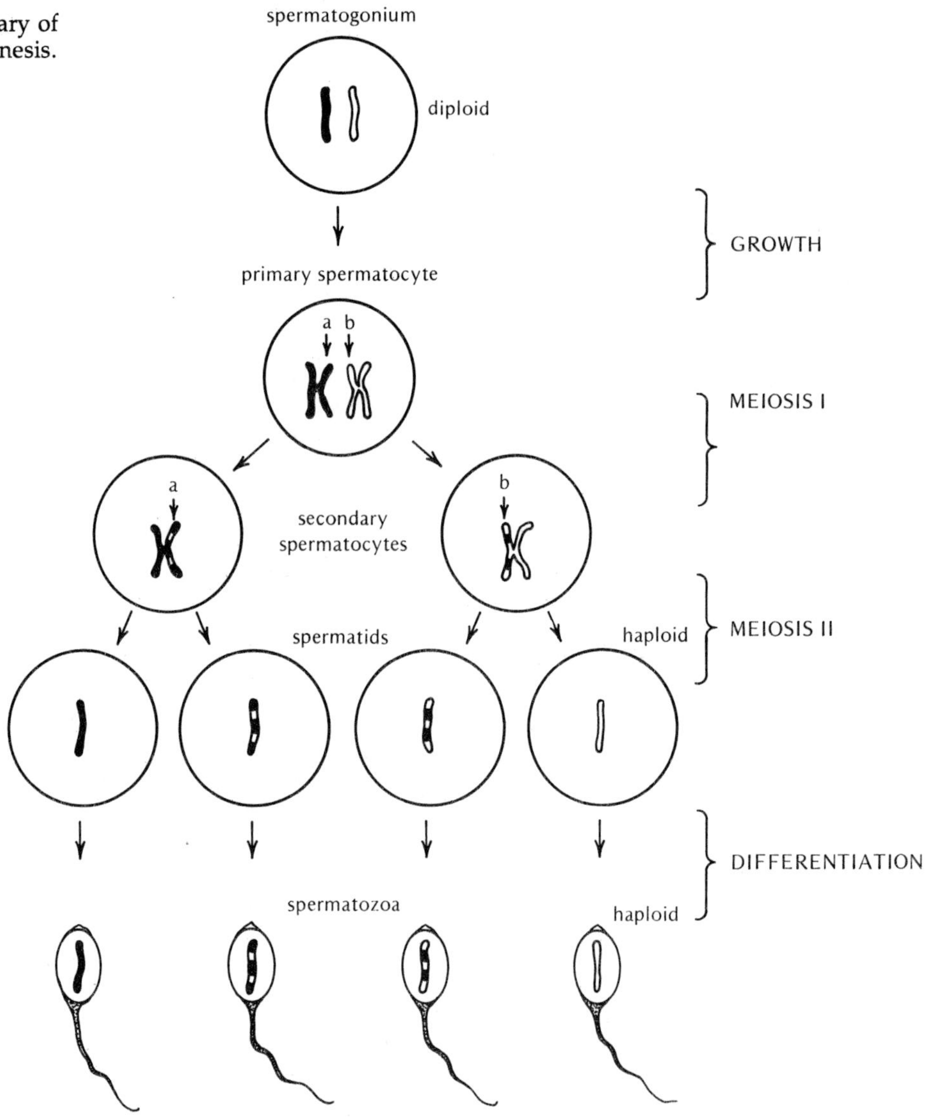

spermatogonium

diploid

GROWTH

primary spermatocyte

a b

MEIOSIS I

a

b

secondary
spermatocytes

spermatids

haploid   MEIOSIS II

DIFFERENTIATION

spermatozoa

haploid

Opp, Fig. 1-3

spermatids

spermatozoan

of chromosomes during telophase. Cell division occurs and each daughter cell now has half the number of chromosomes present in the spermatogonia. These haploid cells are called spermatids. The spermatids undergo a differentiation phase in which all the components of the mature sperm are formed. Thus, each spermatid differentiates, without dividing, into a mature sperm cell, called a spermatozoan.

To sum up, one diploid spermatogonium grows to form the primary spermatocyte. The primary spermatocyte divides in meiosis I to form two secondary spermatocytes. Each of these cells divides in meiosis II to form two spermatids that differentiate into spermatozoa. Thus, one diploid spermatogonium forms four haploid mature sperm cells. Spermatogenesis is summarized in Figure 1–3.

Mechanisms of Spermatogenesis

Now that we have an understanding of the basic steps in formation of sperm cells, let us examine some interesting experimental results that shed light upon the mechanisms involved in spermatogenesis. In addition, we will look at some promising results that may lead to the development of safe and effective male contraceptives. Basic developmental biology, as will be seen, leads to many medically and socially promising applications. We will concentrate on what occurs in the mammal, the system of direct importance to man.

luteinizing hormone

Spermatogenesis is controlled by hormones. Two hormones produced by the anterior pituitary gland, luteinizing hormone (LH) and follicle stimulating hormone (FSH), appear to regulate spermatogenesis. These hormones, especially luteinizing hormone, stimulate interstitial cells in the testis (Leydig cells) to synthesize male sex hormones called androgens. These hormones appear to be among the factors needed for sperm maturation.

follicle stimulating hormone

Leydig cells

androgens

Over 100 years ago La Valette noticed something of interest when teasing apart testes. He observed that the developing male germ cells appeared to be connected to each other. Fawcett and co-workers, using the electron microscope, demonstrated that the spermatocytes were in fact connected to each other by cytoplasmic bridges. These connections seem to result from incomplete cell division of spermatogonia, resulting in the occurrence of channels between daughter cells (Figures 1–1 and 1–4). Many developing spermatocytes are thus connected together. These connections may be an important factor in allowing molecules to pass from cell to cell so that differentiation of many sperm can occur at one time. Many sperm must be mature at specific times to insure proper numbers of sperm for successful fertilization to occur. Sertoli cells may provide molecules that stimulate sperm differentiation.

At this point let's digress for a moment to think about some phases in spermatognesis that may be vulnerable to interference with specific drugs that could be potential male contraceptives. Fawcett suggests that the hormonal stimulation of gametogenesis by follicle stimulating hormone and the separation of joined sperm may be steps that eventually could be interrupted by specific drugs. Also, mature sperm traveling in the collecting ducts of the testis must be vulnerable to drug action. In fact, several drugs are being studied that directly effect mature sperm. These include α-chlorohydrin and 1-amino, 3-chlor, 2-propanol hydrochloride. The site of action of these drugs on the sperm is not as well understood as that of other new substances. Several studies have recently indicated that both sperm and egg cells have sugar-containing molecules on their surfaces. Certain substances called lectins (carbohydrate binding proteins) will specifically attach to the sugars on the cell surfaces of the sperm and eggs. It is possible that sperm or eggs could be treated with lectins that bind to surface sugars in order to prevent fertilization. Oikawa, Yanigamachi and Nicolson have indeed found that one

lectins

Figure 1–4.
Interconnections of mammalian sperm. From D. Fawcett, "Gametogenesis in the Male: Prospects for its Control," in C. L. Markert and J. Papaconstantinou, eds., *The Developmental Biology of Reproduction,* Academic Press, 1975, p. 38.

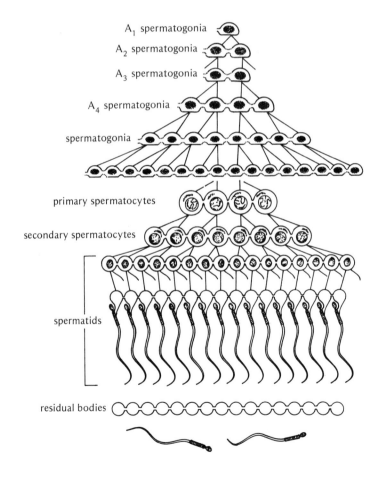

A₁ spermatogonia
A₂ spermatogonia
A₃ spermatogonia
A₄ spermatogonia
spermatogonia
primary spermatocytes
secondary spermatocytes
spermatids
residual bodies

of these lectins (wheat germ agglutinin) will block fertilization of mammalian eggs by sperm. It is believed that lectins might be developed into useful contraceptives because they specifically bind to the surfaces of sperm and egg cells. Such binding might inhibit sperm motility or prevent sperm attachment to egg cell surfaces.

Mechanisms of Sperm Differentiation

We have briefly examined meiosis but have stopped short of describing the transformation of spermatids into mature sperm. We call this transformation phase of spermatogenesis spermiogenesis. In the next chapter on fertilization we will investigate sperm structure and the function of the various sperm parts. Here, let's look at the interesting process of the transformation of round, non-descript spermatids into highly differentiated mature sperm cells.

spermiogenesis

During the primary spermatocyte stage, before any sperm differentiation occurs, the chromosome that confers "maleness" (for example, the Y chromosome in humans) appears to be active in synthesizing messenger RNA. This was demonstrated by observing the appearance of RNA on the loop of the Y chromosome using the radioactive label ^{3}H-uridine that is specifically incorporated into newly formed RNA, by using a technique called autoradiography. This technique involves placing a photographic emulsion over the specimen. Radioactivity in specific regions of the specimen (such as the Y chromosome) is observed as black dots. The researchers used organisms such as the fruit fly, *Drosophila*, but the results probably are of general importance in many systems. What is the importance of the synthesis of RNA on the Y chromosome in the primary spermatocyte? First of all, this specific Y chromosome RNA presumably codes for specific proteins that are responsible for "maleness". Some of these proteins may be needed for the differentiation of spermatids into mature sperm, but remember that spermatids contain the haploid chromosome number. Each spermatid contains *either* an X or a Y chromosome, not both. Thus any messenger RNA formed by the Y chromosome for spermatid differentiation must be made before the first meiotic division when the X and Y chromosomes are separated into different cells. The key to the puzzle seems to be that the important Y chromosome messenger RNA is produced in the primary spermatocytes. This messenger RNA is distributed to all cells and thus all secondary spermatocytes and spermatids have the genetic information from the

Y chromosome to allow synthesis of specific proteins that may be needed for formation of mature spermatozoa.

acrosome   The differentiation of spermatids into spermatozoa involves transformation of the Golgi apparatus into a structure called the acrosome. This structure will be described in the next chapter. The main function of the acrosome is to contact the egg surface and aid the sperm in penetration of the outer egg coats during fertilization. Not all sperm possess acrosomes. Those that do not have to penetrate complex egg coats may not need acrosomes. Acrosomes contain a variety of substances including enzymes that help to break down the outer egg coats.

Transformation of spermatids into sperm also involves compaction of the nucleus and association of nuclear DNA with basic proteins. The basic proteins may help compress the DNA into the small space it must occupy in the sperm head. Other changes occurring in the spermatid include formation of a sperm tail by assembly of proteins called micro-

Figure 1–5.
Transformation of spermatid into mature sperm, chronological sequence. After Gatenby and Beams, 1935.

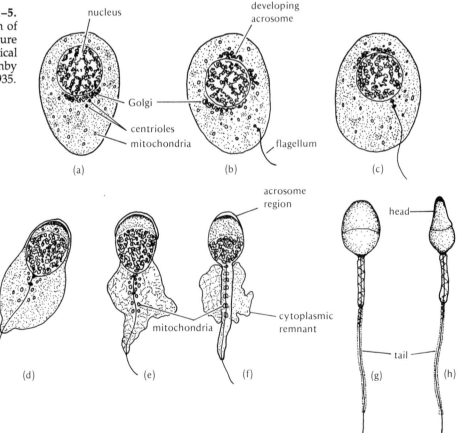

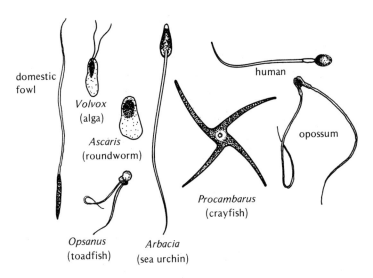

Figure 1–6. Variety of sperm forms.

domestic fowl

Volvox (alga)

Ascaris (roundworm)

human

opossum

Procambarus (crayfish)

Opsanus (toadfish)

Arbacia (sea urchin)

tubules, beginning at a centriole of the spermatid. Sperm structure will be examined in the next chapter. A summary of the transformation events of spermiogenesis is given in Figure 1–5. Some of the varying forms that mature sperm take in different organisms are shown in Figure 1–6.

Oogenesis

Summary of Events

Before looking at some intriguing aspects of the development of female gametes, let us summarize the process of oogenesis. In this way, we will have a framework for discussing the findings that follow.

1. Primordial germ cells, as indicated in our discussion on spermatogenesis, do not originate in the gonad but migrate or are carried to the gonad from other regions of the body. This will be discussed in more detail in Chapter 7.

2. The primordial germ cells proliferate and give rise to the oogonia, the cells that develop into eggs.

oogonia

Figure 1–7. Summary of oogenesis.

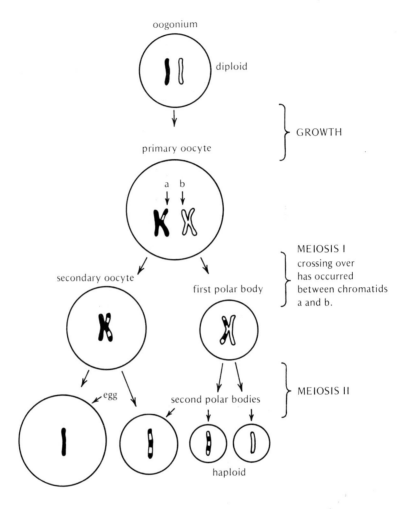

3. Some of the oogonia begin to grow. This growth in some species could be striking. The volume might increase to 100,000 times the original volume. Obviously, this extensive growth provides the egg cell with materials needed for the embryo to develop.

primary oocytes

4. The DNA of the chromosomes duplicates. These growing cells are called primary oocytes. While oogonia have the same amount of chromosomal DNA as normal body cells, the primary oocytes, like primary spermatocytes, possess double the amount of chromosomal DNA found in normal body cells.

5. The primary oocytes, like primary spermatocytes, undergo two meiotic division cycles. By the end of these divisions the cells will have only half the number of chromosomes found in normal body

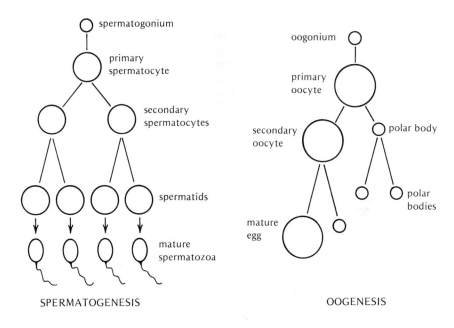

Figure 1–8. Comparison of spermatogenesis and oogenesis.

SPERMATOGENESIS   OOGENESIS

cells. Thus, haploid eggs can combine with haploid sperm to form diploid fertilized eggs to begin embryogenesis.

In oogenesis and spermatogenesis the meiotic stages are exactly the same in terms of crossing over and distribution of the chromosomes to daughter cells (Figure 1–7). One major difference between oogenesis and spermatogenesis involves cytoplasmic division. One primary oocyte divides, in meiosis I, to form one large secondary oocyte and one tiny polar body (that may or not divide again). The secondary oocyte, in the second meiotic division, divides to form one large mature egg and another tiny polar body (Figures 1–7 and 1–8). (Mechanisms that cause such unequal cytoplasmic division are discussed in Chapter 3.) Such unequal cytoplasmic division preserves the vast stores of materials that are produced in the oocyte growth phase. If division was equal, each cell would have much less in the way of stored material needed for the successful development of the embryo. The only other major difference between oogenesis and spermatogenesis is that spermatids must go through a differentiation phase to form mature sperm. The egg formed after the second meiotic division has already prepared itself, during the earlier growth phase, for its role in supporting the new embryo (Figure 1–8).

As in spermatogenesis, hormones exert a controlling influence on oogenesis in most, if not all, organisms. Different hormones act in different systems. Some of these hormones appear to act at the oocyte cell

secondary oocyte

polar body

mature egg

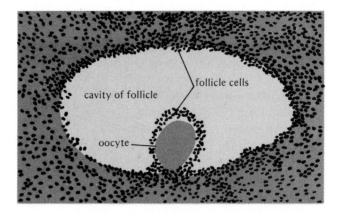

Figure 1–9. Drawing of mammalian oocyte in its follicle, surrounded by follicle cells. From J. D. Ebert and I. M. Sussex, *Interacting Systems in Development*, Holt, Rhinehart and Winston, 1970, p. 17.

surface to somehow stimulate completion of meiosis. In mammals such as human beings, a variety of hormones influence oocyte maturation, development of the follicle that surrounds the oocyte, and rupture of the follicle, releasing the egg (Figure 1–9). The nature of the hormones that control oocyte development in mammals is rather well understood. The development of the oral contraceptive, or birth control pill, is an example of how knowledge of the hormonal control of oogenesis has been applied to an important social problem. These pills contain hormones (such as estrogen and progesterone) involved in controlling oocyte and follicle development. The concentration levels of these hormones in the pills are such that development and release of eggs are prevented.

Oocyte Growth and Development

Unlike sperm cells, egg cells must possess extensive stores of materials to nourish and control the development of the embryo. So, one key aspect of oogenesis is the accumulation of the necessary stores of nucleic acids, proteins, and other substances needed for embryonic development. Let us now examine the events that are involved in this rather remarkable process of oocyte growth.

Nucleic Acids

The primary oocytes enter the first meiotic prophase and go through the same phases as described previously for spermatocytes. The primary oocytes go from leptotene to zygotene to pachytene to diplotene (Figure 1–10). In diplotene, during spermatogenesis, the chromosomes condense and begin to prepare for separating from each other. In oogenesis, however, during diplotene the chromosomes may become greatly extended, forming the so-called lamp brush structure. These chromosomes resemble lamp brushes because thin threads or loops develop perpendicular to the long axis of the chromosomes proper (Figure 1–11). Lamp brush chromosomes are actively involved in synthesizing the RNA needed for development of the egg. All RNA is formed from DNA templates. There are several classes of RNA, including messenger RNA that codes for the amino acid sequence in proteins; ribosomal RNA that is required as a structural component of ribosomes; and transfer RNA that is involved in bringing amino acids to their proper places along the messenger RNA molecules during protein synthesis.

lamp brush structure

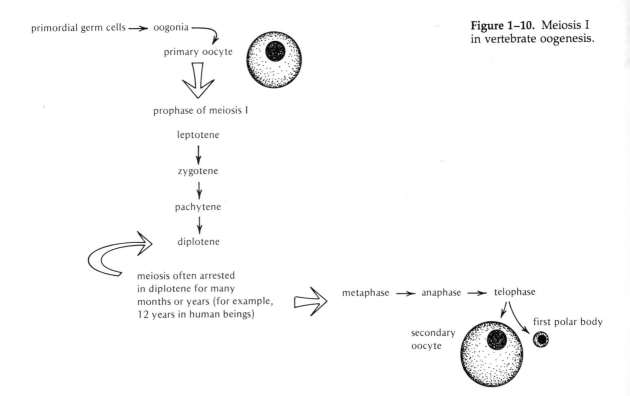

Figure 1–10. Meiosis I in vertebrate oogenesis.

Figure 1–11.
Photomicrograph
showing lampbrush
chromosomes. Courtesy
of J. G. Gall.

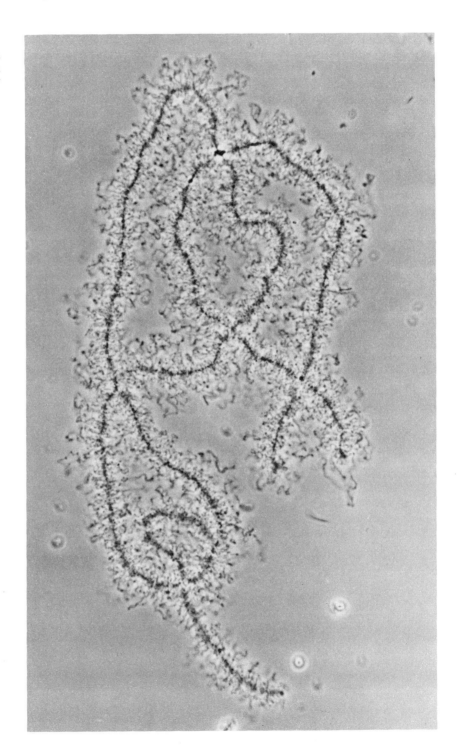

The primary oocytes often remain in diplotene for months or even many years. In the human female, for example, some oocytes may remain in the diplotene stage for up to 50 years. During this time usually only one oocyte matures each month, in response to hormonal stimulation. It may be that a major reason why birth defects are more likely to occur in children born to older women is that oocytes that give rise to these children have remained in diplotene for many decades. This could increase the frequency of chromosomal damage.

What do the lamp brush chromosomes do for the developing oocytes? The loops of the lamp brush chromosomes actively synthesize RNA. This can be shown by incubating oocyte chromosomes with radioactive uridine (a substance from which RNA is formed). Many of these experiments are done using amphibian oocytes because of their large size, large nuclei, and easily obtainable lamp brush chromosomes. Isolated oocyte nuclei can be broken up on a slide, releasing the lamp brush chromosomes. The chromosomes are incubated with radioactive uridine and then washed to remove the uridine. Any newly synthesized RNA incorporates the radioactive uridine, so that only newly synthesized RNA is radioactively labeled. A photographic emulsion is placed over the specimen. Radioactive RNA can then be located as black dots in the emulsion over the regions of the specimen containing radioactivity. Using this technique of autoradiography, it was shown that the loops of the lamp brush chromosomes are actively engaged in synthesizing RNA. Only about five percent of the total genome is in the form of loops at any one time, and only about five percent of the genome is transcribed at the lamp brush stage.

What is the RNA synthesized on lamp brush chromosomes used for? Some of the RNA synthesized during the lamp brush stage is messenger RNA. This RNA codes for proteins that are needed during early embryonic development. Thus, the egg is pre-programmed with messenger RNA that will be needed later on.

In addition to messenger RNA, a very large amount of ribosomal RNA is sythesized in oocytes. This RNA is needed to produce the many ribosomes that are the seats of protein synthesis. The production of ribosomal RNA in many oocytes occurs by a process called selective gene amplification. The DNA segments of the chromosomes that code for ribosomal RNA are selectively copied, while other genes are not. That is, only the ribosomal RNA genes are reproduced thousands of times. These DNA segments form nucleoli. Ribosomal RNA is then formed (transcribed) in these nucleoli from the DNA templates. A single oocyte may contain over 1,000 nucleoli. By again using radioactive uridine it can be shown that the nucleoli are actively sythesizing ribosomal RNA. Estimates have been made that suggest that if all of the ribosomal RNA needed to produce the numerous ribosomes required for early develop-

selective gene amplification

Figure 1–12. Passage of nucleic acids from nurse cells to oocyte of the fly. (a) Follicle (oocyte and nurse cells) one hour after injection of H^3 cytidine. (b) Five hours after injection. Arrows show radioactive nucleic acid streaming from nurse cells into oocyte. After K. Bier, 1963, *Roux Arch.* 154, 552-575.

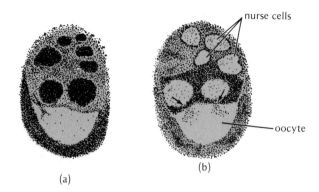

(a)   (b)

nurse cells

ment was made without gene amplification it would take about 500 years. Instead, as a result of gene amplification, the period required to synthesize the needed ribosomal RNA for a frog is reduced to several months.

Not all species possess a lamp brush stage. In species that have no lamp brush chromosomes, the oocytes receive RNA and sometimes DNA from cells that surround the oocytes. In the case of the fruit fly, *Drosophila*, an oogonium divides four times, giving rise to 16 cells. Only one of these 16 cells becomes the oocyte. The others become nurse cells that function to nourish the oocyte. Thus, the nurse cells and oocytes are derived from oogonia and are very closely associated. In fact there appear to be direct cytoplasmic bridges between the oocyte and the nurse cells. The nurse cells synthesize large amounts of RNA, ribosomes, and proteins, and this material literally pours into the oocyte through the connecting cytoplasmic bridges. Autoradiography, mentioned previously, has been used to show that radioactively labeled nucleic acid pours from the nurse cells into the cytoplasm of the oocyte in the common fly, *Musca domestica* (Figure 1–12).

Protein and Other Molecules

Two key aspects of oocyte growth include synthesis of large amounts of RNA and the accumulation of proteins. We have already outlined the RNA story, and now we will move on to proteins. Many of the proteins needed for embryonic development, which accumulate in oocytes, may be produced outside of the ovary and brought to it by the bloodstream.

Experiments have shown that identifiable proteins (either radioactively labeled or detected with specific antibodies) do indeed enter oocytes from the outside.

How do protein molecules get to the oocytes? What is the nature of the molecules and what is their function? One important substance that accumulates in eggs is yolk. Yolk is a major food reserve for some developing embryos. There are many types of yolk, and many methods of yolk production. Some yolk is mainly protein with some lipid (protein yolk). Other types of yolk consists mainly of phospholipid and fat and possibly some protein (fatty yolk). In vertebrates, yolk is synthesized in the liver and carried in soluble form by the bloodstream to the ovaries. Once in the ovaries, it appears to be picked up and transferred to the oocyte by follicle cells that surround the oocyte. The follicle cells, unlike nurse cells, are not formed from oogonia but are formed from the ovary epithelium, the ovary surface layer. Follicle cells are not connected by cytoplasmic channels to the oocytes. Instead, many fine projections called microvilli from the follicle cell surface intertwine with microvilli on the oocyte cell surface (Figure 1–13). The microvilli of the oocyte may absorb yolk by pinching off tiny portions of the membranes of the follicle cell microvilli and engulfing the yolk material in these vesicles, or they may absorb yolk previously released by the follicle cell microvilli. This process of cell drinking is termed pinocytosis or micropinocytosis. In some animals without circulatory systems, yolk may be synthesized in the oocyte itself on ribosomes on the endoplasmic reticulum that is associated with Golgi apparatus.

We should also mention that even in organisms in which yolk is transported to the oocytes from the liver, the yolk often still undergoes packaging into platelets in the oocyte. Some yolk platelets are probably formed inside mitochondria in fish, amphibian, and snail oocytes (Figure 1–14). A mitochondrial enzyme called protein kinase may play an important role in yolk platelet formation. This enzyme adds a phosphate group to one type of soluble yolk (phosvitin), causing the yolk to become insoluble. This may directly cause the yolk to crystalize out of solution to form platelets, yolk granules needed to nourish an embryo. Yolk platelets are also formed in the cytoplasm within the vesicles formed by micropinocytosis.

In addition to yolk, many other molecules needed for early embryonic development accumulate in oocytes. These substances include glycogen, an important energy-rich carbohydrate storage molecule, lipid for membrane synthesis and energy supply, and a variety of proteins, some of which are subunits for the cytoplasmic contractile system (Chapter 9) and others of which are enzymes.

In summary, in many organisms yolk is synthesized in the liver and brought by the bloodstream to the follicle cells surrounding oocytes. The

yolk

follicle cells

microvilli

pinocytosis

protein kinase

glycogen

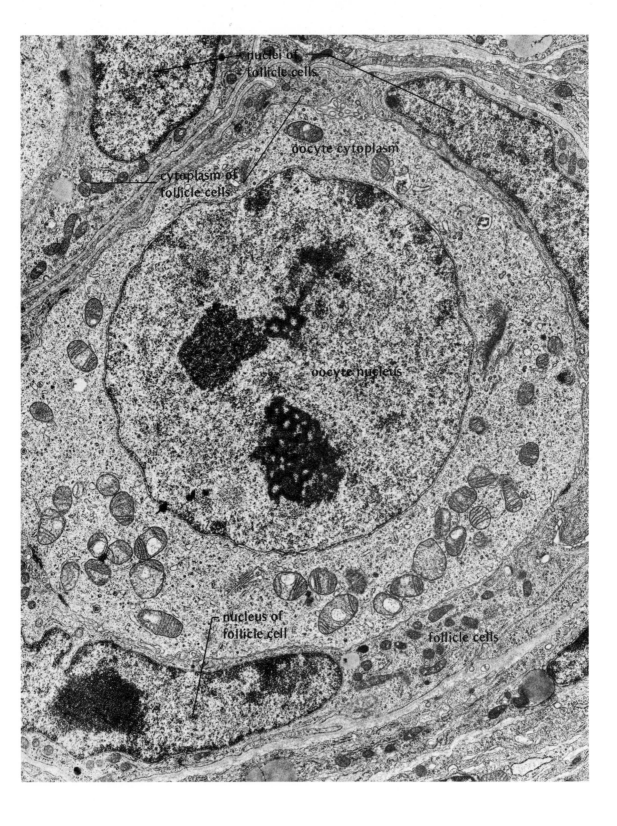

nuclei of
follicle cells

cytoplasm of
follicle cells

oocyte cytoplasm

oocyte nucleus

nucleus of
follicle cell

follicle cells

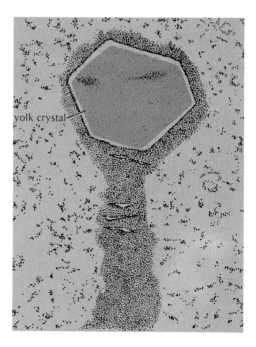

Figure 1–14. Yolk crystal inside mitochondrion of frog oocyte. After R. T. Ward, *J. Cell Bio.* 14:309-341 (1962).

follicle cells transfer this material to the oocytes where it is packaged into yolk platelets. Glycogen, lipid, and non-yolk proteins also accumulate in oocytes to help prepare these cells for embryonic development. Let us turn next to the development of larger structures in oocytes that are also essential components of these highly specialized cells.

Higher Orders of Structure

We have examined how oocytes accumulate large stores of nucleic acid, proteins, and other molecules that are required for the development of the embryo. Oocytes do not become prepared for fertilization and development only by accumulating and synthesizing molecules. They also develop new cellular organelles (cell components with specific func-

Figure 1–13 (opposite). Young mouse oocyte surrounded by follicle cells. Courtesy of Dr. E. Anderson, Harvard Medical School.

tions) that are needed for fertilization and development, and that are composed of many, many molecules that fit together to form more complex structures.

In the next chapter we will see that many eggs contain structures called cortical granules, located in the surface cytoplasm just below the plasma membrane of the egg. These structures play a major role in the fertilization reaction. Cortical granules form from the oocyte Golgi membrane complex and move to the periphery of the oocyte. These structures contain glycoproteins (proteins with attached sugar chains) that play a role in the fertilization reaction described in the next chapter.

cortical granules

Oocytes develop a variety of surface coats. These coats protect the cell and may play a role in assuring that only the right type of sperm will stick to the egg. The coats also often help form the fertilization membrane (see Chapter 2). Many eggs develop special coats in the tight space between the oocyte cell membrane and follicle cell membranes. The follicle cells (and possibly the oocyte itself) secrete mucoproteins and fibrous proteins into this space. The coat formed in this way is called the vitelline membrane in molluscs, insects, amphibians, and birds, the chorion in fishes and tunicates, and the zona pellucida in mammals. In addition to these coats produced between oocytes and follicle cells, other

vitelline membrane

chorion

zona pellucida

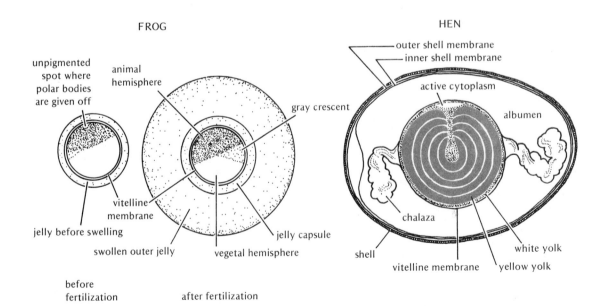

Figure 1–15. Frog and hen's eggs. From I. B. Balinsky, *An Introduction to Embryology*, 4th ed. © 1975 by W. B. Saunders Co., p. 65.

coats that surround some eggs are formed by various glands as the eggs pass through the oviducts. These coats include the frog egg jelly coat and the egg white and shell layers of the bird egg (Figure 1–15).

Summary

To sum up, gametogenesis results in the formation of haploid gametes. Genetic information is mixed in gametes as a result of crossing over and by random distribution of maternal and paternal chromosomes during meiosis. Male gametes become highly specialized as a result of the differentiation of spermatids. Female gametes accumulate vast stores of RNA, protein, and other substances essential for development during the growth phase of the first meiotic prophase. The first and second meiotic divisions preserve these vast stores of materials by unequal cytoplasmic divisions. Thus, only one large, rich egg is formed from each primary oocyte.

The next step in our story of embryology is fertilization. In this process, male and female gametes join to form the fertilized egg or zygote. We will see how the relatively inactive egg suddenly becomes "turned on" to produce the multitude of materials required for the rapid development of the newly formed being.

Readings

Fawcett, D. W., Gametogenesis in the Male: Prospects for Its Control. In *The Developmental Biology of Reproduction*, C. L. Markert and J. Papaconstantinou, eds., Academic Press, 25–53 (1975).

Fawcett, D. W., The Structure of the Mammalian Spermatozoon. *Int. Rev. Cyt.* 7:195–234 (1958).

Fawcett, D. W., A Comparative View of Sperm Ultrastructure. *Biol. Reprod.* supplement 2:90–127 (1970).

Gall, J. G., and H. G. Callan, ³H-Uridine Incorporation in Lampbrush Chromosomes. *Proc. Natl. Acad. Sci. U.S.* 48:562–570 (1962).

Hadek, R., The Structure of the Mammalian Egg. *Int. Rev. Cyt.* 18:29–71 (1965).

MacGregor, H. C., The Nucleolus and Its Genes in Amphibian Oogenesis. *Biol. Rev.* 47:177–210 (1972).

Oppenheimer, S. B., Interactions of Lectins with Embryonic Cell Surfaces. In *Current Topics in Developmental Biology,* A. A. Moscona and A. Monroy, eds. Vol II, pp. 1–16 (1977).

Raven, C. P., *Oogenesis: The Storage of Developmental Information.* Pergamon, (1961).

Roth, T. F., and K. R. Porter, Yolk Protein Uptake in the Oocyte of the Mosquito, *Aedes aegypti L. J. Cell. Biol.* 20:313 (1964).

Wallace, R. A., and J. N. Dumont, The Induced Synthesis and Transport of Yolk Proteins and Their Accumulation by the Oocyte in *Xenopus laevis. J. Cell. Physiol.* 72(supplement I):73 (1968).

CHAPTER 2

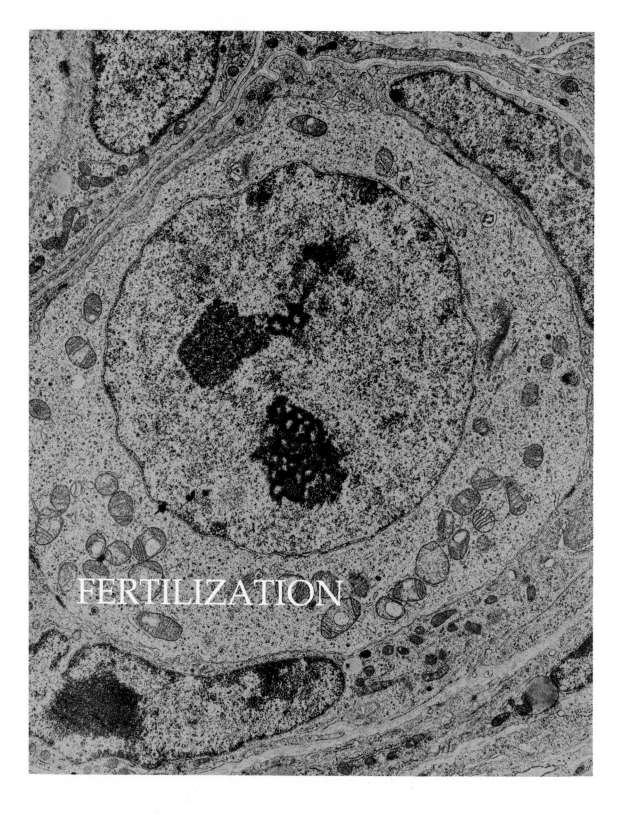

FERTILIZATION

WHAT TURNS ON the complex series of metabolic reactions that occur in eggs immediately after fertilization? How can a tiny sperm that fuses with only 0.0002 percent of the egg surface trigger the multitude of changes that occur in the new zygote? The answers to these questions are becoming better understood and the program of events occurring during the fertilization process is becoming well established. Let us examine the fertilization reaction from a variety of approaches so that by the end of the chapter the questions raised above will, in part, be answered. We will look at: the ultrastructural aspects of fertilization—those aspects we can visualize with the electron microscope; the biochemical and physiological program of events occuring during fertilization; and the molecular aspects of sperm-egg recognition, or how a sperm gets to and sticks to an egg. No attempt will be made to examine these aspects of the fertilization reaction in all types of organisms. Instead, we will discuss representative systems which for one reason or another have yielded unusually important results in the field.

zygote

Ultrastructural Aspects of Fertilization

With the aid of the electron microscope we can examine the structural basis of fertilization by examining the sperm, the eggs, and sperm-egg interaction and the resulting changes at the egg surface.

Sperm Structure

A typical sperm cell is shown in Figure 2–1. This is a diagram of an electron micrograph of the sperm of an annelid worm, *Hydroides.* The

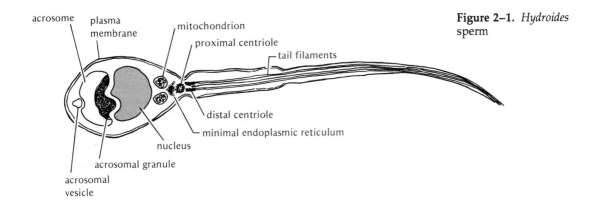

Figure 2–1. *Hydroides* sperm

head of the sperm includes a nucleus that contains the genetic information, and components of the acrosome, the anterior tip of the sperm head that functions to aid the sperm in penetrating through the outer egg coats and in establishing connection with the egg cytoplasm.

acrosome

Vacquier and his colleagues have recently isolated a molecule from the acrosomal granule of sea urchin sperm. It is a protein called bindin that appears to be of major importance in allowing the sperm to recognize and adhere to the egg surface. In addition to bindin, the acrosomal granule contains enzymes called lysins that aid the sperm in penetrating the outer egg coats.

bindin

lysins

Directly behind the nucleus, in the neck region of the sperm, mitochondria and centrioles can be identified. The mitochondria help supply the sperm with energy and the centrioles are involved in forming asters involved in the division of the fertilized egg (proximal centriole) and for attachment of the tail fibers (distal centriole). The tail allows the sperm to move and contains fibers composed of microtubules (rod shaped proteins) that contract and propel the sperm cell.

Egg Surface Structure

Before examining the ultrastructural aspects of the fertilization reaction, it is important to briefly look at some representative eggs and their surfaces. The cell surface plays a major role in sperm-egg interaction. Three types of eggs are shown in Figure 2–2.

Different types of eggs have different kinds of surface coats. They all, however, contain a similar true membrane or plasma membrane

Figure 2–2. Surfaces of three eggs.

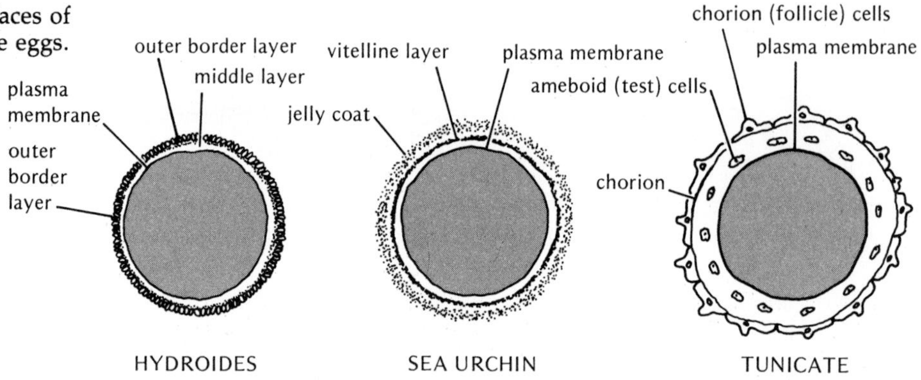

plasma membrane

outer border layer

middle layer

outer border layer

vitelline layer

jelly coat

plasma membrane

ameboid (test) cells

chorion (follicle) cells

plasma membrane

chorion

HYDROIDES   SEA URCHIN   TUNICATE

bounding the cytoplasm. The coats can be cellular in nature (tunicate) or proteinaceous (sea urchin), but the true plasma membrane has a very well defined structure. The sperm plasma membrane also uses this structure. It is generally agreed that the plasma membrane consists of a lipid bilayer in which chunks of protein or glycoprotein are imbedded as shown in Figure 2–3.

fluid-mosaic model

This model of the cell membrane was developed by Singer and Nicolson and is called the fluid-mosaic model. The chunks of proteins and glycoproteins can move in the plane of the membrane like floating protein islands in a sea of lipid. This movement may be restricted by rod-shaped protein elements such as microtubules or microfilaments attached to the inner membrane surface. The control of membrane protein mobility by these rod-shaped proteins is not well understood but is the topic of a great deal of recent research. As can be seen in Figure 2–3, the outermost boundary of the plasma membrane consists of sugar chains attached to the protein or lipid. These sugars may play an important role in mediating initial contact between sperm and egg plasma membranes during fertilization.

Figure 2–3. Current model of plasma membrane structure.

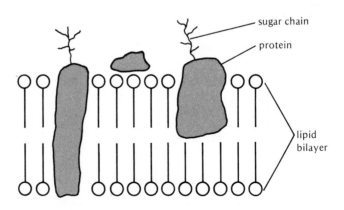

sugar chain

protein

lipid bilayer

Sperm-Egg Interaction

Let us now consider the ultrastructural aspects of sperm-egg interaction. How does the fertilization process look using the electron microscope? Figures 2–4 and 2–5 are elegant photos of sperm attaching to the egg surface, as seen with the scanning electron microscope.

In order, however, to understand exactly what is occuring between sperm and egg membranes during fertilization, the transmission electron microscope must be used. The transmission electron microscope allows viewing of ultrathin sections through the sperm and eggs so that the interior parts of the membranes, coats, and cytoplasm can be visualized.

A classic study of fertilization in *Hydroides* was performed by Colwin and Colwin using the transmission electron microscope. Figure 2–6 gives a diagramatic summary of their results.

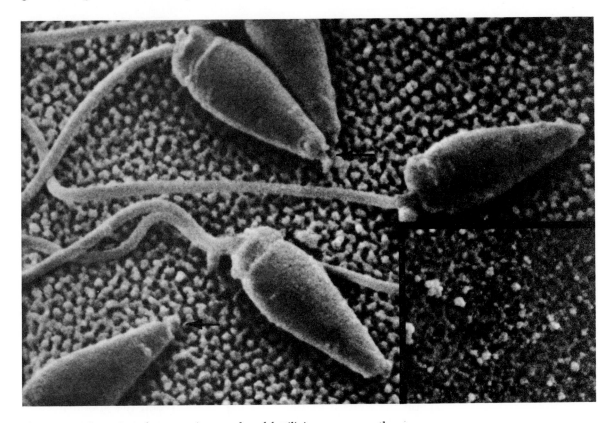

Figure 2–4. Scanning electron micrographs of fertilizing sperm on the egg surface. Inset shows inner vitelline layer. Courtesy of Dr. Charles Glabe and Victor Vacquier.

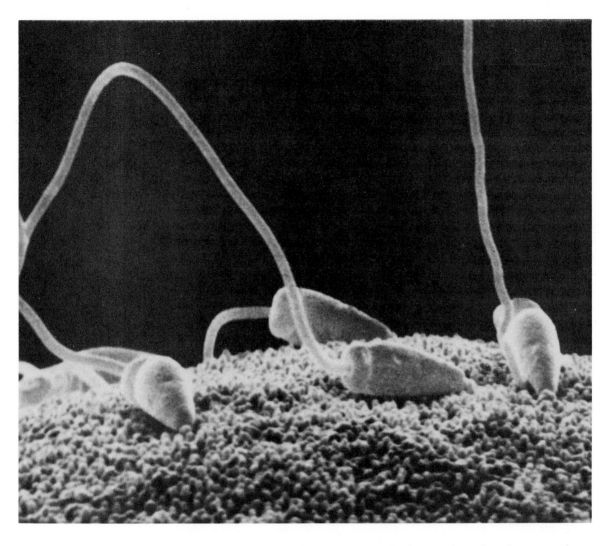

Figure 2–5. Scanning electron micrograph of sperm bound to the egg surface. Courtesy of David Epel.

As can be seen in Figure 2–6, when the sperm head hits the outer border layer of the egg, the acrosomal vesicle bursts and the acrosomal membrane becomes continuous with the sperm plasma membrane. The acrosomal granule dumps lysins onto the egg. These enzymes aid in sperm penetration of the outer egg coats. When the sperm head is approximately halfway through the outer egg layers, microvilli form at the base of the acrosome in the sperm, and from the egg plasma membrane. Finally sperm and egg microvilli fuse, the sperm plasma

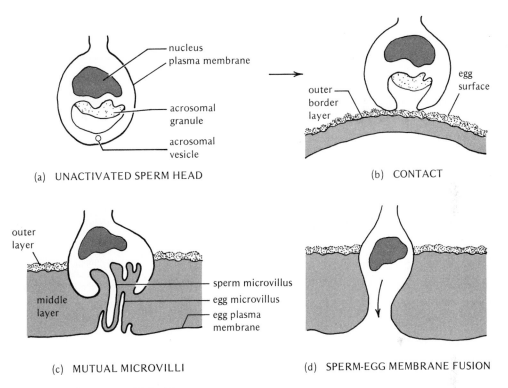

(a) UNACTIVATED SPERM HEAD

(b) CONTACT

(c) MUTUAL MICROVILLI

(d) SPERM-EGG MEMBRANE FUSION

Figure 2–6. Fertilization in *Hydroides*.

membrane and egg plasma membranes become continuous, and the sperm nucleus and other contents move into the egg cytoplasm. Thus fertilization here represents an apparent fusion between sperm and egg.

Egg Cortical Reaction

What is happening, at an ultrastructural level, in the egg at fertilization? Let us examine what happens at the surface of the sea urchin egg at fertilization. The sea urchin has become a model system in the study of fertilization because gametes can be extracted from them in massive numbers, and the entire fertilization reaction can be easily observed in

Figure 2–7. Fertilization membrane formation in the sea urchin, showing the cortical reaction.

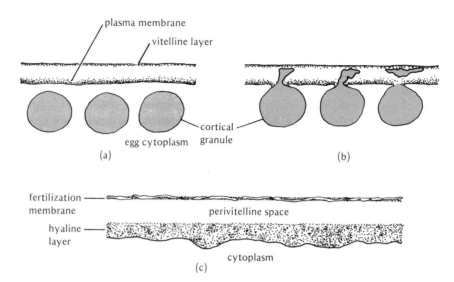

the laboratory in plain sea water. Figure 2–7 shows that upon fertilization in the sea urchin egg, directly below the plasma membrane, tiny structures called cortical granules begin to fuse with the plasma membrane. Each cortical granule is about one micrometer (10^{-3} millimeter) in diameter, and each egg contains about 15,000 of these tiny structures. The cortical granules release some of their contents into the space between the plasma membrane and vitelline layer as seen in Figure 2–7. Cortical granules contain enzymes, structural proteins, and sugar-protein complexes. Carroll and Epel have shown that one of the cortical granule enzymes alters sperm receptor proteins on the vitelline layer, preventing additional sperm from attaching. Thus, this enzyme is an important factor involved in the so-called block to polyspermy, the prevention of fertilization by more than one sperm. This is not the whole story, however, because even earlier in the fertilization reaction, about one second after sperm attachment, sodium ions flow into the cell, causing a brief voltage change that appears to prevent additional sperm from entering the egg.

cortical granules

block to polyspermy

Another cortical granule enzyme serves to disconnect the vitelline layer from the plasma membrane, allowing the vitelline layer to lift away from the surface of the egg to form the fertilization membrane. The fertilization membrane, therefore, is a combination of structural proteins released from the cortical granules and the vitelline layer. The fertilization membrane is an effective additional means to block polyspermy. Other material derived from the cortical granules sticks to the plasma membrane, forming a clear surface coat called the hyaline layer (Figure 2–7). This layer will help the sea urchin embryo cells stay together during the cleavage or cell division stage of development.

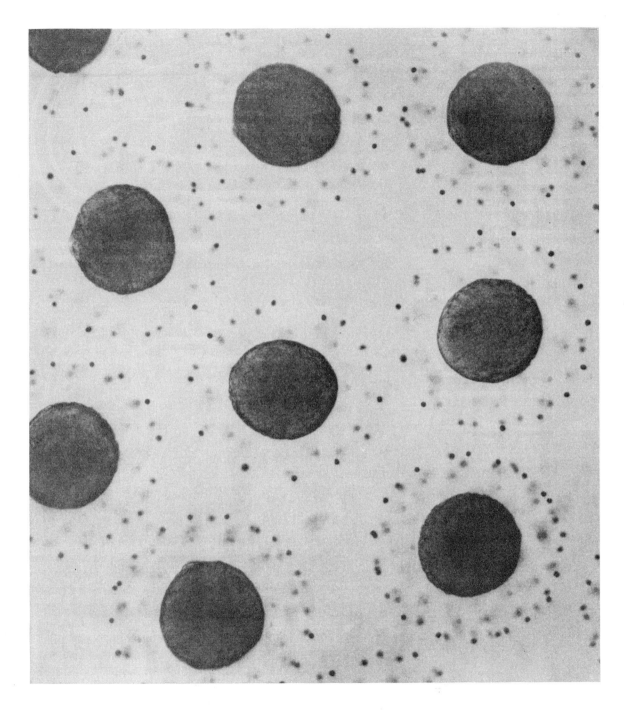

Figure 2–8. Echinoderm unfertilized egg. Courtesy of Victor Vacquier.

Figure 2–9 (right, and opposite page). Sea urchin fertilization. Note progressive formation of the fertilization membrane. Courtesy of Victor Vacquier.

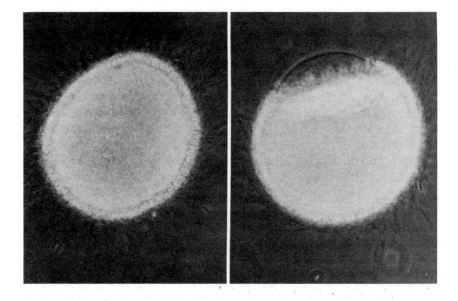

About twenty minutes after the sperm has made initial contact with the sea urchin egg, the sperm nucleus fuses with the egg nucleus. The first division occurs shortly thereafter. Fertilization is complete and the embryo begins to develop.

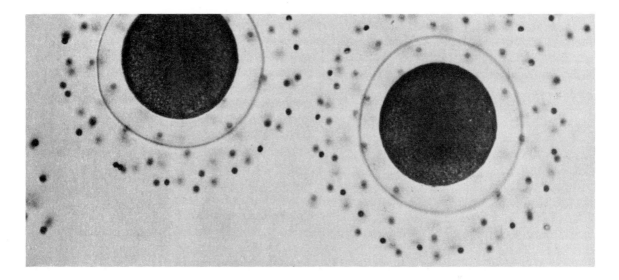

Figure 2–10. Sea urchin fertilized egg. Fertilization membrane is complete. Courtesy of Victor Vacquier.

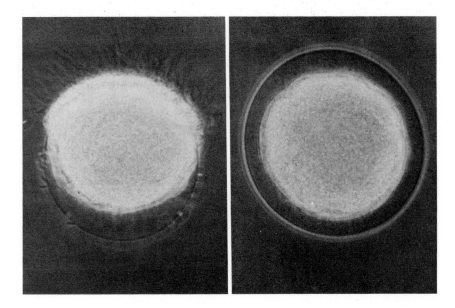

Fate of Sperm Mitochondria

Before turning to the biochemical and physiological program of events occurring during fertilization, let's briefly consider one additional question. What happens to the sperm mitochondria at fertilization? Mitochondria supply cells with energy and also contain DNA and therefore deserve consideration here. The fate of sperm mitochondria appears to be different in different organisms. For example, in insects sperm mitochondria enter the egg. In the rat they also enter but disintegrate in about 30 minutes. In *Hydroides*, mitochondria from the sperm also enter the egg at fertilization and remain intact at least through the fifth cleavage. In tunicates, however, an elegant study by Ursprung and Schabtach utilizing the electron microscope showed that the sperm mitochondrion did not enter the egg but instead was knocked off the sperm as the sperm began to enter the outer egg coats. Figure 2–11 illustrates what appears to happen in this system. As the sperm squeezes between the chorion cells at the outer boundary of the egg, the mitochondrion can not make it through and separates from the rest of the sperm.

Not only is the fate of mitochondria different in different species, but the number of sperm that enter the egg also varies. We mentioned that a block to polyspermy occurs in the eggs described that prevents more than one sperm from entering the egg. This, however, is not the

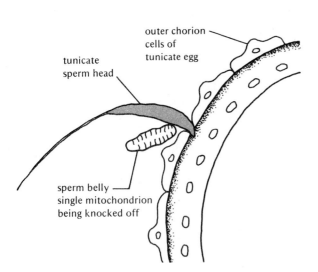

outer chorion
cells of
tunicate egg

tunicate
sperm head

sperm belly
single mitochondrion
being knocked off

Figure 2–11 (above). Fate of the sperm mitochondrion in the tunicate. Based on experiments by H. Ursprung and E. Schabtach, *J. Exp. Zool.* 159: 379–384 (1965).

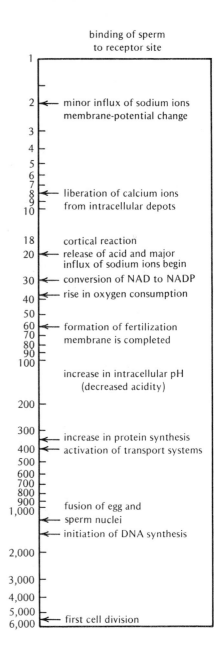

binding of sperm
to receptor site

1

2 ← minor influx of sodium ions
membrane-potential change

3
4
5
6
7
8 ← liberation of calcium ions
9 from intracellular depots
10

18   cortical reaction
20 ← release of acid and major
influx of sodium ions begin
30 ← conversion of NAD to NADP
40 ← rise in oxygen consumption
50
60 ← formation of fertilization
70   membrane is completed
80
90
100

increase in intracellular pH
(decreased acidity)

200

300
increase in protein synthesis
400 ← activation of transport systems
500
600
700
800
900
1,000   fusion of egg and
← sperm nuclei
← initiation of DNA synthesis

2,000

3,000

4,000

5,000
6,000 ← first cell division

Figure 2–12 (right). Timetable of events following fertilization of the sea urchin egg. The time scale shows seconds elapsed after sperm binds to egg. From D. Epel, "The Program of Fertilization," *Scientific American* 237:129. Copyright © 1977 by Scientific American, Inc. All rights reserved. Used with permission.

case in all organisms. For example, in eggs such as those of some molluscs, selachians, birds, reptiles, and urodeles, many sperm can enter the egg. All but one of the sperm, however, eventually degenerate in the egg cytoplasm. Thus, these eggs allow entry of more than one sperm but possess a mechanism that eliminates all but one of these entering sperm. This type of fertilization in which many sperm enter the egg but all but one disintegrate is termed physiological polyspermy.

physiological polyspermy

Summary

Different things occur during fertilization in different systems, and seldom does one find a mechanism that is exactly the same in all organisms. Many successful components of individual fertilization processes have therefore evolved. In summary, in this section we have examined: sperm structure, egg surface structure, sperm-egg interaction at the structural level, the egg cortical reaction and the block to polyspermy, and the fate of sperm mitochondria. Now that we have this background in the larger structural aspects of fertilization, we can turn to examining the fertilization reaction at the molecular level. The story at the molecular level, as you will see, is still incomplete. A body of fascinating information, however, is beginning to emerge.

The Biochemical and Physiological Events of Fertilization

At the start of this chapter we asked: What turns on the complex series of events that occur in eggs after initial sperm contact? Some of the mysteries that have clouded the answer to this question have recently been solved. Let us turn to an examination of the physiological and biochemical events occuring during fertilization and egg activation now that we have an understanding of what occurs at the structural level.

Timetable of Events Following Fertilization

Figure 2–12 gives the timetable of events following the fertilization of the egg of the sea urchin *Strongylocentrotus purpuratus* as described by David Epel. As mentioned previously, much of the work in this area has been done with the sea urchin, because massive numbers of eggs and sperm can be removed from these animals in the laboratory and because all of the early events of fertilization can easily be observed in the laboratory using plain sea water medium.

The sequence of events following sperm contact with the egg can be listed as follows:

1. influx of sodium ions

2. liberation of calcium ions from intracellular depots

3. cortical granule reaction, release of acid and another major influx of sodium ions

4. conversion of NAD to NADP

5. rise in oxygen consumption

6. completion of fertilization membrane

7. increase in intracellular pH (decrease in acidity)

8. increase in protein synthesis

9. activation of transport systems

10. fusion of egg and sperm nuclei

11. initiation of DNA synthesis

12. first cell division

Numerous experiments from many laboratories have provided us with the sequence of events shown in Figure 2–12 (reviewed by Epel, 1977). An example of the type of data that has helped to form this picture is the elegant set of experiments that indicated that calcium ions are released into the egg cytoplasm early in the fertilization process, as first suggested by Mazia in 1937. Ridgway, Gilkey, and Jaffe used a luminescent protein extracted from jellyfish, called aequorin, that glows in the presence of free calcium ions. Aequorin was injected into the large unfertilized eggs of a fish, the Japanese medaka (*Oryzias latipes*). The eggs only glowed slightly. Upon fertilization, however, there was a ten thousand-fold increase in luminescence, indicating that a large amount of calcium was released from a bound to a free state in the egg. In order to determine if the rise in calcium plays a key role in egg activation, Steinhardt, Epel, Chambers, Pressman, and Rose used a substance called ionophore A23187, which causes the release of calcium ions in cells. In the absence of sperm, this ionophore caused an activation response resembling actual fertilization in sea urchin eggs. Epel, Carroll, Yanagimachi, and Steinhardt found that the ionophore activated many types of eggs including those of amphibians, mammals, tunicates, and mollusks. Thus the release of calcium seems to be an important early event in the fertilization reaction that may directly trigger the occurrence of some of the other events that occur in the activation of the egg.

As can be seen in Figure 2–12, at about 30 seconds after the sperm has bound to the egg, an enzyme, NAD kinase, is activated. This enzyme catalyzes the transfer of a phosphate group from ATP to NAD, forming NADP. NAD and NADP are coenzymes and an increase in the amount of NADP in the freshly fertilized egg is one important part of the process of preparing the zygote for the many synthetic reactions that will shortly follow.

We can also see from Figure 2–12 that there is a rise in oxygen consumption at about 35 to 45 seconds after sperm binding to the egg. At about the same time as this is taking place another enzyme becomes activated. This enzyme, glucose-6-phosphate dehydrogenase, is very important in initiating metabolic reactions that are involved in sugar metabolism. Some of these synthetic reactions that suddenly begin after fertilization appear to be initiated as a result of the activation of certain enzymes that are needed for these reactions to occur.

Modern technology makes it relatively easy to obtain evidence regarding the sequence of events in the fertilization process. For example, to show that protein synthesis increases at about 350 seconds after sperm binding simply involves incubating eggs with radioactive amino acid at various times after sperm binding. New protein synthesis is measured by examining the amount of amino acid incorporated into protein with a scintillation counter. Many such experiments have been done in different systems by different investigators. The studies described in this chapter represent a tiny sampling of the numerous volumes of data in this area. They by no means represent a complete story and are only intended to introduce the reader to the types of findings that have been made in this field.

Protein synthesis in sea urchins increases rapidly after fertilization, as shown in Figure 2–13. Protein synthesis is measured by counting the amount of radioactive amino acid (in this case methionine) incorporated into sea urchin protein that can be precipitated from sea urchin embryo homogenates with acid. As seen, there is a rapid increase in protein synthesis during the first couple of hours after fertilization. By four hours, protein synthesis decreases, and begins increasing again at about ten hours. These data suggest that upon fertilization, there is a need for a large variety of proteins that are required for the numerous reactions and syntheses that occur in the new embryo. Of immediate importance is the need for spindle proteins and chromosomal proteins required for chromosome replication and cell division. New membrane proteins are needed for the synthesis of new membranes required as cells divide.

The reasons for the sudden stimulation of protein synthesis upon fertilization are not well understood. Factors that may be involved in this activation include activation of inactive ribosomes; appearance of new cofactors, (small molecules that may be needed to activate an enzyme or

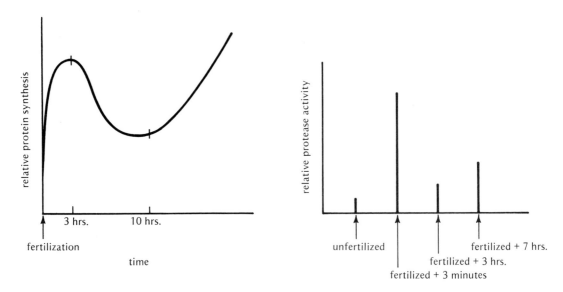

Figure 2–13 (left). Protein synthesis in the sea urchin embryo. Based on experiments by H. Ursprung and K. D. Smith, *Brookhaven Symp Biol.* 18:1–13 (1965), Figure 1.

Figure 2–14 (right). Protease activity in the sea urchin egg. Based on experiments by Lundblad and Lundblad.

reaction), ions, or enzymes; activation of messenger RNA templates, or transfer RNA, or activating enzymes. The ribosomes of unfertilized eggs can act as the seats of protein synthesis if synthetic messenger RNA is added to these ribosomes in an *in vitro* protein synthesizing system. Thus it does not appear that ribosomes in unfertilized eggs are the reason for reduced protein synthesis before fertilization, since these ribosomes can translate added messages. It may be that the messenger RNA in unfertilized eggs is inactive, or that the message is in some way masked, possibly by a protein coat. It has been shown that upon fertilization there is a rapid increase (maximum at 3 minutes after fertilization) in the activity of enzymes (proteases) that catalyze the hydrolysis (breaking down) of proteins (Figure 2–14). We can speculate that these enzymes may digest away some of the protein that may be masking messenger RNA molecules, in this way activating the messages.

Protein synthesis may be activated by other mechanisms. Addition of polyadenylic acid residues to messenger RNA appears to be important in activating the message so that it can function in protein synthesis. At fertilization, adenylation of messenger RNA occurs in the cytoplasm. This may activate the messages. It is clear that the total picture of what activates protein synthesis at fertilization is, as yet, not well understood.

The experiments described in Figure 2–13 give the picture for total cell protein synthesis. Although the rate of total cell protein synthesis

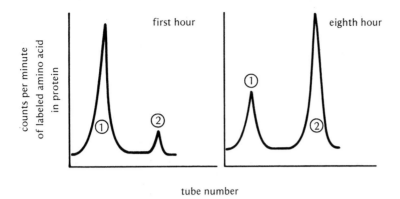

Figure 2–15. Synthesis of individual proteins after fertilization in the sea urchin. Each peak represents a specific protein. The results indicate that more of protein 1 than protein 2 is synthesized during the first hour. During the eighth hour, the opposite is true. Based on results of Ellis, *J. Exp. Zool.* 163:1–22, Figures 3, 4, 6).

may increase at a given time, this does not mean that synthesis of all of the proteins in the cell is increasing. Quite the contrary, certain individual proteins do not follow the pattern shown in Figure 2–13. Figure 2–15 shows that the levels of synthesis of two specific proteins are quite different at different times following fertilization in the sea urchin. In Chapter 11, we will examine in detail the methods by which specific protein synthesis is determined. Figure 2–15 shows the results of an experiment where sea urchin embryos were labeled with radioactive amino acids during the first hour after fertilization and during the eighth hour after fertilization. The embryos were homogenized and placed on chromatography columns that separate specific proteins according to molecular size and charge. As can be seen, more of protein 1 than protein 2 is synthesized during the first hour after fertilization. During the eighth hour, however, more of protein 2 than protein 1 is synthesized. Different amounts of the same protein are synthesized at specific times after fertilization. This finding is consistent with the contention that developing embryos require specific amounts of different proteins at given times during early development. It is not the case that all proteins in the cell follow the pattern of total cell protein synthesis shown in Figure 2–13.

Hypothetical Relations among the Events in Fertilization

Figure 2–16 outlines a largely hypothetical scheme of the relationships of the events occurring during fertilization, as proposed by Epel and based upon considerable experimental evidence. It should be stressed that this

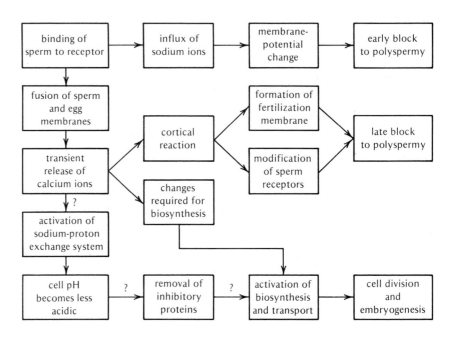

Figure 2–16. Flow chart outlines the largely hypothetical relations among the early calcium-dependent changes upon fertilization (such as the cortical reaction) and the late pH-dependent changes (such as DNA synthesis). The mechanism by which calcium turns on the sodium-proton exchange system to result in a reduction of the acidity of the egg cytoplasm (an increase in pH) is not yet understood. How this pH shift initiates protein and DNA synthesis is also not known, but it may involve the removal of inhibitory proteins from functional proteins such as those of ribosomes or from structural proteins such as actin or tubulin subunits. In the latter case, removal of an inhibitory protein might enable the subunits to polymerize into filaments, resulting in major changes in cell structure that could initiate a variety of biochemical events. From D. Epel, Scientific American 237:129 (1977).

scheme is tentative and will surely be altered over the next few years as additional experiments are performed. A few pieces to the puzzle of what triggers each event are beginning to emerge. For example, the release of calcium ions in the egg cell following sperm-egg contact seems to trigger the cortical reaction and other early changes. Vacquier performed an elegant experiment to support this notion (Figure 2–17). He isolated sea urchin egg plasma membranes with their associated cortical granules on a glass slide. Upon addition of calcium, the cortical granules released their contents. Thus the release of calcium ions in the egg (from a bound to a free state) appears to directly trigger the cortical reaction.

What is the significance of the reduction of acidity of the egg cytoplasm, another important event shown in Fig. 2–12 and 2–16? Experiments by Nishioka, Epel, Shen, and Steinhardt suggest that the pH need only be raised for a ten minute period for development to begin. Since many of the synthetic changes occur at a later time, after the transient pH shift has occurred, it is likely that the pH increase initiates a cellular change that allows metabolic activation to occur. Such a change may involve a pH-induced dissociation of an inhibitor from enzymes or ribosomes, enabling synthetic reactions to proceed.

We may also note that some of the early changes occurring during fertilization involve the release or influx of ions such as calcium and sodium. Synthetic reactions necessary for development to proceed are

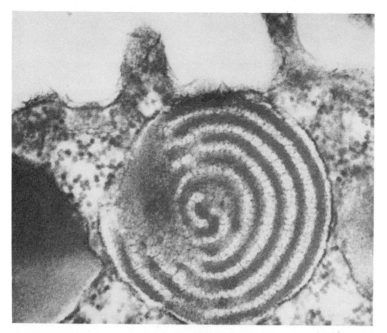

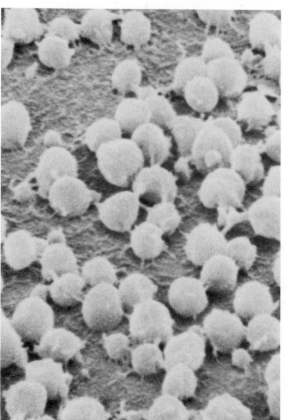

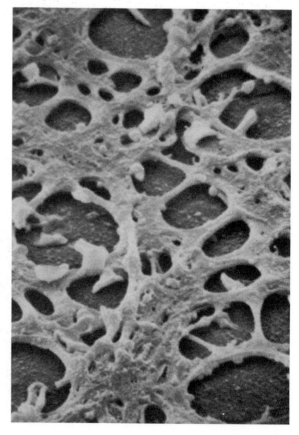

Figure 2–17. Calcium initiated discharge of isolated sea urchin cortical granules. (Left) Top sectioned cortical granule (transmission electron micrograph). (Below, left) Intact cortical granules on inside-out plasma membrane (scanning electron micrograph). (Below, right) Bottom after addition of calcium, cortical granules fuse (scanning electron micrograph). Courtesy of Victor Vacquier.

controlled by enzymes. Enzymes are often activated by specific ions. Thus, changes in cytoplasmic ion concentrations may activate certain synthetic reactions by activating the enzymes that catalyze these reactions.

Summary

In this section we have examined the sequence of events that occurs during fertilization, trying to look at these events in a causal manner. Which event triggers the other events? We cannot fully answer this question as yet. We did see, however, that release of specific ions such as calcium and alteration in cellular pH appear to be important in causing some of the other events to occur.

Molecular Aspects of Sperm-Egg Recognition

Now that we have surveyed fertilization from the ultrastructural and physiological viewpoints, we will briefly consider the important question of how a sperm cell recognizes an egg cell. Again we will center our attention on the sea urchin system.

In the ocean, sea urchins release up to 400 million eggs per female and 100 billion sperm per male during their three to eight month breeding season. How does a sea urchin sperm know that it has reached a sea urchin egg of the same species? In certain organisms, such as the fern and moss, male gametes are attracted to female gametes by specific chemicals given off by the female gametes. In the fern, L-malic acid seems to be the chemoattractant, while sucrose appears to serve a similar function in moss. In the water mold, male gametes are attracted to a specific organic compound, L-sirenin, released by the female gametes. The male cells respond to levels of L-sirenin as low as 0.5×10^{-10} M in solution! In most (but not all) animals, however, female gametes do not

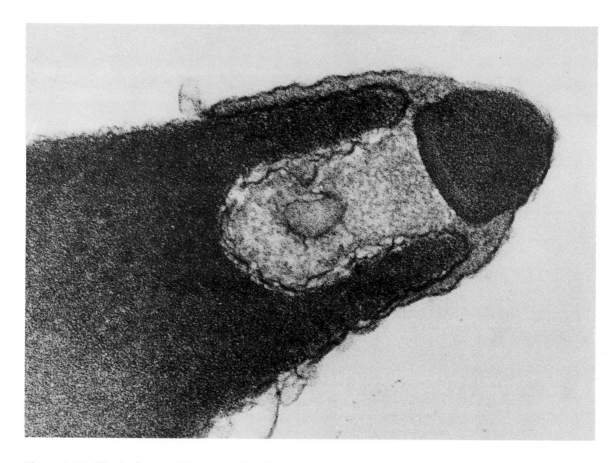

Figure 2–18. Head of sea urchin sperm showing acrosomal granule at tip. Courtesy of Victor Vacquier.

appear to chemically attract male gametes over long distances. Instead, most of the movement of animal sperm appears to be random and the key aspect of sperm-egg recognition appears to be the specific sticking of sperm to egg. What is the nature of the molecules involved in this important process of sperm-egg recognition? To answer this question we must consider two processes: the attachment of sperm to the outer egg coats, and the recognition of the underlying egg plasma membrane by the sperm.

The jelly coat of sea urchin eggs contains a substance named by Lillie fertilizin. This substance has been found to be an acid mucopolysaccharide (a protein containing polysaccharide and sulfate ions). Fertilizin appears to bind sperm. If fertilizin is acid extracted from sea urchin egg jelly coats, it will clump sperm, suggesting that fertilizin is a molecule with several receptor sites for the sperm. Fertilizin, or a molecule associated with it, also appears to activate sperm by increasing their respiration and motility and by causing the acrosomal reaction to occur.

fertilizin

Figure 2–19. Isolated sea urchin sperm acrosomal granules. Courtesy of Victor Vacquier.

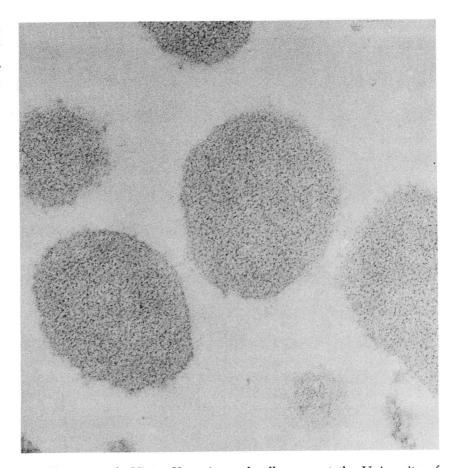

bindin

Very recently Victor Vacquier and colleagues at the University of California at Davis have isolated a protein from the acrosomal granules of sea urchin *sperm* that appears to mediate species specific recognition of sperm to the egg surface (Figures 2–18 and 2–19). This protein, termed bindin, only agglutinates (causes to clump together) eggs from the same species of sea urchin. Eggs whose surfaces have been treated with carbohydrate-destroying agents are not agglutinated with bindin. This suggests that bindin may recognize specific carbohydrate portions of egg surface glycoproteins. Vacquier has also isolated a molecule, which appears to be glycoprotein in nature, from the *egg* surface. This molecule has a species-specific affinity for bindin and may be the egg surface receptor site for bindin.

Many questions still remain unanswered. Is the bindin receptor site on the egg surface fertilizin? What is the relationship between bindin and a molecule on the sperm cell surface that binds to fertilizin, postulated to exist long ago and termed antifertilizin? What are the functional

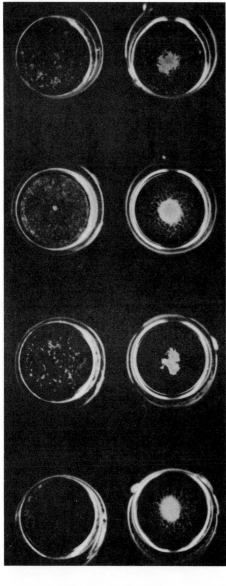

S. purpuratus eggs   +
S. purpuratus bindin

S. purpuratus eggs   +
S. franciscanus bindin

S. franciscanus eggs   +
S. franciscanus bindin

S. franciscanus eggs   +
S. purpuratus bindin

Figure 2–20. Specificity of sea urchin sperm bindin in agglutinating sea urchin eggs. Figure shows that the best agglutination of eggs occurs with bindin isolated from sperm of the same species. Courtesy of Victor Vacquier.

groups of bindin and of the egg surface bindin receptor sites that are responsible for interaction of these molecules? The very preliminary evidence cited above suggests that perhaps we are dealing with the interaction of protein (bindin) with sugar portions of the egg surface receptor sites (Figure 2–22). As will be seen later on, protein-carbohydrate interaction appears to play an important role in many cell-cell recognition phenomena other than sperm-egg interaction.

cell-cell recognition

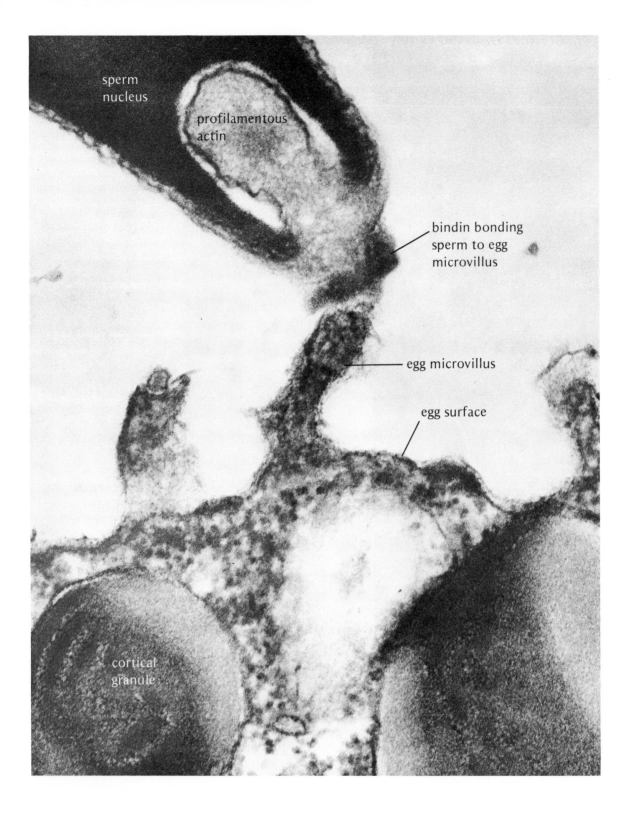

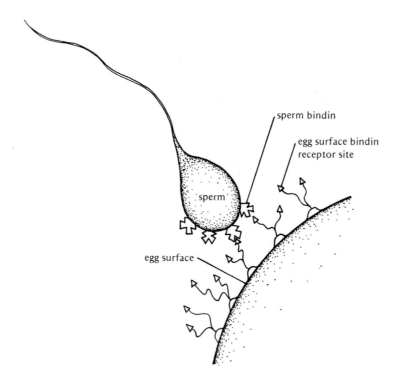

sperm bindin

egg surface bindin
receptor site

sperm

egg surface

Figure 2–22.
Hypothetical model of
the molecular basis of
sperm-egg recognition.
The lock and key
scheme may involve a
protein (bindin) on the
sperm interacting with
sugar groups on the egg
surface. Size
relationships of these
hypothetical molecules
are greatly exaggerated.

Summary

The fertilization process is important to us all. It marks the beginning of
the new organism. It provides for a mixing of genetic information from
the mother and father to form a new, unique being. In this chapter we
began to get a glimpse of the complexities involved in fertilization. We
examined how the sperm penetrates the outer egg coats and finally how
it fuses with the egg plasma membrane. We observed the egg cortical
reaction and how it often provides a block to polyspermy. At the molecular level, we noted that many biochemical reactions become activated
upon fertilization and that events such as release of ions and change in
cellular pH may trigger the occurrence of some of the other events.
Finally we looked at the molecular basis of sperm-egg recognition and

Figure 2–21 (opposite). Sea urchin sperm bound to egg microvillus by bindin.
Courtesy of Victor Vacquier.

examined molecules such as fertilizin and bindin that have been implicated in mediating gamete contact. We began to get a glimpse of some of the answers to the many questions involved in the fertilization reaction. The exploration of fertilization, however, has barely begun. It is likely that exciting new work during the next several years will provide us with a much clearer view of the presently little understood aspects of the fertilization reaction. Perhaps the question posed at the beginning of the chapter: "How can a tiny sperm that fuses with only 0.0002 percent of the egg surface trigger the multitude of changes that occur in the new zygote?" will be fully answered during the next few years.

Readings

Key Recent Review

Epel, D., The Program of Fertilization, *Scientific American* 237:129–139 (1977). This recent review is unusually stimulating because it describes possible causal relationships among the events of fertilization and provides up-to-date evidence for these relationships.

Selected Additional Readings

Carroll, E. J., and D. Epel, Isolation and Biological Activity of the Proteases Released by Sea Urchin Eggs Following Fertilization. *Develop. Biol.*, 44:22–32 (1975).

Colwin, L. H., and A.L. Colwin, Role of Gamete Membranes in Fertilization, 22nd Symposium of The Society for the Study of Developmental Growth. Academic Press, Chicago (1964).

Mazia, D., The Release of Calcium in Arbacea Eggs upon Fertilization. *J. Cell Comp. Physiol.* 10:291–304 (1937).

Metz, C. B., and A. Monroy, *Fertilization*. Chicago, Academic Press (1967).

Singer, S. V., and G. L. Nicolson, The Fluid Mosaic Model of the Structure of the Cell Membrane. *Science* 175:720–731 (1972).

Steinhardt, R. A. and D. Epel, Activation of Sea Urchin Eggs by Calcium Ionophore, *Proc. Natl. Acad. Sci. U.S.* 71:1915 (1974).

Tegner, M. J., and D. Epel, Sea Urchin Sperm-Egg Interactions Studied with Scanning Electron Microscopy. *Science* 179:685–688 (1973).

Tyler, A., Gametogenesis, Fertilization and Parthenogenesis. In B. H. Willier, P. Weiss, and V. Hamburger (eds.), *Analysis of Development*, Saunders, 1955.

Ursprung, H., and E. Schabtach, Fertilization in the Tunicate: Loss of the Paternal Mitochondrion Prior to Sperm Entry. *J. Exp. Zool.* 159:379–384 (1965).

Vacquier, V. D., and G. W. Moy, Isolation of Bindin: The Protein Responsible for Adhesion of Sperm to Sea Urchin Eggs. *Proc. Natl. Acad. Sci. U.S.* 74:2456–2460 (1977).

CHAPTER 3

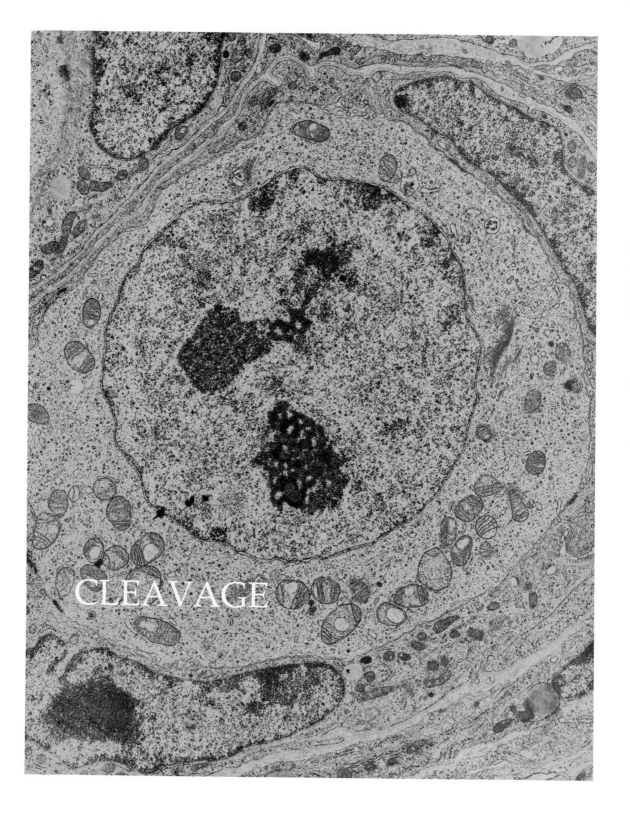

CLEAVAGE

cleavage

blastomeres

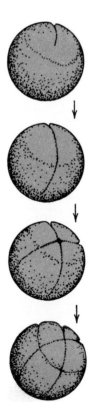

FERTILIZATION HAS OCCURRED. The zygote has become "activated". Many synthetic reactions have begun. What next? When we think of an embryo we think of a multicellular organism. At this point all that exists is a single cell, the fertilized egg. The process of division of the fertilized egg, called cleavage, transforms the single cell into many cells, called blastomeres in early embryos. Cleavage is a visually dramatic process and can easily be observed in many embryos in the laboratory. Such observations lead to some interesting conclusions. One of these is that the pattern of cleavage is not the same in all organisms. In this chapter we will briefly examine some of the different ways in which embryos cleave. In addition, we'll look at some of the mechanisms and factors that appear to influence cleavage patterns and the cleavage process itself. By the end of this chapter our embryos will be multicellular and ready to twist and turn to form something that begins to resemble a "real" being.

Cleavage Patterns and Mechanisms

Let us look at various patterns of cleavage and describe some of the factors that appear to influence these patterns. We will select some representative embryos that illustrate the concepts of interest.

Figure 3–1. Cleavage in frog.

Total Cleavage

Total or holoblastic cleavage simply means that the divisions pass through the entire fertilized egg. This type of cleavage occurs in embryos with a small or moderate amount of yolk, such as sea urchin, *Amphioxus*, frog, and most mammalian embryos. Dense accumulations of yolk retard cleavages.

holoblastic (total) cleavage

Frog Embryo and the Influence of Yolk on Cleavage. The influence of yolk on cleavage is nicely shown in the frog egg, which is moderately telolecithal. That is, yolk is substantially more concentrated in the vegetal hemisphere of the egg than in the animal hemisphere. The animal hemisphere is defined as that region of the egg where the nucleus resides. The vegetal hemisphere is the other half. When eggs are free to rotate, the animal hemisphere is usually up. The more dense, more yolky vegetal region is down. By peering through a low power lens, we can easily see that the first cleavage division in the frog begins in the pigmented animal hemisphere and slowly moves down vertically (a meridonal cleavage) through the more yolky vegetal hemisphere. The yolk appears to retard movement of the cleavage through the vegetal region. Before the first division is completed, the second division often begins in the animal region, also vertically, at right angles to the first cleavage (Figure 3–1). So, yolk retards the movement of the division plane into the vegetal region. The first two cleavages are total, however, and eventually they are able to divide completely through the yolky vegetal area. The third cleavage in the frog embryo is horizontal (equatorial). This cleavage does not separate the embryo exactly in half. Instead it is displaced towards the animal region so that the upper cells formed are smaller than the lower, more yolky cells. Thus, we can classify the cleavage pattern of the frog as total and unequal. Continued division results in more numerous, smaller, animal cells and larger, yolky, vegetal cells (Figure 3–1).

moderately telolecithal egg

animal hemisphere

vegetal hemisphere

unequal cleavage

Yolk obviously plays an important role in frog cleavage. The nucleus lies at the top of the dense yolk. Since the mitotic spindle forms in the area of the nucleus, the cleavage furrows start at the animal pole region and move toward the vegetal area, being retarded by the dense yolk. The third cleavage also is influenced by the yolk because of the position of the blastomere nuclei. The position is above the equator, above the concentrated yolky cytoplasm. So, since the mitotic spindles form in conjunction with the nuclei, they are displaced above the equator. In this way yolk directly causes the third cleavage to occur closer to the animal pole of the blastomeres. So now, smaller animal cells (micromeres) are separated from larger vegetal cells (macromeres). When the animal and vegetal cells separate, the nuclei divide and the vegetal nuclei sit atop the dense yolk of the vegetal cells. The smaller, less yolky, animal cells continue to cleave more rapidly than the yolky vegetal cells. In this way the frog embryo takes form with yolk playing a major role in determining the size of the cleavage blastomeres.

In vertebrates and certain other organisms (cephalopods, nematodes, and tunicates), at some time during cleavage, the arrangement of blastomeres becomes such that a plane of bilateral symmetry is established. In other words, a right and left side of the embryo that are mirror images of each other become established. When this occurs the pattern is called bilateral cleavage.

bilateral cleavage

Amphioxus. In the primitive chordate *Amphioxus*, which possesses a small amount of evenly distributed yolk (oligolecithal or isolecithal egg), the first three cleavages are much like those in the frog embryo. But probably because of the smaller amount of yolk, the third cleavage separates the animal and vegetal regions more equally. Thus, the animal and vegetal cells formed are nearly equal in size, with the vegetal blastomeres being only slightly larger.

oligolecithal egg

isolecithal egg

Sea Cucumber. In many other embryos, such as the sea cucumber, that have only a small amount of evenly distributed yolk (oligolecithal egg), cleavages are total and also equal. As shown in Figure 3–2, in the sea cucumber the first cleavage is vertical, and the second cleavage is also vertical at right angles to the first. The third cleavage is horizontal. Unlike the frog embryo, note that the third cleavage in the sea cucumber separates the embryo into upper and lower cells that are about equal in size. The upper cells lie exactly over the lower cells. Such an arrangement is called radial cleavage because the pattern of cells is radially symmetrical around the polar axis of the egg. These patterns appear to be influenced, in part, by yolk distribution.

radial cleavage

Figure 3–2. Radial cleavage in the sea cucumber.

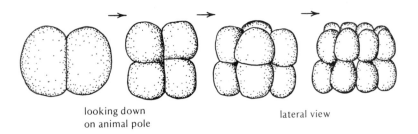

looking down
on animal pole

lateral view

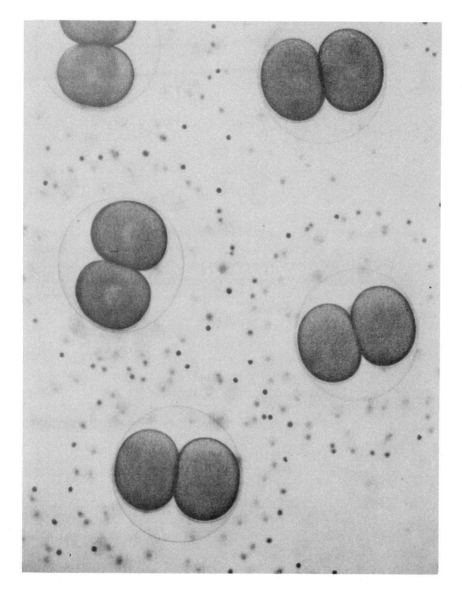

Figure 3–3. First cleavage in the sea urchin. Courtesy of Victor Vacquier.

Unequal Cleavage in the Sea Urchin Embryo. The sea urchin, which also exhibits a radial cleavage pattern and has an oligolecithal egg, goes along cleaving rather routinely and then gives us a real surprise. The first three cleavages are similar to those described previously. As shown in Figure 3–3, the first is vertical. The second cleavage is vertical at right angles to the first, and the third is equatorial. Up to now, the cleavage pattern in the sea urchin has been similar to that in *Amphioxus* and the

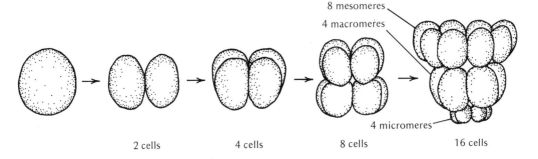

Figure 3–4. Cleavage in the sea urchin.

mesomeres

macromeres

micromeres

sea cucumber. The fourth cleavage in the sea urchin, however, is the big surprise. The four animal cells divide vertically and equally (forming eight middle sized cells called mesomeres). The four vegetal cells, however, cleave horizontally and unequally producing four larger cells (macromeres) and four tiny cells (micromeres) at the extreme tip of the vegetal region (vegetal pole) as shown in Figures 3–4 and 3–5. The 16-cell stage, therefore, consists of eight animal mesomeres, four large vegetal macromeres, and four tiny vegetal micromeres.

The mechanisms that cause the unequal fourth cleavage in the sea urchin embryo are not well understood. Ikeda has found that during cleavage cycles in the sea urchin embryo, the quantity of certain proteins that contain sulfhydryl groups varies in a cyclic way. A variety of experiments suggest that these so called sulfhydryl cycles may, in part, control cleavage. For example, if sulfhydryl cycles are blocked with ether and the embryos are then returned to ether-free sea water, the fourth cleavage is not unequal and micromeres are not formed. Also, a cleavage can be blocked with ultraviolet light or with the metabolic inhibitor, 2, 4-dinitrophenol. These treatments do not inhibit the embryo's sulfhydryl cycle, but have caused embryos to lose a cleavage. When the third cleavage occurs in these treated embryos, it is unequal, producing micromeres at the vegetal pole. Since the sulfhydryl cycles are not blocked, at the time of the third cleavage in the treated embryos, the fourth sulfhydryl cycle occurs. Thus, perhaps the fourth sulfhydryl cycle somehow causes an unequal cleavage. Future work will undoubtedly shed more light upon this problem.

The causes of the specific cleavage patterns are, as seen, not well understood. The fourth cleavage in the sea urchin embryo does, however, uncover another interesting problem. The eight medium sized

ectoderm

endoderm

mesomeres give rise to most of the ectoderm or outer layer of the embryo. The four large macromeres give rise to some ectoderm and all of the endoderm, the innermost layer of the embryo. The four small micro-

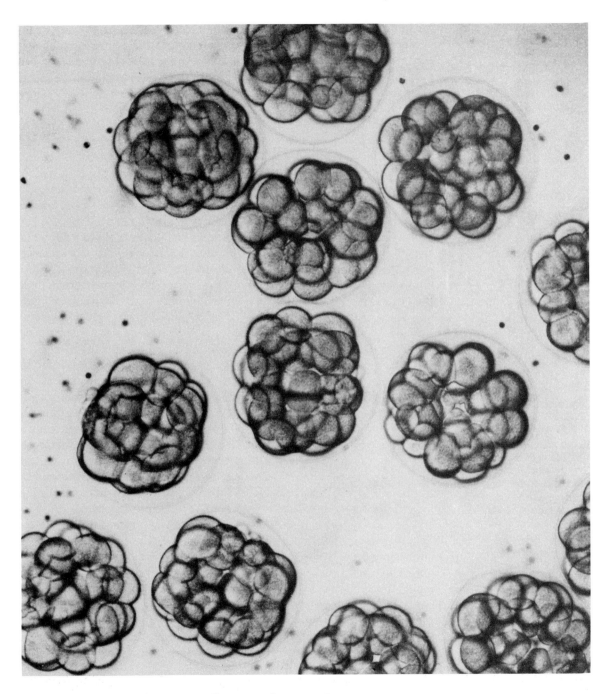

Figure 3–5. Sixteen to thirty-two–cell stage in the sea urchin. Courtesy of Victor Vacquier.

primary
mesenchyme cells

meres become the primary mesenchyme cells that form the skeletal elements of the embryo. The size differences of these cell types makes it possible to determine the fates of these cells by watching what each cell type forms. Also, in some species, some of the cell types contain pigment granules that aid the observer in following the cells during development.

At the 16-cell stage, the mesomeres, macromeres and micromeres can be separated from one another and then recombined in specific ways. These experiments show that combinations of mesomeres and micromeres can give rise to a normal embryo. Mesomeres plus macromeres can form a nearly normal embryo. Mesomeres or micromeres or macromeres alone do not develop into a normal embryo. These experiments by Hörstadius indicate that in order to form a normal embryo, one must combine some cells from the animal half of the embryo with some cells from the vegetal half of the embryo. These results have been interpreted to suggest that two basic kinds of cells might exist in the embryo. One is most concentrated in the animal half of the embryo; the other in the vegetal region. In order for a normal embryo to develop, it has been suggested that some animal "stuff" and some vegetal "stuff" must be present.

What is the nature of the animal and vegetal "stuff"? This is the $100,000 question. Before we produce the illusion that we really understand what is going on here, let's flatly say—who knows what the "stuff" is. Many interesting experiments have been done, however, that shed a little light upon this problem. It is unlikely that the gradients involve cytoplasmic materials that can easily be displaced by centrifugation. This conclusion is based upon experiments in which the egg cytoplasm is rearranged by centrifugation. The blastomeres then contain different materials than they would have under normal conditions. A normal embryo can develop from "mixed up" cytoplasm. Many other experiments have been done with similar inconclusive results.

A variety of investigations suggest that important differences among mesomeres, macromeres, and micromeres may reside in the surface region of the cells. The surface region includes the cell membrane and the cortex, the region just below the cell membrane. Experiments using the large squid egg suggest that the cell surface may indeed contain key information that can control development. These experiments by Arnold involved use of an ultraviolet light microbeam to irradiate small regions of the egg surface. These treatments wiped out formation of specific organs or tissues that normally would form at each irradiated site.

But what about the sea urchin embryo? Is there any indication that the surfaces of micromeres, mesomeres, and macromeres are different? The answer to this question is yes. Roberson, Neri and Oppenheimer recently showed, in an article in the journal *Science,* that there are differ-

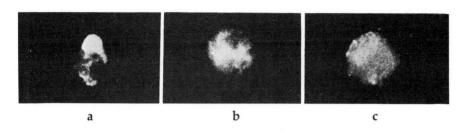

a   b   c

Figure 3–6. Cell surface differences of sea urchin micromere (a), mesomere (b), and macromere (c). Only the micromere exhibits mobile cell surface fluorescent concanavalin A receptor sites This is shown by the accumulation of fluorescence to one tip of the micromere. Other experiments have shown that fluorescence moves from the cell body to the tip. From M. Roberson, A. Neri, and S.B. Oppenheimer, *Science* 189:639 (1975).

ences in the surface properties of micromeres, and those of macromeres and mesomeres. These workers used a lectin (a carbohydrate binding protein—see Chapter 9) called concanavalin A to study the surfaces of micromeres, mesomeres, and macromeres. Concanavalin A binds to certain sugar-containing receptor sites on the cell surface. This binding can be visualized by coupling a flourescent dye to the concanavalin A. In this way, if one adds the fluorescent concanavalin A to the cells, one can, with a fluorescence microscope, observe the binding of this material to the cell surface as a fluorescent glow in the region of binding. Using this technique, it was shown that only on the micromere cell surface did the fluorescence move to one pole of the cell. No movement of fluorescence was observed on the macromeres and mesomeres (Figure 3–6). These results suggest that certain sugar-containing sites on the micromere cell surface can move to accumulate at one end of the cell. On the macromeres and mesomeres, the sites do not move and are more rigidly embedded in the cell surface. What does all this mean? It simply suggests that the surfaces of micromeres are clearly different from the surfaces of macromeres and mesomeres. Whether or not these surface differences are important in helping the cell types achieve their fates remains to be determined. It is clear, however, that micromeres do become migratory mesenchyme cells that play an important role in sea urchin embryonic development (Chapter 4). Perhaps their "mobile" surfaces assist these cells in migration.

Spiral Cleavage and Special Cytoplasm. Many embryos, such as those of annelids, mollusks, some flatworms, and nemereans, exhibit a spiral cleavage pattern. In this type of cleavage, as a result of oblique orientations of the mitotic spindle, daughter cells do not lie directly above or below one another but instead are shifted so that each upper cell lies over the junction of two lower cells as in Figure 3–7.

There are two important developmental concepts that can be learned from studying spirally cleaving embryos. They will be developed

spiral cleavage

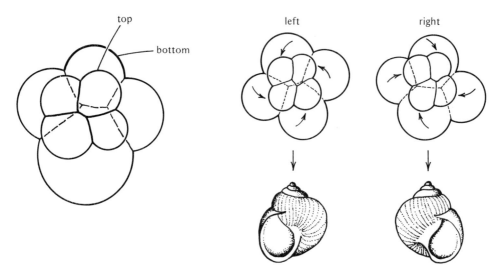

Figure 3–7 (left). Diagram of spiral cleavage in the mollusc *Unio*. Each top cell is directly above the junction of two lower cells. After Lillie, in Kellicott, *General Embryology*, 1914.

Figure 3–8 (right). Diagram of left-handed (counterclockwise) and right-handed (clockwise) spiral cleavage and shell coiling in the snail. After Conklin, in T.H. Morgan, *Experimental Embryology*, Columbia University Press, 1927.

more fully in other sections of the text. It would be useful, however, to briefly consider them here in the context of cleavage.

The first concept deals with causes of the direction of cleavage. What causes spiral cleavages to be left-handed or right-handed? In other words, when a cleavage occurs the top cell will form to the left or right of the bottom cells. What causes the different spindle orientations that result in left or right-handed cleavages? A partial answer to this question comes from a study of the snail *Limnaea*. These snails have their shells and organs coiled in either a left or right-handed manner. Such coiling results from whether early cleavages were left-handed (counterclockwise) or right-handed (clockwise) as shown in Figure 3–8. A single gene appears to determine whether the cleavage is left or right-handed. Right-handedness is dominant over left-handedness. So, you would expect if you mate a pure left-handed coiled (and cleaving) female with a pure right handed coiled (and cleaving) male, all offspring would possess right-handed cleavage and coiling because it is the dominant characteristic. Instead, just the opposite occurs. All first generation offspring exhibit left-handed cleavage and coiling. The control of cleavage pattern appears to reside in the genes of the mother's body. If the mother has pure genes for left-handed cleavages, her body appears to produce some sort of substance that gets into the cytoplasm of the oocytes, causing the

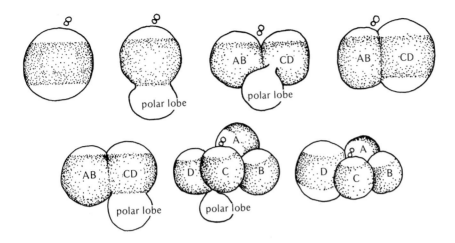

Figure 3–9. Cleavage of the mollusc *Dentalium.* After E.B. Wilson, *J. Exp. Zool.* 1: 1–72, 1904.

fertilized egg to cleave in the left-handed manner. If the mother has mixed right and left genes, her offspring cleave in a right-handed spiral because right is dominant, and the maternal right factor supercedes the genetic characteristics of the offspring. Thus, the genes of the mother form a cytoplasmic substance for the oocytes that will later control their cleavage pattern. The genes of the fertilized egg do not seem to be able to control the pattern. It is the genes of the mother that, in this case, appear to be of major importance in producing the substance that controls the direction of cleavage spindle orientation. Thus, it is seen that cytoplasmic substances appear to be of major importance in controlling a significant developmental event.

What is the second important concept that can be learned from studying spirally cleaving embryos? Not only do cytoplasmic substances appear to be important in controlling the pattern of cleavage, but they also play a key role in controlling the fate of the blastomeres. This is nicely shown in the spirally cleaving embryo of the mollusc *Dentalium.* The cytoplasm of the egg of *Dentalium* is divided into three very distinct regions: a clear region at the animal pole, a broad equatorial region of granular cytoplasm, and another clear region of cytoplasm at the vegetal pole. As shown in Figure 3–9, right before the first cleavage the clear vegetal cytoplasm becomes extruded into a polar lobe. As the first division becomes completed, the clear vegetal cytoplasm withdraws back into the embryo, but only into one of the two daughter cells (called the CD cell). The other cell (AB cell) only has two of the original cytoplasmic regions, while the CD cell has all three. At the second cleavage a similar thing occurs and the clear vegetal cytoplasm becomes segregated in only one of the four blastomeres (the D cell). The A, B, and C cells do not possess the clear vegetal material. Thus, certain cells receive certain cytoplasms during cleavage. One can separate the blastomeres at the two-cell

polar lobe

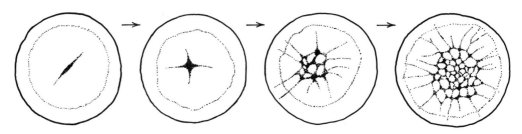

Figure 3–10. Incomplete cleavage of chick egg. Surface view. Looking down onto blastodisc at animal pole. (Modified from J.T. Patterson, *J. Morph.* 21: 101–134, 1910).

and four-cell stage. Only blastomeres containing the clear vegetal cytoplasm (CD cell or D cell) could give rise to a normal (yet smaller) larva. The other cells give rise to larvae without a middle germ layer (mesoderm layer). Thus it seems that the clear vegetal cytoplasm is important for mesoderm formation.

That the special polar lobe material is the important factor in mesoderm formation is supported by the following experiment. If the polar lobe, which contains only the special cytoplasm and no nucleus, is nipped off before the completion of the first or second cleavage, the remaining embryo forms a larva without mesoderm. Thus it appears that it is the cytoplasm (or surface region) and not the nucleus that makes blastomere D different and able to give rise to mesoderm.

There are many other examples of the importance of cytoplasmic materials in embryonic development. We will see later (in Chapter 8) that, in all probability, in all embryos, certain cytoplasms become segregated during cleavage and that these cytoplasms play a key role in determining the fate of cells. The example given above illustrates this principle because of the clear visual differences in the cytoplasmic regions. In many other embryos such cytoplasms may not be as dramatically visible and therefore cytoplasmic influence on development may not be as apparent.

Incomplete Cleavage

strongly telolecithal egg

In some organisms such as sharks and rays, bony fish, reptiles, and birds, the egg is strongly telolecithal. That is, the yolk is so concentrated in the vegetal region that all of the non-yolky cytoplasm floats atop the

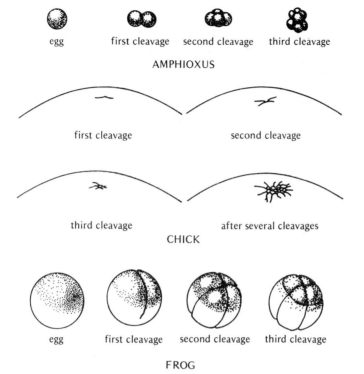

egg   first cleavage   second cleavage   third cleavage

AMPHIOXUS

first cleavage   second cleavage

third cleavage   after several cleavages

CHICK

egg   first cleavage   second cleavage   third cleavage

FROG

Figure 3–11. Comparison of *Amphioxus*, chick and frog cleavage.

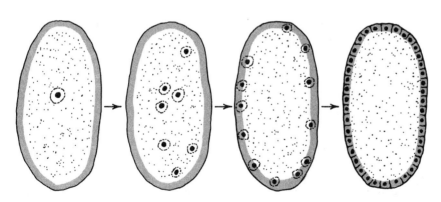

Figure 3–12. Superficial cleavage in centrolecithal egg (insect egg).

yolk as a little cap. We already saw that even a moderate amount of yolk retards cleavage. What happens here, where almost the entire egg is dense yolk? In this case, the cleavages divide up only the non-yolky cytoplasmic cap, and do not divide up the dense yolk at all. This is called incomplete cleavage. (See Figures 3–10, 3–11).

Insects and many other arthropods have centrolecithal eggs. These eggs have their yolk concentrated in the interior of the egg. The non-

centrolecithal egg

yolky cytoplasm forms a thin surface layer around the yolk, and a patch of non-yolky cytoplasm in the center of the egg. This cytoplasm contains the nucleus (Figure 3–12). In these eggs the nucleus divides before any cytoplasm divides. The divided nuclei surrounded by small amounts of central cytoplasm move outward toward the egg surface. The surface now consists of many nuclei in an uncleaved layer of cytoplasm. The cytoplasm becomes subdivided, with the compartments still connected to the central yolk. In time the cells separate from the yolk and use the yolk for nourishment. This type of incomplete cleavage is called superficial cleavage.

superficial cleavage

Thus we see that in eggs with a small or moderate amount of yolk, cleavage is usually total. In eggs with highly concentrated regions of yolk, cleavage occurs only in the non-yolky cytoplasm. Yolk, therefore, is one important factor in controlling the cleavage pattern, along with other factors such as cycles of certain proteins and other cytoplasmic factors that are genetically controlled.

Mechanics of Cleavage

Cleavage involves division of the cell cytoplasm and division of the nuclear components. What are the mechanics of these processes? A ring of microfilaments that are 30–70 angstroms in diameter, probably composed of the protein actin, has been observed just below the cell surface of many eggs. As discussed in detail in Chapter 12, these filaments are composed of globular subunits that can polymerize and depolymerize by alteration of cytoplasmic factors such as ionic strength. These filaments appear to form a contractile ring that causes the cytoplasm to separate in a way similar to pulling purse strings to decrease the size of the purse opening. A drug, cytochalasin B, that disrupts cytoplasmic microfilaments, also interferes with cytoplasmic division of cultured mammalian cells. If the drug is removed, cell division resumes as the microfilaments reappear. Such experiments suggest that microfilaments are involved in cytoplasmic division. It should be stressed, however, that the drug cytocholasin B also affects other cell processes such as protein synthesis, sugar uptake, respiration, etc. Thus, interpretation of drug studies must be made with caution. It may be that cytoplasmic division is controlled by microfilaments. Cytochalasin B studies, however, cannot be cited as definitive evidence for such a conclusion.

The spindle, or more correctly the asters, may initiate cytoplasmic division by producing some sort of diffusible factor that acts on the contractile ring. This has been shown by experiments in which asters were removed or separated. Asters must be present in the area where the cleavage furrow occurs. Asters, therefore, may set up conditions that are required for the organization of a constriction mechanism.

With respect to nuclear division, the mitotic spindle apparatus is composed mainly of tubule-like protein, tubulin A and B (see Chapter 12). This protein is present in the egg and blastomeres even when the mitotic spindle is not present. The protein appears to reside in the cytoplasm in subunit form. At the proper time for spindle formation, these subunits probably become stimulated to come together (polymerize) to form the visible mitotic apparatus. We will return to the interesting problem of polymerization of cellular organelles from pre-formed subunits in Chapter 12.

What are some of the metabolic events that occur during cleavage? Large amounts of DNA are synthesized to provide each blastomere with a set of chromosomes. Many proteins are also required for cleavage. Some of these proteins are stored in the oocyte and need not be synthesized, but others are synthesized during cleavage. Inhibition of protein synthesis with the drug puromycin brings cleavage to a complete halt. This suggests that protein synthesis is essential for cleavage to occur. We should keep in mind, however, that inhibitor studies should be approached with caution. Some inhibitors may also be interfering with other processes than those that are known to be affected by the inhibitor.

Large amounts of RNA synthesis, however, are not needed during early cleavage. RNA synthesis can be blocked with the drug actinomycin D. If embryos are treated in this way, protein synthesis and cleavage still occur. This experiment suggests that synthesis of RNA is not needed during early cleavage because most of the RNA required to synthesize proteins, ribosomes, etc., is stored in the oocyte. Again, it should be stressed that drug studies are difficult to interpret. It may be that in this case a tiny amount of RNA is synthesized in the presence of inhibitor. We will return to a more detailed look at such studies in Chapter 10.

Let us move on to the next stage of development. We now, through cleavage, have a ball or cap of cells. These embryos still do not resemble a 'real' being, as we would recognize it. In the next chapters we will see how the embryo twists and turns to form something that begins to resemble an organism. Let us introduce these events by briefly looking at how the cleaved cell mass readies itself for the events that will take place next.

Figure 3–13. Possible mechanism of blastocoel formation. Dividing cells push outward forming central cavity. After Gustafson and Wolpert, *Biol. Rev.* 42: 442–498, 1967

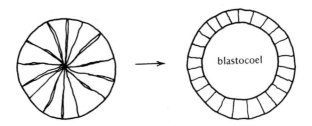

Blastula

morula

blastula

Many embryos become transformed from a solid ball (often termed a morula) or cap of cells into a hollow ball called a blastula. This transformation prepares the embryo for the numerous cellular rearrangements that shape the embryo in the next stage of development.

The forces involved in causing the solid ball of cells to become hollow in the sea urchin and similar embryos may include osmotic pressure exerted from the influx of water into the ball. A buildup of macromolecules occurs in the center of the ball. This causes an influx of water, and the pressure of the water may push the cells outward so that the central cavity enlarges. Other forces involved may include cell-cell adhesiveness and surface tension, and the fact that cells are attached to an outer elastic layer called the hyaline layer. If the cells are attached to each other as they divide, they will tend to push the elastic hyaline layer outward to form a central cavity (Figure 3–13).

blastocoel

Blastulas vary in appearance in different species. Many are hollow balls of cells with the cavity called the blastocoel located centrally. Such blastulas are often found in organisms with oligolecithal eggs such as the sea urchin and *Amphioxus* (Figures 3–14 and 3–15). In embryos with a more uneven yolk distribution, such as the frog, the blastocoel is displaced toward the animal pole (Figure 3–14).

In very yolky eggs where the embryo proper develops as a cap atop an uncleaved mass of yolk, as in birds, bony fish, and reptiles, a blastula-like stage also occurs. In these embryos a cavity does not form in the yolk

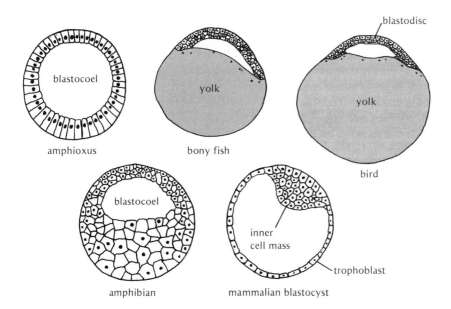

Figure 3–14.
Comparison of
blastulas.

at all. Instead, a cavity forms under the cap or disc of cells (blastodisc). **blastodisc**
Thus the cavity is located between the blastodisc and the uncleaved yolk.
Some researchers do not consider such a cavity a true blastocoel because
it lies outside (below) the embryo proper. Other cavities also appear
within the blastodisc (within the embryo proper) as described in
Chapter 4.

In organisms with centrolecithal eggs and superficial cleavage such
as insects, no cavity appears at all. The cells form at the periphery of the
embryo while the center remains filled with yolk. In a way, this is still a
blastula if you imagine that a central cavity does exist, but is filled with
yolk.

In most mammals, cleavage forms a solid ball of cells (morula). The
outer layer of this ball will form membranes that surround the embryo **inner cell mass**
proper, while the embryo proper itself forms from the inner part of the
morula (inner cell mass). The outer layer (trophoblast) lifts away from **trophoblast**
one side of the inner cell mass, forming a blastula-like stage called the
blastocyst (Figure 3–14). It is at the blastocyst stage that the human **blastocyst**
embryo implants itself in the wall of the uterus, where the embryo
remains and is nourished until birth. The mammalian embryo resembles
the chick embryo, but the mammalian blastocyst is not filled with any
yolk. Instead, nourishment is supplied by the mother. Since mammals
have evolved from yolky-egged ancestors, it is reasonable to assume that
at one time the blastocyst cavity was filled with yolk. Development of a
system of nourishing the embryo in the mother probably resulted in the
loss of yolk. The early stages of the embryo and blastocyst, however,

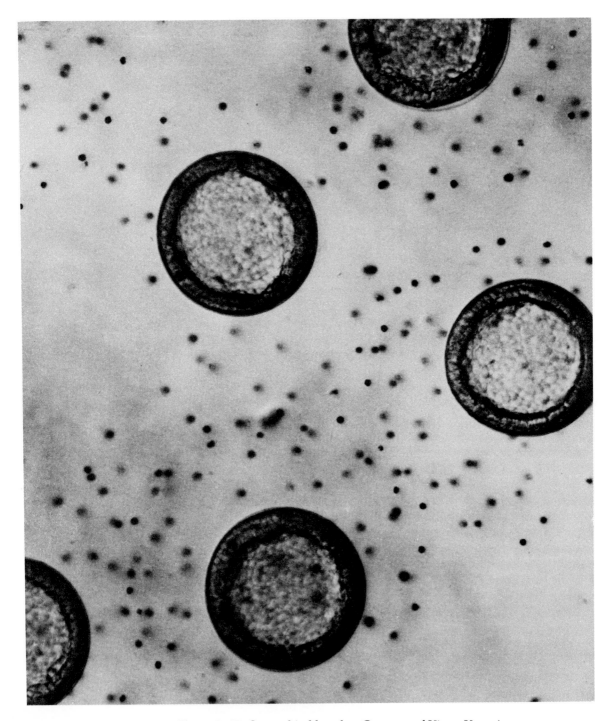

Figure 3–15. Sea urchin blastulas. Courtesy of Victor Vacquier.

probably did not need to change. In this way we can explain similarities of mammalian development to that of the reptile and bird.

We now have a ball or cap of cells. In the next chapter we will see how these rather nondescript embryos undergo a remarkable process that ends in the development of embryo forms that begin to resemble what we think of as "real" organisms. We will see how embryos become transformed by twists, turns, and cellular rearrangement.

Readings

Gustafson, T., and L. Wolpert, Cellular Movement and Contact in Sea Urchin Morphogenesis. *Biol. Rev.* 42:442–498 (1967).

Morgan, T. H., *Experimental Embryology.* Columbia University Press, New York (1927).

Neri, A., M. Roberson, D. T. Connolly, and S. B. Oppenheimer, Quantitative Evaluation of Concanavalin A Receptor Site Distributions on the Surfaces of Specific Populations of Embryonic Cells. *Nature* 258:342–344 (1975).

Patterson, J. T., Studies on the Early Development of the Hen's Egg. I. History of the Early Cleavage and Accessory Cleavage. *J. Morph.* 21:101–134 (1910).

Rappaport, R., Cleavage. In *Concepts of Development,* J. Lash and J. R. Whittaker, eds. Sinauer, Sunderland, Mass. (1974).

Roberson, M., A. Neri and S. B. Oppenheimer, Distribution of Concanavalin A Receptor Sites on Specific Populations of Embryonic Cells. *Science* 189:639–640 (1975).

Schroeder, T. E., Dynamics of the Contractile Ring. In *Molecules and Cell Movement,* S. Inoue and R. E. Stephens, eds., Raven Press, New York p. 304 (1975).

Wilson, E. B., Experimental Studies on Germinal Localization. I. The Germ Regions in the Egg of *Dentalium.* II. Experiments on the Cleavage-Mosaic in *Patella* and *Dentalium. J. Exp. Zool.* 1:1–72 (1904).

Wolpert, L., The Mechanics and Mechanism of Cleavage. *Intl. Rev. Cytol.* 10:164 (1960).

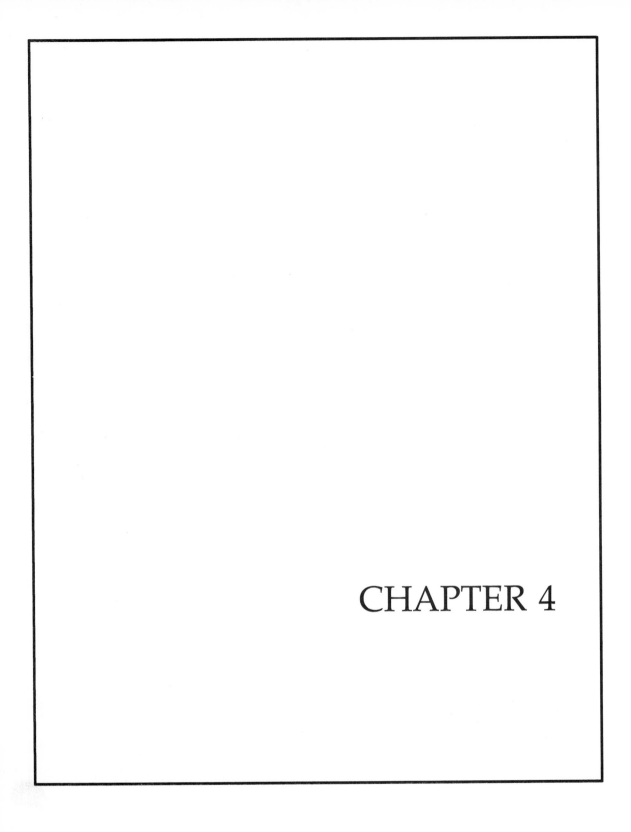

CHAPTER 4

GASTRULATION

WE HAVE REACHED the point in embryonic development at which the embryo is a ball or cap of cells and does not even slightly resemble the adult organism. How does an early conglomeration of cells become transformed into a layered embryo that begins to look like a familiar organism? Now we will examine the important process of gastrulation in which the hollow ball of cells, the blastula, becomes transformed into a layered embryo, the gastrula, by mechanisms such as cell movements and selective adhesion. Before we examine gastrulation in some representative systems let us briefly consider the topic of fate maps. These projections of how specific areas of early embryos will develop at later stages help us to better understand the movements that occur during gastrulation.

gastrulation

gastrula

fate maps

Fate Maps

Fate maps are constructed by marking specific areas on early embryos and following these marks to see where they finally reside in later stages. The technique of marking embryos was developed by Vogt. By pressing agar impregnated with vital stains such as Nile blue sulfate or neutral red against specific areas on embryos such as amphibian blastulas, Vogt was able to mark various regions. By observing the fate of these marks in the older embryos, Vogt was able to construct the fate maps which showed the fate of each embryonic region. Fate maps of *Amphioxus*, amphibian, and bird embryos are given in Figure 4–1. In the fate maps, and also in the gastrula-stage embryos that will be described, the term *prospective* (or *presumptive*) is used to denote what the specific region will become in the more advanced embryo.

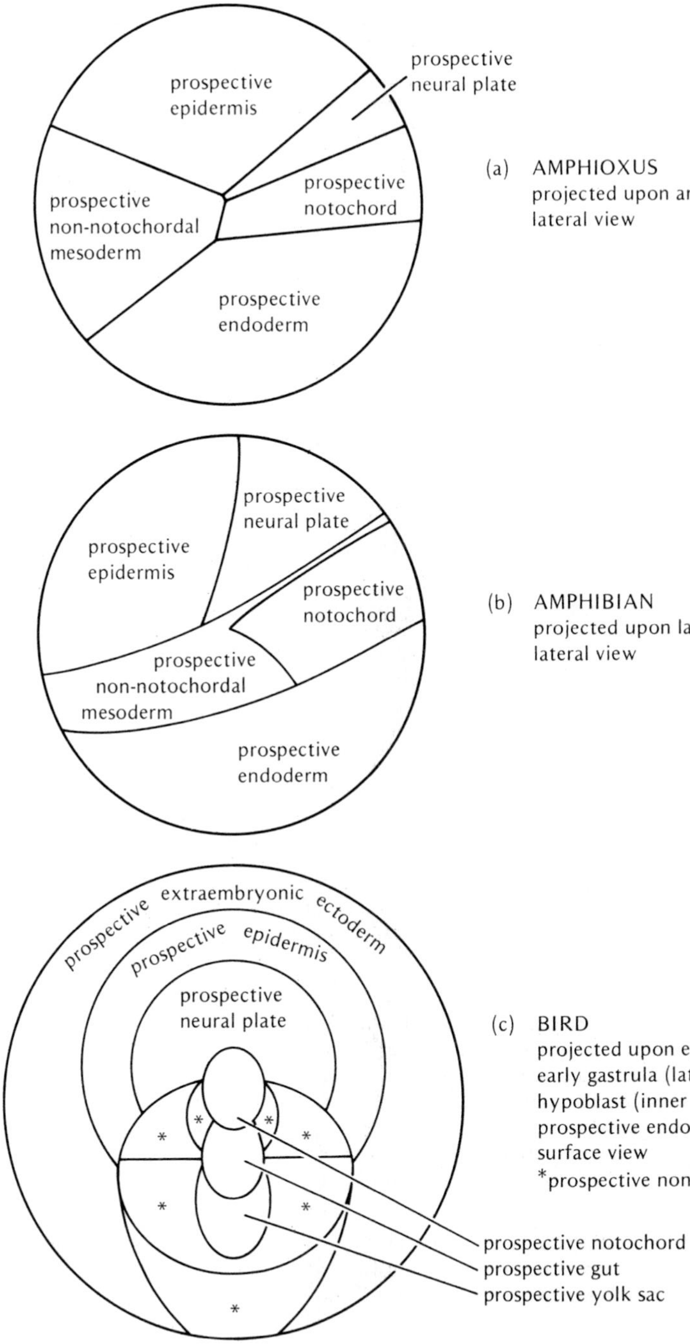

Figure 4–1. Fate maps of major areas in three embryos. Diagrammatic.

(a)   AMPHIOXUS
projected upon an uncleaved zygote
lateral view

(b)   AMPHIBIAN
projected upon late blastula
lateral view

(c)   BIRD
projected upon epiblast (top) layer of
early gastrula (late blastula). The
hypoblast (inner layer) is mostly
prospective endoderm.
surface view
*prospective non-notochordal mesoderm

prospective notochord
prospective gut
prospective yolk sac

Figure 4–2. Gastrulation in the sea urchin. After T. Gustafson, *Experimental Cell Research*, 32, Academic Press.

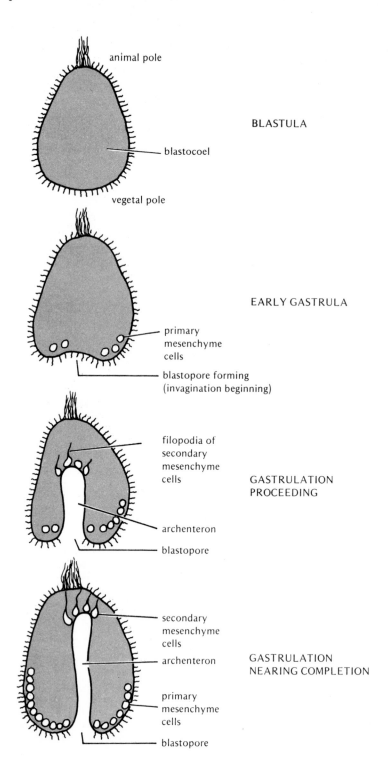

animal pole

blastocoel

vegetal pole

BLASTULA

primary mesenchyme cells

blastopore forming (invagination beginning)

EARLY GASTRULA

filopodia of secondary mesenchyme cells

archenteron

blastopore

GASTRULATION PROCEEDING

secondary mesenchyme cells

archenteron

primary mesenchyme cells

blastopore

GASTRULATION NEARING COMPLETION

Amphioxus and Amphibian Fate Maps

The fate maps in Figure 4–1 of *Amphioxus* and the amphibian are similar.
As can be seen, the animal region of each embryo gives rise to epidermis
and the neural plate (which becomes the nervous system and associated
structures). The prospective epidermis and prospective neural plate will
make up the outer layer of the gastrula, the ectoderm. The middle or
marginal zone of each embryo gives rise to the middle region of the
gastrula, the mesoderm, which is composed of prospective notochord
and prospective non-notochordal mesoderm. The notochord is a rod that
develops below the neural tube, that may help support the developing
embryo. It usually disappears in the adult. The non-notochordal mesoderm develops into many structures including limbs, heart, muscles,
kidneys, and gonads. The vegetal region of the *Amphioxus* and amphibian embryos gives rise to the endoderm, the inner embryonic layer, that
forms the gut and derivatives of the gut tube. The structures that develop from these regions will be described in detail in Chapters 6
and 7.

epidermis

neural plate

ectoderm

mesoderm

notochord

endoderm

Bird Fate Map

The bird fate map shown in Figure 4–1 is different from the other two.
This is because, as described in the previous chapter, the bird blastula is
not a hollow ball but instead is a layered cap sitting atop the yolk. The
fate map of the top (epiblast) layer is shown in Figure 4–1. As can be
seen, this layer gives rise to the extraembryonic ectoderm, the membranes that are outside the embryo itself; the epidermis; the neural plate;
notochord and non-notochordal mesoderm; the yolk sac; and the gut
endoderm. The lower (hypoblast) region gives rise to endoderm and
some of the notochord. Thus, epiblast and hypoblast are not at all equivalent to ectoderm and endoderm. Details of the fate of these regions are
given in Chapters 5, 6 and 7.

Fate maps describe the fate of specific regions of the early embryo.
Now we will see how these specific regions begin to get to their respec-
tive destinations to achieve their fate. Gastrulation is the important
process that facilitates these key cellular rearrangements.

Figure 4–3. Early sea urchin gastrula. Courtesy of Victor Vacquier.

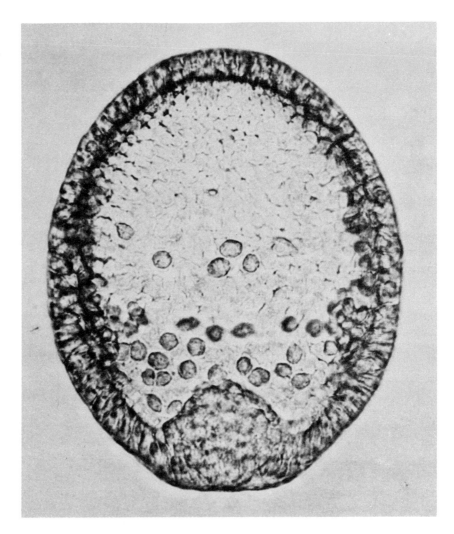

Gastrulation in the Sea Urchin

In previous chapters we have noted that the sea urchin embryo is ideal for a variety of studies in developmental biology. Because of the ease of maintaining these embryos in the laboratory, and the ease of obtaining large numbers of them that develop beautifully under the microscope, sea urchin embryos have been widely used to study the gastrulation process. The embryos are relatively transparent, allowing the observer to easily analyze the components of the gastrulation process. Thus we will examine this process in the sea urchin and then move on to looking at gastrulation in other interesting organisms that accomplish the same end by somewhat different mechanisms.

Sequence of Events

Figure 4–2 shows the sequence of events occurring during gastrulation in the sea urchin, as studied by time lapse cinematography. The sea urchin egg, as described previously, is isolecithal, having a rather even distribution of a relatively small amount of yolk. There is not, therefore, an excessive quantity of yolk that could interfere with the buckling in or invagination of the blastula that is seen to occur in Figure 4–2. Gastru- **invagination** lation in the sea urchin begins with cellular shape changes that occur in the vegetal plate. Some of the cells lose adhesion with their neighbors and are forced into the blastocoel, as shown in Figure 4–3. These are the **primary mesenchyme** primary mesenchyme cells. As a partial result of this loss of vegetal plate **cells** cells, the vegetal area begins to buckle in or indent. This indentation forms a cavity or tube in the gastrula, the archenteron, or primitive gut. **archenteron** The opening of this tube to the exterior, at the vegetal end, is the blas- topore. The tip of the archenteron advances toward the animal pole. At **blastopore** the tip of this advancing tube, cells extend projections called filopodia. The cells that extend these filopodia are the secondary mesenchyme **filopodia** cells, that become the mesoderm. The tips of the filopodia appear to "feel" the inner surface of the gastrula and finally stick to the animal **secondary** end. They then contract, pulling the archenteron tube toward the animal **mesenchyme cells** end (see Figure 4–4).

Mechanisms of Gastrulation

What are the mechanisms involved in gastrulation in the sea urchin? Cell movement or cell motility obviously is one important mechanism, as **cell motility** seen from mesenchyme cell activity. A second important gastrulation

selective adhesiveness

contractility

mechanism in this system is selective adhesiveness. The primary mesenchyme cells were seen to lose adhesiveness with their neighbors in the vegetal plate while the secondary mesenchyme cells appeared to preferentially stick (via their filopodia) to the animal end. Finally, contractility of the filopodia seemed to play a role in pulling the gut tube towards the animal pole. Gastrulation has transformed the sea urchin blastula, a hollow ball of cells, into a gastrula with a gut tube, which begins to resemble the body of the adult sea urchin. We see that during gastrulation, mechanisms such as cell motility, selective adhesiveness and contractility appear to control the cellular rearrangements needed for the embryo to begin to take the form of the adult. We will examine these mechanisms in detail in Chapter 9.

Figure 4–4. Sea urchin gastrula, gut forming. Courtesy of Victor Vacquier.

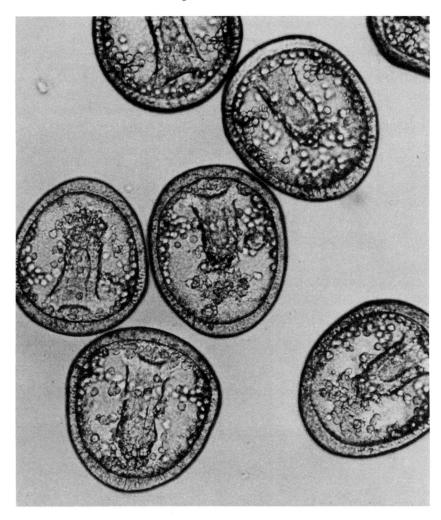

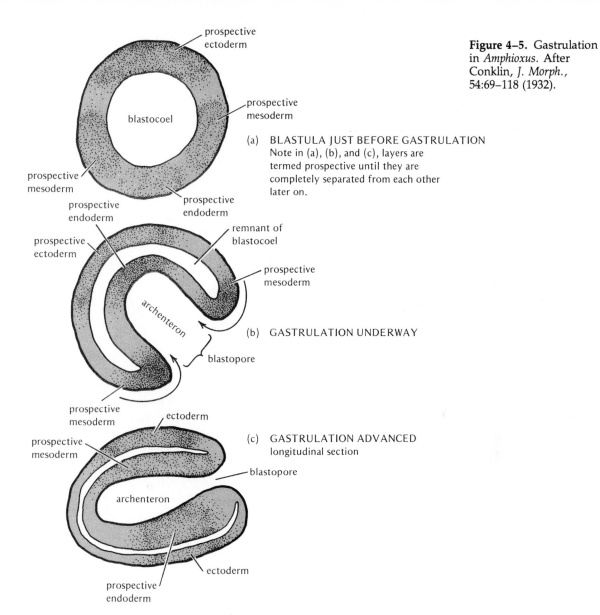

Figure 4–5. Gastrulation in *Amphioxus*. After Conklin, *J. Morph.*, 54:69–118 (1932).

(a) BLASTULA JUST BEFORE GASTRULATION
Note in (a), (b), and (c), layers are termed prospective until they are completely separated from each other later on.

(b) GASTRULATION UNDERWAY

(c) GASTRULATION ADVANCED
longitudinal section

Gastrulation in *Amphioxus*

Like the sea urchin egg, the *Amphioxus* egg is isolecithal. There is no excessive amount of yolk to hinder invagination. Let us examine gastrulation in *Amphioxus* and compare the process with that of the sea urchin.

Gastrulation in *Amphioxus* also occurs by invagination but not with the aid of secondary mesenchyme cells. Instead, as can be seen in Figure 4–5, the vegetal area flattens and bends inward or invaginates. The embryo begins to resemble a punched-in ball. The outer portion of the

Figure 4–6. Cross
section through
Amphioxus gastrula.
Note at this point the
mesoderm and
endoderm are not yet
separated. Thus the
inner layer is termed
mesendoderm. The outer
layer is ectoderm.
Modified from Conklin,
J. Morph., 54: 69–118
(1932).

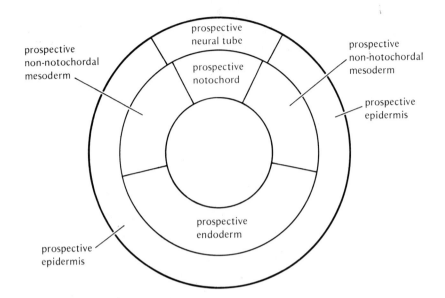

Figure 4–6. Cross section through *Amphioxus* gastrula. Note at this point the mesoderm and endoderm are not yet separated. Thus the inner layer is termed *mesendoderm*. The outer layer is ectoderm. Modified from Conklin, *J. Morph.*, 54: 69–118 (1932).

embryo consists of prospective epidermis and prospective neural
system. Together these make up the ectoderm. The inner part of the cup,
where your fist would touch as the ball is pushed in, consists mainly of
the endoderm, the prospective gut and gut derivatives. The mesoderm
(prospective notochord and non-notochordal mesoderm) moves into the
cup, from the rim into the pocket.

Like the sea urchin, the *Amphioxus* gastrula now possesses a primitive gut, or archenteron. The opening of the archenteron to the outside is
the blastopore. After examining gastrulation in the amphibian and bird
embryo, we will come back to *Amphioxus* to investigate the next phases of
development, namely, separation of mesoderm and endoderm and
development of the nerve tube (see Figure 4–6).

Gastulation in Amphibian

We have seen that both the sea urchin and *Amphioxus* embryos possess a
relatively even distribution of a small amount of yolk. We also noted in
the previous chapter that the amphibian egg (and embryo) is moderately
telolecithal. Yolk is unevenly distributed and the vegetal region contains
much more yolk than the animal region. We saw that cleavage in the

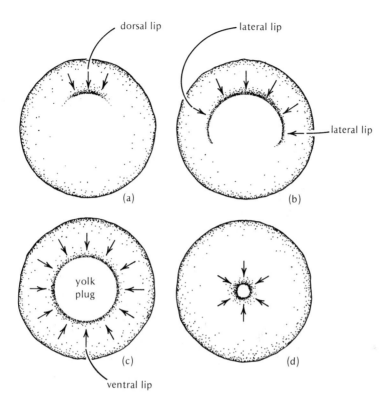

Figure 4–7. Surface views of amphibian gastrulation.

amphibian embryo was unequal—the yolk appeared to retard cleavage in the vegetal region. Gastrulation also appears to be influenced by yolk. In the amphibian, unlike *Amphioxus* and the sea urchin, invagination or "punching in" of the embryo does not occur. Instead, cells move from the exterior to the interior of the embryo by active migration of the cells through a groove that begins to form at the surface of the embryo. The vegetal region of the amphibian is too thick and too yolk-laden to allow the type of invagination found in the sea urchin and *Amphioxus.*

Sequence of Events—Surface View

Figure 4–7 shows surface views of gastrulation in the amphibian. A slit forms just below the equator. This slit is the blastopore. Cells from the surface of the embryo move inside the embryo through the blastopore. This migration first occurs only in a small region below the equator

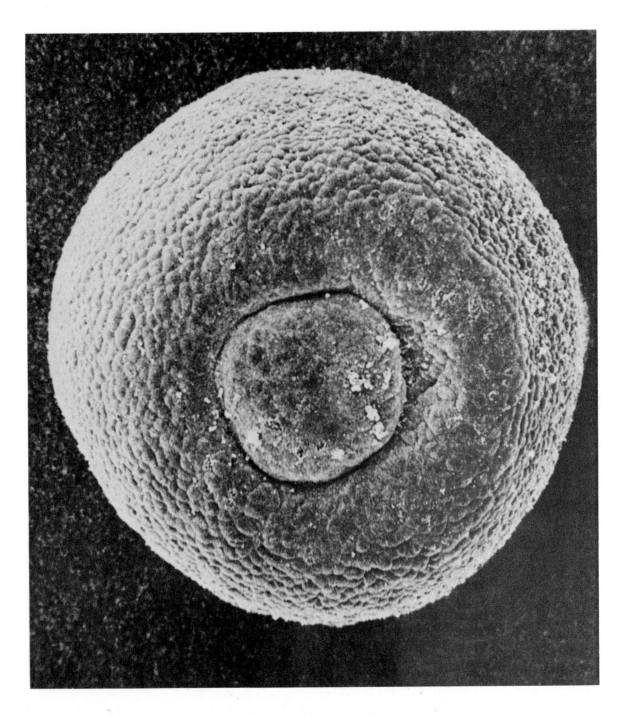

Figure 4–8. Frog yolk plug. Courtesy of Peter Armstrong.

between animal and vegetal hemispheres. The area just dorsal to this beginning cleft is termed the dorsal lip of the blastopore. The dorsal lip forms at the site of the grey crescent described in Chapter 3. Cells migrate over this lip, through the blastopore, and into the embryo.

    As can be seen in Figure 4–7, the blastopore lengthens and becomes crescent-shaped, then semicircular, and finally it forms a full circle. This results from cells on the surface of the embryo moving inward. The first cells move in from the dorsal area of the embryo; as the blastopore becomes crescent shaped, cells from lateral regions of the embryo move in. Finally ventral cells move in, completing the circular blastopore. The lateral lips and ventral lip of the blastopore are simply those regions, lateral and ventral respectively, over which cells migrate into the embryo through the blastopore. When the blastopore is complete, forming a full circle, the center of the circle is filled with yolky endoderm cells. This plug of yolky endoderm is called the yolk plug (Figure 4–8).

**dorsal lip
of the blastopore**

lateral lips

ventral lip

yolk plug

Mechanisms of Gastrulation

What are the mechanisms that cause surface cells of the amphibian embryo to move inward? Cellular shape changes first occur at the dorsal lip region and cellular movement begins. Cells expand and contract, reminiscent of the changes in the secondary mesenchyme cells of the sea urchin embryo. Such cellular expansion and contraction appear to play important roles in the inward movement of the active cells as well as those cells attached to the active cells. The nature of the forces involved in amphibian gastrulation is not well understood. They do, however, act even if the dorsal lip is cut out of the embryo and transplanted to a different part of the embryo. In this case, the dorsal lip cells will begin to migrate inward as if they were in the proper position in the embryo.

Sequence of Events—Cross-Section

Let us now turn from the surface of the embryo, by slicing it in half, to examine what is occurring inside the amphibian embryo during gastrulation. As shown in Figure 4–9, the notochord forms from cells that move from the dorsal surface of the embryo over the dorsal lip region, through the blastopore, to the interior of the embryo below the prospective neural tube. The notochord forms the roof of the archenteron. The prospective epidermis and prospective nervous system (together comprising the ectoderm) expand over the entire surface of the embryo. The blastopore becomes circular, the notochordal mesoderm enters dorsally and the non-notochordal mesoderm enters laterally and ventrally. The yolk plug endoderm disappears from the surface of the embryo, as the rim of the blastopore contracts, pulling the yolk plug inside.

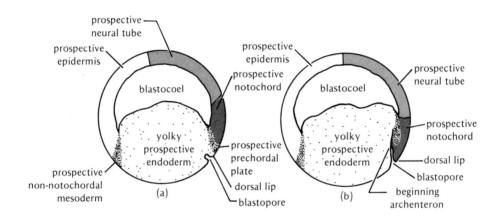

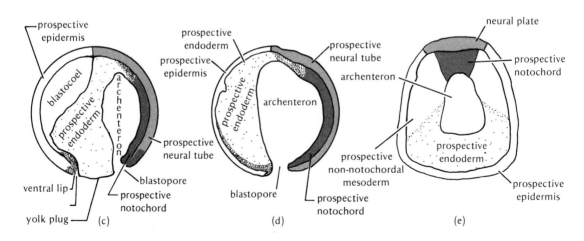

Figure 4–9. Gastrulation in amphibian. (a) to (d) Sagittal section. (e) Cross section. After Vogt, W., *Roux Arch. 120*, 385–706 (1929).

Gastrulation in the Bird Embryo

We have examined gastrulation in the sea urchin, *Amphioxus,* and the amphibian. We noted that in embryos consisting of cells with relatively small amounts of yolk (sea urchin and *Amphioxus*), gastrulation occurs by in-pocketing or invagination. In the amphibian embryo, the large, yolky vegetal cells appear to prevent in-pocketing, and instead cells migrate into the embryo through a narrow slit. As discussed in Chapter 3, the bird embryo is very different from the sea urchin, *Amphioxus,* and the amphibian embryos. All of the cleavage occurs only in the little cytoplasmic cap that sits atop the massive quantity of yolk in the bird egg. At the blastula stage of the bird embryo, the cytoplasmic cap (blastodisc) has cleaved to form a multilayered embryo (the blastoderm) that sits atop a space, called the subgerminal space, that separates most of the embryo from the yolk below it.

Gastrulation in the bird embryo begins when the blastoderm separates into two layers, the top epiblast layer and bottom hypoblast layer (see Figure 4–10). The space between these layers is the cleft space.

epiblast

hypoblast

The second step in bird gastrulation is the formation of a thickening at one end of the blastodisc by movement of lateral cells towards the center. This thickening is called the primitive streak (Figure 4–10).

primitive streak

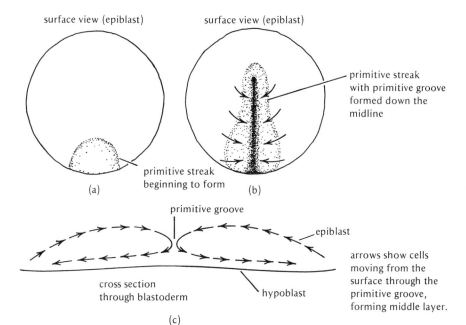

surface view (epiblast)

surface view (epiblast)

primitive streak with primitive groove formed down the midline

primitive streak beginning to form

(a)

(b)

primitive groove

epiblast

cross section through blastoderm

hypoblast

arrows show cells moving from the surface through the primitive groove, forming middle layer.

(c)

Figure 4–10. Bird gastrulation.

primitive groove   An indentation called the primitive groove forms down the midline
of the primitive streak. The primitive groove serves the same function as
the blastopore of the other embryos that we have studied. The primitive
streak elongates, and cells move from the epiblast surface through the
primitive groove and into the cleft space. Recall the bird fate map (Figure
4–1) and note that some of the epiblast cells that move in form endoderm. These cells enter the hypoblast. The other epiblast cells that move
in form the middle mesoderm layer. In the bird, therefore, gastrulation
occurs only in the non-yolky cap of cells. The yolk remains uncleaved
and uninvolved in the gastrulation process.

Genesis of Early Embryonic Structures

How does a flat embryo, such as the bird gastrula, become transformed
into a three dimensional entity? How are the structures that surround
developing embryos formed? Let us begin with the bird gastrula and
examine how this embryo begins to take shape.

Body Folds of the Bird Embryo

The bird embryo becomes transformed into a three-dimensional organism as a result of a series of folds that occur in the embryo proper. Recall
that the bird embryo sits atop the yolk as a layered sheet of cells. The
embryo begins to lift itself off of the yolk by folds that occur anteriorly,
posteriorly, and laterally. These folds are shown in Figure 4–11 Similar
foldings occur in reptile and mammalian embryos, such as man.

As can be seen in Figure 4–11 the head fold occurs at the anterior
end of the embryo, the tail fold occurs at the posterior end and the lateral
body folds occur at the sides of the embryo. Thus via these folds, the
embryo becomes a three-dimensional cylinder that has lifted itself above
the yolk. The midgut region of the embryo remains open to the yolk.
Neural folds also occur, as described in Chapter 5.

Extraembryonic Folds of the Bird Embryo

The bird embryo is beginning to resemble a real three-dimensional organism. The body folds have transformed a flat sheet into a cylinder that sits atop the yolk. Another series of foldings occurs in the area of cells surrounding the embryo proper. These folds are the extraembryonic folds and result in the formation of membranes that function to protect the embryo and store its wastes.

As shown in Figure 4–12, the extraembryonic somatopleure, which is the combination of ectoderm and the mesoderm that is in contact with it anterior to the head fold of the embryo proper, begins to fold over the head of the embryo. Similarly the posterior extraembryonic somatopleure also folds over the tail fold of the embryo proper. Laterally, this same extraembryonic somatopleure also begins to fold over the sides of the embryo proper. These extraembryonic folds are called the head, tail and lateral folds of the amnion respectively. These folds cover the embryo anteriorly, posteriorly and laterally and fuse together to form a double "helmet" that covers the embryo as shown in Figure 4–12. The

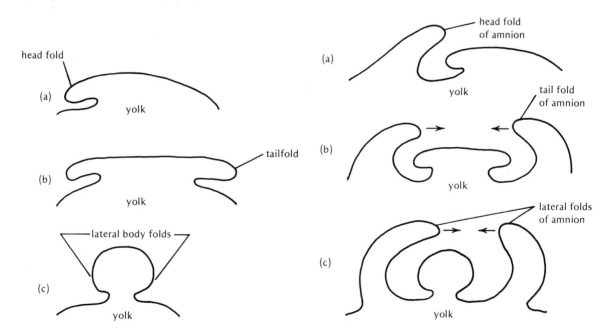

Figure 4–11 (left). Body folds of the chick embryo. (a) and (b) are longitudinal sections. (c) is a cross section.

Figure 4–12 (right). Extraembryonic folds of the chick embryo.

chorion

amnion

extraembryonic coelom

amniotic cavity

allantois

outermost membrane of the double helmet is called the chorion. The innermost membrane closest to the embryo is called the amnion. Each consists of somatopleure. The space between the amnion and chorion is extraembryonic coelom and the space between the amnion and the embryo proper is the amniotic cavity. This cavity contains amniotic fluid that protects the embryo from drying out and serves as a cushion against mechanical injury.

Another important sac develops as an outpocketing of the hindgut and consists of endoderm and splanchnic mesoderm—that mesoderm that touches the endoderm (that is, splanchnopleure). This structure is the allantois and is shown in Figure 4–13. This sac extends into the extraembryonic coelom and serves as an embryonic excretory organ in which uric acid is deposited. Also, as seen in Figure 4–13, the combination of the allantoic membrane and chorion is called the chorioallantoic membrane. This membrane becomes full of blood vessels and functions as a respiratory organ until hatching. Oxygen from the outside passes through the shell and into the chorioallantoic vessels while carbon dioxide passes out of the vessels.

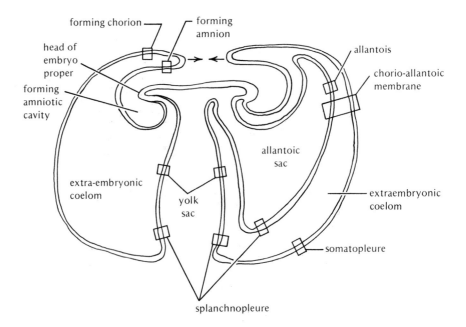

Figure 4–13. Summary diagram of early chick embryo with its forming extraembryonic membranes. Longitudinal view.

Mammals

Gastrulation in mammals, including humans, resembles that of the bird. Little or no yolk is present in many mammalian eggs (except in the egg-laying mammals). If yolk quantity and distribution influence gastrulation, why is human gastrulation similar to that of the bird? Birds' eggs have massive amounts of yolk while mammal eggs have little or none. It may be that although yolk disappeared from most mammalian eggs as development inside the mother came about, the pattern of gastrulation did not become significantly altered with the loss of yolk and still resembled that of its yolky-egged ancestors.

The mammalian blastodisc consists of an epiblast and hypoblast layer like the bird's. A primitive streak forms. Cells from the primitive streak migrate to form a middle layer between epiblast and hypoblast.

It should be noted that it is much more difficult to study embryonic development in mammals than in sea urchins, frogs, and birds. This is because mammal embryos must be in the mother to develop normally. Sea urchin, frog, and bird embryos develop nicely outside of the mother under simple conditions. Human eggs and early embryos, for example, are occasionally available for study from medical operations, or by using hormones in volunteers to induce oocyte development. There is some controversy surrounding this sort of experiment. Mammalian eggs can be removed from the mother and fertilized in a test tube. Early development can be observed outside of the mother but development soon deteriorates unless the embryo is implanted into the uterus of a receptive female. Recently, a human egg was removed from the ovary of a female with blocked oviducts. The egg was fertilized outside the mother, in a "test tube," with the husband's sperm, and implanted into the mother's uterus. Normal development and birth of a normal baby girl took place. This has also been accomplished in other mammals, such as mice. At this point, however, the lack of detailed information about some aspects of embryonic development in many mammals can be attributed to the complex conditions needed to study mammalian embryos and the lack of sufficient quantities of such embryos.

Let us stop here and review some of the mechanisms that appear to be important in gastrulation. In the sea urchin, invagination appeared to occur as a result of mechanisms such as cell motility, selective adhesiveness, and contractility. The vegetal plate of the sea urchin embryo invaginated as a result of active movement of the primary mesenchyme cells and contraction of the filopodia of the secondary mesenchyme cells that specifically adhered to the animal end. We observed that yolk quantity and distribution also appears important in the gastrulation process. In the sea urchin and in *Amphioxus*, the relatively small amount of yolk permits invagination or in-pocketing of the embryo to occur. As the

Figure 4–14. Sea urchin larvae (early pluteus). This stage follows gastrulation. Basic body plan is nearly complete. Courtesy of Victor Vacquier.

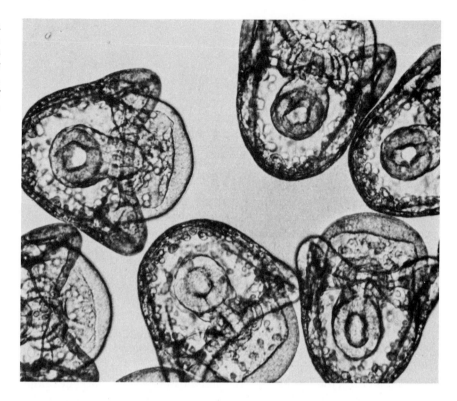

quantity of yolk increases, the pattern of gastrulation changes. In the amphibian, cells migrate into the embryo through a narrow slit. Invagination of the embryo does not occur. The bird blastodisc sits atop the yolk. The entire gastrulation process, like the entire cleavage process, occurs only in the non-yolky cap of cytoplasm.

Although the patterns of gastrulation in various organisms differ widely, the end is the same. In all cases gastrulation results in embryonic cellular rearrangements that begin to give form and structure to a nondescript ball or cap of cells. In Chapter 9, we will investigate the molecular nature of some of the mechanisms (such as selective cell adhesiveness) that appear to control the rearrangements of cells in embryos during gastrulation and other stages.

In the next chapter, we will focus on how the embryo achieves a three-layered state, and how the nervous system begins to form. In the next chapter, as in this one, no attempt will be made to review these processes in all or even most types of organisms. Instead, certain systems are selected that provide basic examples of mechanisms that appear to influence embryonic development. We began our study of embryology with sperm and egg cells. We have progressed to the point at which the embryo is beginning to resemble something more than a blob of cells. It still does not resemble the adult organism, but the foundations of the basic body plan are almost complete (see Figure 4–14). In the next chapter, the basic plan will become fully established.

Readings

Guidice, G., *Developmental Biology of the Sea Urchin Embryo.* Academic Press, Chicago (1973).

Gustafson, T., and L. Wolpert, Cellular Movements and Contact in Sea Urchin Morphogenesis. *Biol. Rev.* 42:442–498 (1967).

Holtfreter, J., A Study in the Mechanics of Gastrulation I, II. *J. Exp. Zool.* 84:261–318 (1943); 95:171–212 (1944).

Moore, A. R., On the Mechanics of Gastrulation in *Dendraster excentricus. J. Exp. Zool.* 87:101–111 (1941).

Nicolet, G., Avian Gastrulation, *Adv. Morphogen.* 9:231–262 (1971).

Spratt, N. J., and H. Haas, Integrative Mechanisms in the Development of the Chick Embryo. I, *J. Exp. Zool.* 145:97–137 (1960).

Trinkaus, J. P., *Cells Into Organs: The Forces that Shape the Embryo.* Prentice Hall, Englewood Cliffs, N.J. (1969).

Vogt, W., Gastrulation and Mesoderm Formation in Urodeles and Anurens, *Roux Arch.* 120:385–706 (1929).

CHAPTER 5

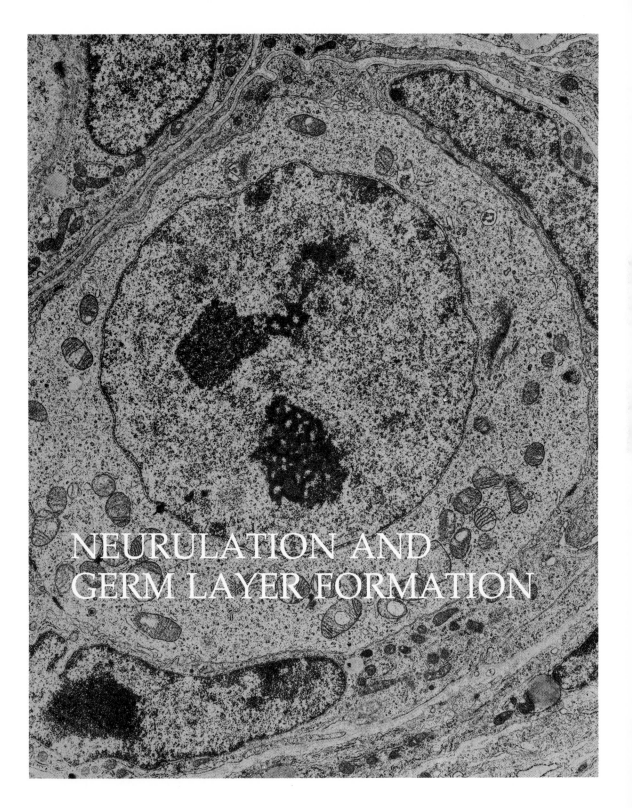

NEURULATION AND
GERM LAYER FORMATION

IN THE LAST CHAPTER we examined the process of gastrulation in several representative systems. We did not, however, reach the point in which the embryo is composed of three distinct layers, ectoderm, mesoderm, and endoderm. How does the three-layered state come about in various embryos? How does the neural tube form in the embryo? What are the mechanisms that cause the three-layered state to form and that cause neurulation, the formation of the nerve tube? The first two questions can be fully answered. The last question that asks about the nature of mechanisms involved in these developmental events will be more difficult to answer because the nature of all of these mechanisms is, as yet, not well understood. We will, however, learn a great deal about the forces that are involved in these events, and Chapters 8 and 9 will deal with these mechanisms in a more thorough manner.

The Three-Layered State

Amphioxus

In the last chapter we left *Amphioxus* in a two-layered state, consisting of an outer ectodermal layer and an inner mesendoderm layer. As can be seen in Figure 5–1, the mesoderm separates from the endoderm as a result of the formation of pouches. The mesoderm separates into three regions. The two lateral mesodermal pouches migrate between the ectoderm and endoderm and fuse ventrally. Thus the middle germ layer or mesoderm forms as a result of pouching of the mesendoderm (prospective mesoderm plus prospective endoderm) and migration of the lateral mesodermal pouches between the endoderm and ectoderm. The middle

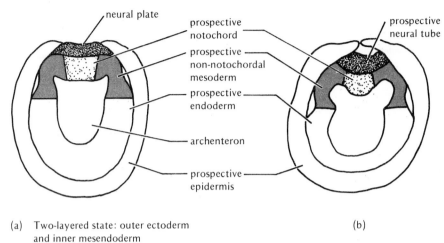

(a) Two-layered state: outer ectoderm
     and inner mesendoderm

(b)

neural plate

prospective
notochord

prospective
non-notochordal
mesoderm

prospective
endoderm

archenteron

prospective
epidermis

prospective
neural tube

Figure 5–1. Formation of the three layered state in *Amphioxus*. Mesoderm and coelom formation occur by pouching. Arrows show movement of mesodermal pouches between epidermis and endoderm. Three layered state nearly complete. Modified and redrawn from Hatohek in Korschelt, 1936, *Vergleichende Entwicklungsgeschichte Der Tiere*, G. Fischer, Jena.

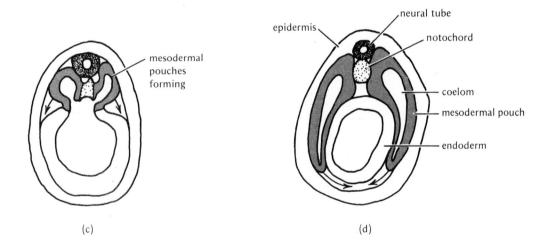

mesodermal
pouches
forming

epidermis

neural tube

notochord

coelom

mesodermal pouch

endoderm

(c)

(d)

mesodermal region forms the notochord. As can be seen in Figure 5–1, a space is present within the mesodermal pouches at the anterior region of the embryo (the space forms by separation of solid mesoderm in the more posterior regions of the embryo). This space is the body cavity, or coelom.

In summary, in the primitive chordate *Amphioxus*, the mesendoderm layer reached the inside of the embryo during gastrulation as a result of the invagination process (Chapter 4). Mesoderm separated from endoderm as a result of a pouching of the mesendoderm layer. The body cavity or coelom developed as a result of this pouching activity in the anterior portion of the embryo, and posteriorly as the result of separation of solid mesoderm.

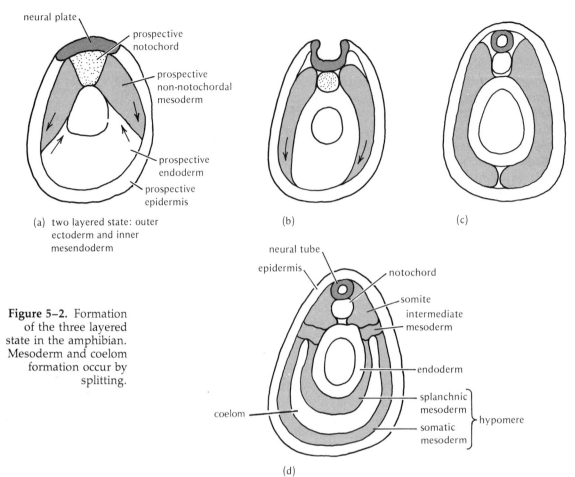

neural plate

prospective notochord

prospective non-notochordal mesoderm

prospective endoderm

prospective epidermis

(a) two layered state: outer ectoderm and inner mesendoderm

(b)

(c)

neural tube

epidermis

notochord

somite

intermediate mesoderm

endoderm

splanchnic mesoderm

somatic mesoderm

hypomere

coelom

(d)

Figure 5–2. Formation of the three layered state in the amphibian. Mesoderm and coelom formation occur by splitting.

Amphibian

Figure 5–2 shows that the amphibian embryo also is in a two-layered state before the mesoderm forms. Just as *Amphioxus*, the amphibian gastrula consists of an outer ectoderm and inner mesendoderm layer. Mesoderm formation in the amphibian, however, does not occur by pouching of the mesendoderm as it did in *Amphioxus*. Instead it forms by separation of the mesendoderm into mesoderm and endoderm. The mesoderm migrates between the ectoderm and endoderm, fusing ventrally.

Coelom formation in the amphibian does not result from pouching, either. Instead, the mesoderm splits, forming somatic (in contact with ectoderm) and splanchnic (in contact with endoderm) mesoderm. The space between these regions is the coelom. The notochord also splits off from the remainder of the mesoderm. The mesoderm forms four major subdivisions: notochord, somite, intermediate mesoderm, and hypomere (somatic and splanchnic mesoderm). We will carefully define these regions and their derivatives in the next chapters.

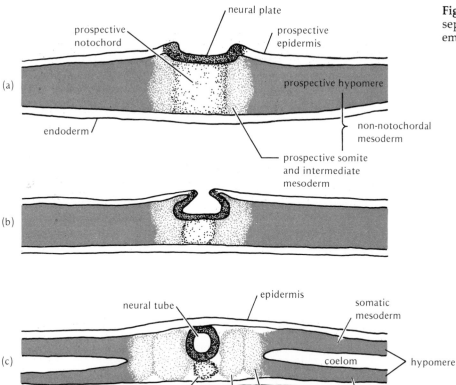

Figure 5–3. Mesodermal separation in chick embryo.

Bird

Development of the three-layered state is similar in birds, reptiles, and mammals, and most of the discussion also applies to the latter two groups. In the last chapter we left the bird embryo at a stage in which the prospective mesoderm and some prospective endoderm from the epiblast were migrating into a central cleft space between the epiblast (top layer) and hypoblast (bottom layer). The prospective endoderm cells originating from the epiblast enter the hypoblast. The prospective mesoderm cells maintain a position between epiblast and hypoblast. It should be noted that endoderm of the gut tube originates in the epiblast, so that the epiblast is not exactly equivalent to amphibian ectoderm. Also, some hypoblast cells finally become located in the notochord so hypoblast is not equivalent to endoderm. These findings resulted from marking regions of the embryo with radioactive tracers and carbon particles and observing the final location of marked cells.

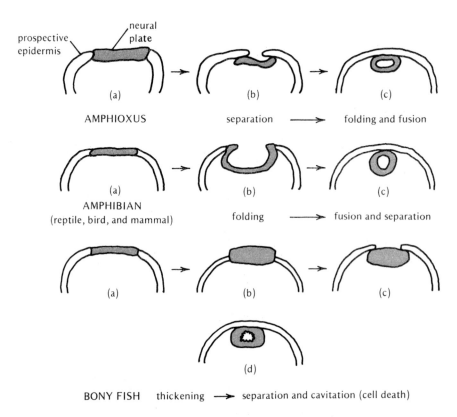

Figure 5–4. Neurulation in *Amphioxus*, Amphibian, and bony fish.

After moving from the epiblast through the primitive groove, the mesoderm comes to occupy the space between epiblast and hypoblast. The presumptive notochordal mesoderm accumulates in an anterior to posterior direction in the cleft space. Cells begin to form the notochord at the anterior end of the embryo. Additional cells move in from the epiblast allowing the notochord to elongate posteriorly. The epiblast cells directly above the prospective notochord are prospective neural plate.

The bulk of the non-notochordal mesoderm also now occupies the space between the epiblast and hypoblast. As will be seen in the next chapter, the parts of this mesoderm adjoining the neural tube become somite and intermediate mesoderm. The remainder of the mesoderm splits, as in the amphibian, into outer somatic and inner splanchnic mesoderm (Figure 5–3). The space between the two is the coelom.

In summary, we have examined the formation of the three-layered state in *Amphioxus*, the amphibian, and the bird. We saw that the mechanisms of coelom formation may differ but the end is the same. In *Amphioxus* the coelom forms from mesodermal pouching and splitting, while in the amphibian and bird the mesodermal tissue separates to form the coelom. Chapter 9, "Mechanisms of Morphogenesis," will deal with some of the factors that may be involved in the development of form in the embryo. We will see, for example, that changes in adhesiveness of cells in a tissue may cause separation of the tissue into two parts. Let us turn to a discussion of neurulation, the formation of the nerve tube, now that we have an idea of how the three-layered state comes about and how the body cavity forms in some representative embryos.

Neurulation

Amphioxus

As shown in Figure 5–4, neurulation in *Amphioxus* occurs in three basic steps: (1) Separation of the neural plate (prospective neural tube) from the prospective epidermis; (2) Folding of the neural plate (prospective neural tube); (3) Completion of neural tube formation and fusion of the overlying epidermis. It seems plausible that, for example, separation of the prospective neural tube from the prospective epidermis may result from adhesive changes in the cells of the ectoderm prior to separation. More will be said about mechanisms of neurulation in the following sections.

Amphibian, Reptile, Bird, and Mammal

The formation of the neural tube is similar in the amphibian, reptile, bird, and mammal. It occurs in a slightly different manner from that of *Amphioxus*. The sequence of events in amphibian, reptile, bird, and mammal neurulation is shown in Figure 5–4 and can be summarized as follows: the neural plate folds, the crests of the neural plate fuse, and the neural tube separates from the epidermis and drops below the surface while the overlying epidermal regions fuse to form an intact outer epidermal layer. Neural crest cells (Chapter 6) also separate from the ectoderm and drop below the surface. Thus in the amphibian, reptile, bird, and mammal, folding of the neural plate is the first step in neurulation, followed by fusion of the neural folds and of the overlying epidermis (Figures 5–5 and 5–6).

Figure 5–5. Amphibian neural plate stage. Courtesy of Peter Armstrong.

Figure 5–6 (opposite page). Amphibian, neural folds closing. Courtesy of Peter Armstrong.

Mechanisms of Neurulation

In Chapter 9 we will examine mechanisms of morphogenesis in some detail. It would be valuable to mention some of the forces, here, that appear to influence the neurulation process. Formation of the neural tube in the systems discussed may involve cell surface adhesive changes in the ectoderm. In addition, contraction of peripheral microfilaments oriented parallel to the short axis of the neural plate cells may cause some folding to occur (Figure 5–7). Indeed, Baker and Schroeder have found bundles of microfilaments just below the surface of folding neural plate cells in certain amphibian embryos (*Hyla* and *Xenopus*). Microtubules may also play a role in neurulation. Waddington and Perry found many microtubules in the neural plate cells of certain amphibian embryos (*Triturus alpestris*). These workers suggest that neurulation may, in part, occur because of cell elongation caused by microtubules (Figure 5–7). The role of microtubules and microfilaments in events such as neural folding will be examined in Chapter 9. Thus, factors such as changes in cell adhesiveness, microfilament contraction, and microtubule elonga-

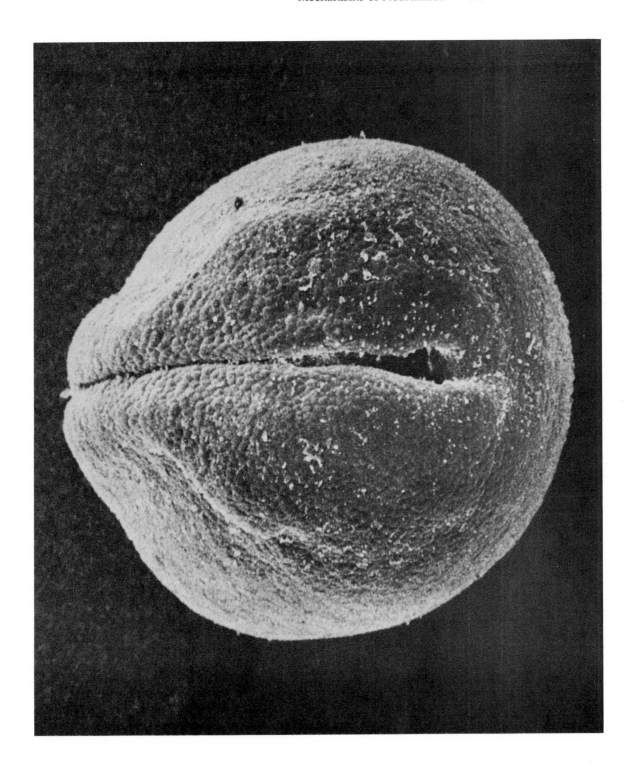

tion may be important in the neurulation process. Before concluding our discussion of neurulation let's look at this process in one other system—that of the bony fish. Neural tube formation in this embryo occurs by a completely different mechanism.

Bony Fish (Teleost) Neurulation

As seen in Figure 5–4, the neural plate in the bony fish thickens but doesn't fold. This thickened neural plate separates from the rest of the ectoderm and begins to sink below the surface. The overlying epidermis fuses and the prospective neural tube drops below the surface. But the prospective neural tube at this point is still a solid rod. How does it become a hollow tube without having folded? Neural tube formation in the bony fish occurs by cavitation. Cells in the solid prospective neural tube rod die. The hollow nerve tube in the bony fish, therefore, forms by differential death of the cells in the center of the prospective neural tube. Cell death plays other important roles in morphogenesis, as will be seen in Chapter 9.

Figure 5–7. Mechanism of neurulation. Two factors that may be involved in causing neural folding in newt embryos are elongation of cytoplasmic microtubules parallel to the long axis of the cells and contraction of a surface layer of cytoplasmic microfilaments parallel to the short axis of the cells. After Burnside, *Devel. Biol.* 26: 434, Academic Press, 1971.

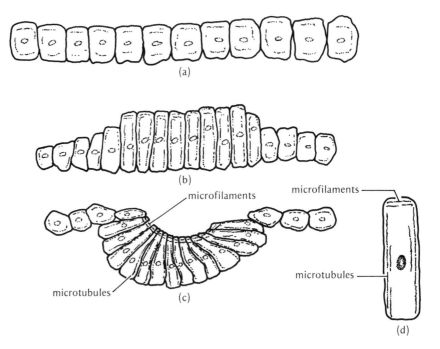

In summary, in this chapter we saw that although the mechanisms differ slightly, the end is the same in all embryos examined, namely the three-layered state and body cavity have formed. Also, regardless of whether the neural tube forms by separation followed by folding, folding followed by separation or by selective cell death in a solid core, the end is the same: the nerve tube has formed. The basic body plan of chordates and vertebrates has been achieved. The embryos we have examined now contain three layers and possess neural tubes.

What structures are derived from each germ layer and what are the mechanisms of their formation? We will begin to answer the question in the next chapter, with an overview of the primary germ layer derivatives. In the following two chapters we will deal with a more detailed examination of organogenesis and the mechanisms controlling the development of embryonic form and structures.

Readings

Baker, P. C., and T. E. Schroeder, Cytoplasmic Filaments and Morphogenetic Movement in the Amphibian Neural Tube. *Devel. Bio* 15:432-450 (1967).

Balinsky, B. I., *An Introduction to Embryology*, 4th. ed. W. B. Saunders, Philadelphia (1975).

Burnside, B., Microtubules and Microfilaments in Amphibian Neurulation. *Am. Zool.* 13:989–1006 (1973).

Karfunkel, P., The Activity of Microtubules and Microfilaments in Neurulation in the Chick. *J. Exp. Zool.* 181–289–302 (1972).

Spemann, H., *Embryonic Development and Induction.* Hafner Publishing, New York, (1938).

Torrey, T. W., *Morphogenesis of the Vertebrates* 2nd ed. John Wiley & Sons, New York (1967).

Waddington, C. H., and M. M. Perry, A Note on the Mechanism of Cell Deformation in the Neural Folds of the Amphibian. *Exp. Cell Res.* 41:691–693 (1969).

Weiss, P., Nervous System (Neurogenesis). In *Analysis of Development*, B. H. Willier, P. Weiss and V. Hamburger, eds., W. B. Saunders, Philadelphia, p. 346 (1955).

CHAPTER 6

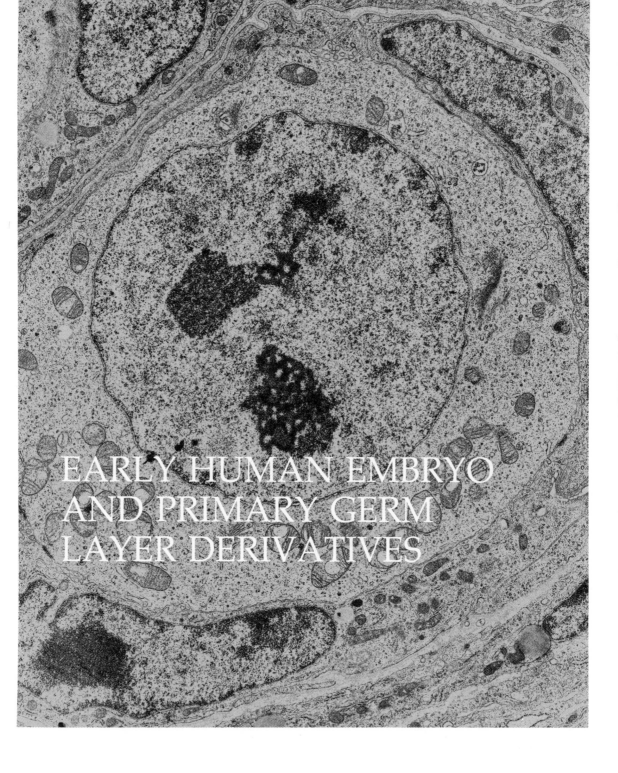

EARLY HUMAN EMBRYO
AND PRIMARY GERM
LAYER DERIVATIVES

THE EARLY HUMAN EMBRYO

WE HAVE EXAMINED early development in several groups of organisms but have made only casual references to development in mammals. We will survey early mammalian development here as a separate section because of the special interest that we have for the development of humans and their relatives in Class Mammalia.

Gametes

Human sperm are formed in the testis from spermatogonia. The primordial germ cells that form the spermatogonia in man appear to originate in the yolk sac endoderm and migrate to the gonad. Mature spermatozoa form in the seminiferous tubules of the testis and are the final product of a growth phase, two meiotic divisions, and a differentiation phase of the original spermatogonia. The steps leading to the formation of mature sperm are:

spermatogonia
  (growth)

primary spermatocytes
  (meiosis 1)

secondary spermatocytes
  (meiosis 2)

spermatids
  (differentiation)

mature spermatozoa

Human sperm have a length of 65 μ. There are about 100 million human sperm per milliliter of semen.

Human eggs are formed in the ovary from oogonia. The primordial germ cells that form human oogonia also appear to originate in the yolk sac endoderm (the inner lining of yolk sac) and migrate to the gonad. In the human female, the outer region of the gonad (cortex) develops into the ovary and those primordial sex cells that develop into oogonia are those that have entered the outer region of the gonad. In the human male, the primordial sex cells that enter the inner (medulla) region of the primitive gonad develop into sperm and it is the inner region of the primitive gonad that develops into the testis in males (see Chapter 7).

Development of human eggs from oogonia, as noted in Chapter 1, involves the following sequence of events:

oogonia
  (growth)

primary oocytes
  (meiosis 1)

secondary oocytes
  (meiosis 2)

mature eggs

As detailed in Chapter 1, we should recall that one oogonium gives rise to only one mature egg, while one spermatogonium gives rise to four mature spermatozoa. This is the case because the cytoplasmic divisions during meiosis in the female are unequal, yielding large cells (secondary oocyte and egg), and tiny cells (polar bodies). In this way the stores of molecules and other components required to take care of embryonic development are not diluted out to four cells. Instead, the unequal cytoplasmic division results in the formation of one large egg with a full supply of stored material. Since mammals usually obtain nourishment from the placenta, their eggs need not be as large as those of some other groups of organisms. Sperm development, on the other hand, results from equal cytoplasmic divisions, so one spermatogonium yields four sperm cells. Large stores are not required for sperm as they are for eggs.

The human egg is about 120–150 μ in diameter. Human polar bodies, which are not fertilizable, measure a maximum of 10 μ in diameter. At birth, in the human female, all oogonia are present. These number between 200,000 and 300,000 in the ovaries. Only 200–300 of these, however, will ever reach maturity. At puberty and thereafter until menopause, maturation of usually one oocyte occurs during each menstrual cycle. Thus, about once a month, one egg matures in the human female. Maturation of these eggs is hormonally controlled.

Maturation of human oocytes in their expanding chambers or follicles is controlled by the combined presence of hormones from the anterior pituitary gland (follicle stimulating hormone and luteinizing hor-

follicle stimulating hormone

Figure 6–1. (a) Human graafian follicle. (b) Section of human ovary. From H. Tuchman-Duplessis et al., *Illustrated Human Embryology*, Vol. I. Springer-Verlag, New York, and Masson, Editeur, Paris (1972).

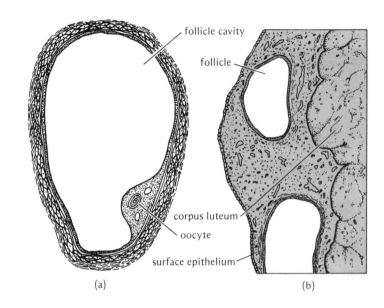

follicle cavity

follicle

corpus luteum

oocyte

surface epithelium

(a)   (b)

luteinizing hormone

estrogen

mone) and a hormone, estrogen, secreted from the follicles themselves. The pituitary hormones are secreted in response to secretions from the hypothalamus that control pituitary action. Ovulation, or the release of the oocyte with a surrounding layer of follicle cells (corona radiata) from the follicle chamber (and ovary), appears to be controlled by a burst of luteinizing hormone secretion, that may directly act by dissolving the surface layer of the follicle.

During maturation, the follicle expands and the follicle cells secrete fluid into the expanding follicle and also serve to nourish the developing oocyte (see Chapter 1). Right before ovulation, the follicle has enlarged and is full of liquid (liquor folliculi). This mature follicle is called a graaf-

graafian follicle

Figure 6–2. Changes in human uterine lining during menstrual cycle. From B. M. Patten, *Human Embryology*. McGraw-Hill (1974).

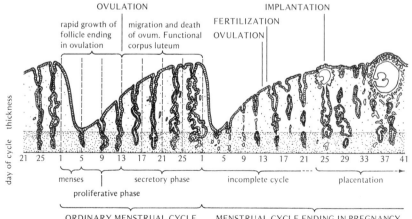

OVULATION

rapid growth of follicle ending in ovulation

migration and death of ovum. Functional corpus luteum

IMPLANTATION

FERTILIZATION

OVULATION

day of cycle   thickness

21 25 1 5 9 13 17 21 25 1 5 9 13 17 21 25 29 33 37 41

menses   secretory phase   incomplete cycle   placentation

proliferative phase

ORDINARY MENSTRUAL CYCLE   MENSTRUAL CYCLE ENDING IN PREGNANCY

ian follicle. After ovulation, the graafian follicle, which no longer contains the oocyte, becomes transformed into a hormone-secreting structure, the corpus luteum. This structure secretes estrogen and an additional hormone, progesterone. These hormones act on the hypothalamus, which sends a signal to the anterior pituitary to shut down synthesis of follicle stimulating hormone and luteinizing hormone.

corpus luteum

progesterone

Estrogen and progesterone from the corpus luteum also serve to prepare the uterus for implantation of the ovulated oocyte if fertilization occurs. The corpus luteum hormones cause the uterine lining (endometrium) to thicken and enrich in blood and gland supply. Note that the human oocyte is ovulated as a secondary oocyte in the second meiotic metaphase (where the chromosomes are ready to separate into two daughter cells). Final maturation occurs upon fertilization. Fertilization stimulates completion of the second meiotic division. If fertilization does not occur, implantation of the oocyte into the uterine lining does not occur, and the corpus luteum, by some unknown means, stops secreting hormones. This causes the rich uterine lining to be sloughed off, resulting in menstruation.

endometrium

Human Gamete Anomalies

Before we discuss fertilization, implantation and early embryonic development in human beings, it would be interesting to briefly describe some of the anomalies that can occur in human gametes. An examination of normal semen (the fluid containing the sperm) reveals that 20

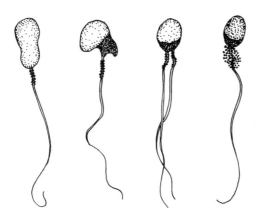

Figure 6–3. Some abnormal human sperm cells.

Figure 6–4. Some abnormal human egg cells. (a) Oocyte with two nuclei. (b) Two oocytes sharing the same follicle.

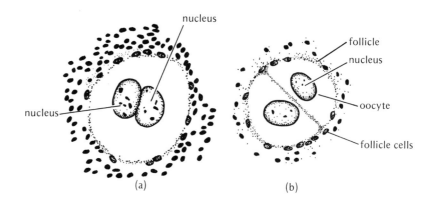

percent or less of the sperm are double. These may form as a result of failure of cell separation during spermatogenesis. In the fetal ovary, unusual oocytes are also occasionally seen. These include oocytes with two nuclei, or the presence of two oocytes in the same follicle. It is likely that an embryo formed from grossly abnormal gametes would be defective and would probably spontaneously abort early.

At the chromosomal level, anomalies can also occur during meiosis. Sometimes one cell gets both chromosomes of a pair, while the other gets none. Gametes with such abnormal chromosome numbers could result in severe problems to any embryo that might develop from such gametes. These chromosomal abnormalities could occur in the somatic chromosomes or the sex chromosomes. So, for example, a gamete may end up with either an X or Y chromosome (normal), or no sex chromosome, or an XY or XX set in a single gamete. When fertilization occurs with one such abnormal gamete plus a normal gamete, a variety of abnormalities may result. These include the XXY karyotype (chromosome pattern) called Klinefelter's syndrome and the XYY syndrome. About one in 500 male births is a Klinefelter's baby. The genitalia are male. Testes are reduced in size, which results in lowered androgen levels, causing reduced body hair and reduced sexual activity. Some Klinefelter's individuals are taller than average, may have normal to

Klinefelter's syndrome

XYY syndrome

Figure 6–5. Distribution of sex chromosomes. (a) Normal distribution. The haploid parent cell, when divided, results in two sperm cells, one carrying the X (female) chromosome, and one carrying the Y (male) chromosome. (b) Abnormal distribution results in one gamete with both sex chromosomes, and one with neither.

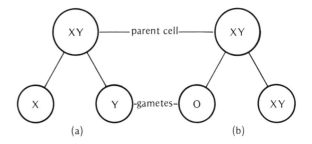

slightly subnormal intelligence, and may be antisocial. XXXY and XXXXY individuals also exhibit the basic characteristics of the XXY individual. With additional X chromosomes, the genitalia become more ambiguous.

The XYY karyotype often results in taller individuals, and may lead to aggressive and antisocial behavior. It should be noted that many individuals may have any of these syndromes and possess normal behavior. It is not known how many of such individuals are rather normal. It may be that only a minority of them (those diagnosed in prisons or mental institutions) are social problems.

The XO karyotype develops into a sterile female. Only about 1 in 1,000 births, however, is an XO individual. This condition is called Turner's syndrome, and the low birth rate is paired with a higher rate of spontaneous XO abortions. Those XO individuals who survive have hormonal deficiencies, short size, and often reduced sexual development.

XXX females often exhibit a slight decrease in mental abilities. They are sometimes fertile, and generally exhibit decreased feminine characteristics even though they have an increased "dose" of X.

The more common autosomal (non-sex) chromosomal anomalies include Trisomy 21 (Down's syndrome or Mongolism). Persons affected possess three instead of two chromosomes in chromosome pair number

Trisomy 21 (Down's syndrome)

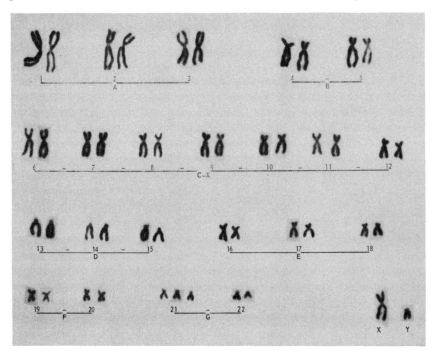

Figure 6–6. Karyotype of male with Down's syndrome (Trisomy 21). Note that there are three chromosomes rather than the normal two in group 21.

21 (Figure 6–6). Individuals with this syndrome are mentally retarded, have flat facial profile, unusual external ears, slow bone growth, and in the case of males are not fertile. At age forty-five, the risk of a woman giving birth to a Down's baby is 1 in 60, while it is 1 in 1,500 for women at age twenty. Trisomy 18 individuals possess three instead of two chromosomes in chromosome pair 18. About 1 in 3,000 babies are born with this anomaly. Only about ten percent survive to the first year. These children are very weak. Trisomy 13 results from 3 instead of 2 chromosomes in chromosome pair 13. This anomaly occurs in 1 of 5,000 births. Fewer than 20 percent survive to the first year. These individuals have improperly developed forebrains and poorly developed optic and olfactory nerves. Facial abnormalities are severe.

Many of these anomalies occur as a result of nondisjunction of chromosomes during meiosis, as indicated earlier in this section. This results in one of the gametes having an extra chromosome and the other a missing chromosome. Nondisjunctions, however, can also occur after fertilization or during cleavage—during any division in which chromosomes segregate to daughter cells. The earlier it occurs, the more cells are involved and the more widespread the effects. Other chromosomal abnormalities can occur as a result of chromosomes that lag behind on the spindle at anaphase in mitosis, resulting in one daughter cell receiving no chromosome while the other is normal. In addition, radiation and a variety of mutagenic chemicals can interfere with chromosome replication and cause chromosomal damage with physiological consequences. The anomalies discussed, and a host of others, occur because certain genes are missing or altered, or present in the wrong dose. The means by which genes lead to normal differentiation of cells are discussed in Chapters 10, 11 and 12.

Figure 6–7. Stages in human embryonic development from ovulation to implantation in the uterus.

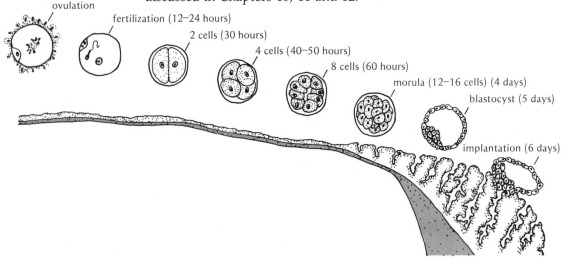

ovulation

fertilization (12–24 hours)

2 cells (30 hours)

4 cells (40–50 hours)

8 cells (60 hours)

morula (12–16 cells) (4 days)

blastocyst (5 days)

implantation (6 days)

Fertilization, Implantation, and Early Development

Fertilization in humans usually occurs inside the distal third of the fallopian tube (the end nearest the ovary). The mature oocyte is ovulated from the graafian follicle and is picked up by the expanded end of the fallopian tube that surrounds the ovary. Fertilization occurs if sperm are

Figure 6–8. Mouse zygote as it appears when magnified by Nomarski optics. Courtesy of Dr. P. Calarco.

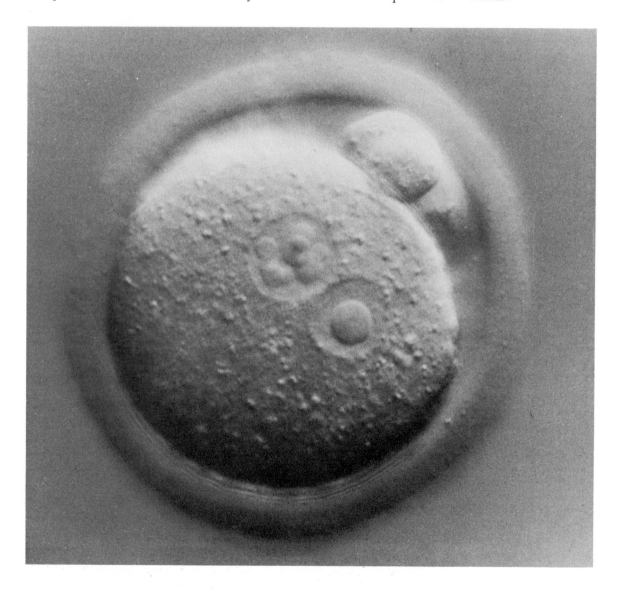

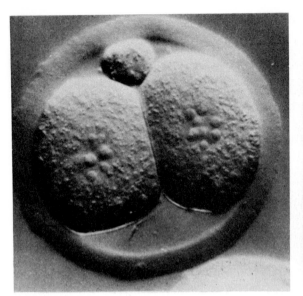 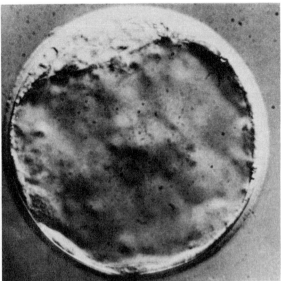

Figure 6–9 (left). Nomarski optics micrograph of 2-cell mouse embryo. Courtesy of Dr. P. Calarco.

Figure 6–10 (right). Mouse blastocyst as it appears by Nomarski optics. Note that this is the same stage as shown in Figure 6–16.

present in the distal third of the fallopian tube. It takes human sperm about ten hours to reach this part of the fallopian tube, so coitus usually occurs hours before fertilization. If the egg is not fertilized within 24 hours of ovulation, it usually dies.

Upon fertilization, the second polar body is extruded and the normal diploid chromosomal number is restored by fusion of the nuclei of the sperm and egg. Sex is determined at fertilization by the sperm. An X chromosome-containing sperm forms an XX zygote that is genetically female. A Y chromosome-containing sperm yields an XY zygote that is genetically male.

In humans, fertilization occurs within 24 hours of ovulation. The oocyte only lasts about 24 hours unless it is fertilized. The embryo begins to develop in transit through the fallopian tube to the uterus. The 2-cell stage occurs at about 30 hours. The 4-cell stage and 8-cell stage occur at 40–50 hours and 60 hours respectively. As the embryo approaches the entrance to the uterus, it is in the 12–16 cell stage, called a morula. This occurs at the fourth day. The morula hatches out of its enclosing coat, the

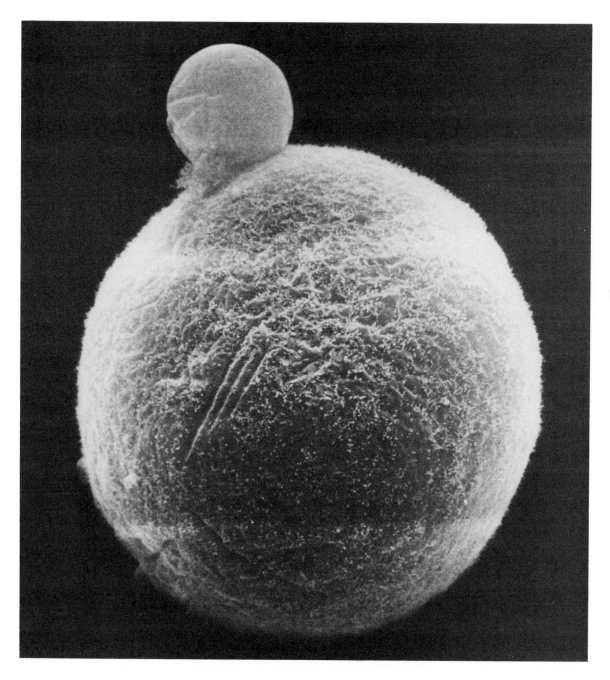

Figure 6-11. Scanning electron micrograph of fertilized mouse zygote. Early developmental stages are similar for most mammals. Courtesy of Dr. P. Calarco.

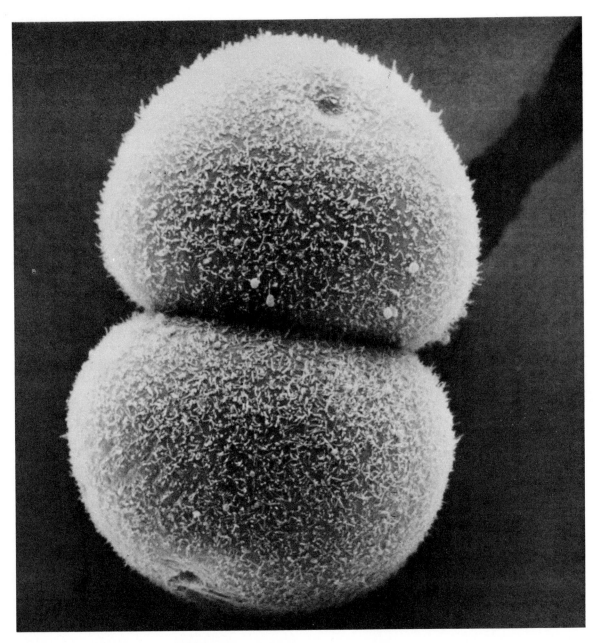

Figure 6–12. Scanning electron micrograph of 2-cell mouse embryo. Courtesy of Dr. P. Calarco.

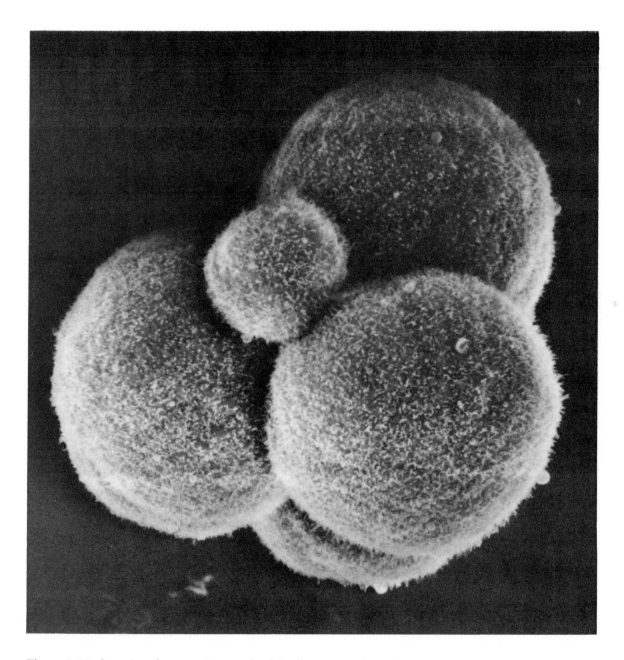

Figure 6–13. Scanning electron micrograph of 4-cell mouse embryo. Courtesy of Dr. P. Calarco.

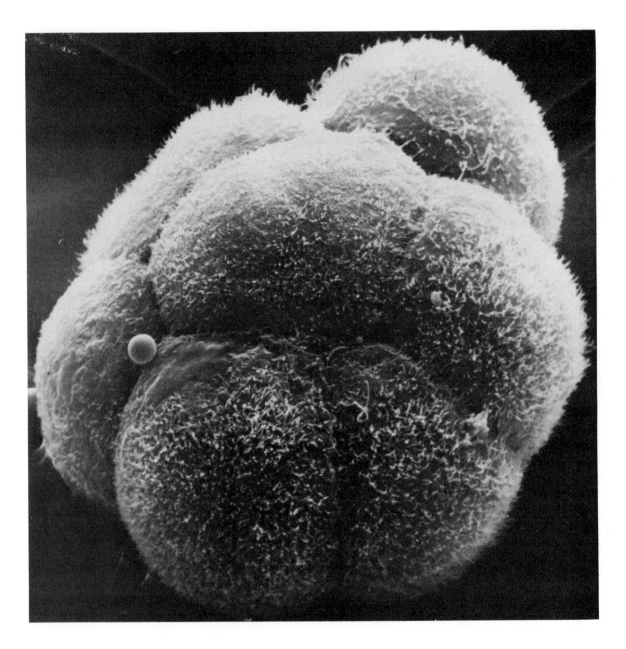

Figure 6–14. Scanning electron micrograph of 8-cell mouse embryo. Courtesy of Dr. P. Calarco.

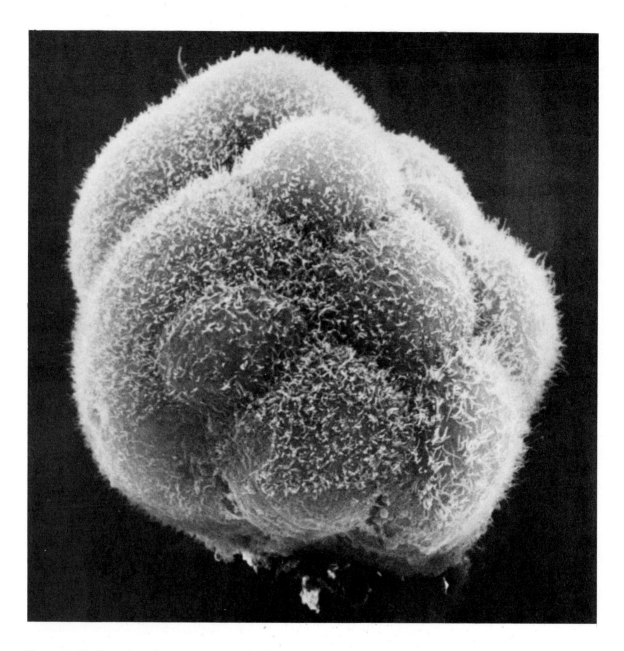

Figure 6–15. Scanning electron micrograph of mouse morula (12–16 cells). Courtesy of Dr. P. Calarco.

Figure 6–16. Scanning electron micrograph of mouse blastocyst. Courtesy of Dr. P. Calarco.

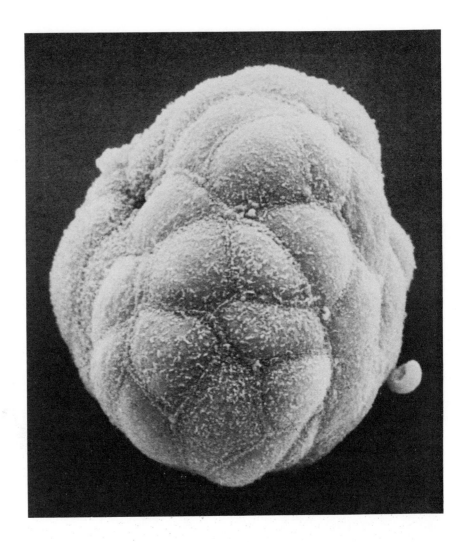

blastocyst

blastocyst cavity

trophoblast

inner cell mass

zona pellucida, during the fifth day. This free embryo is now in the 32–64 cell stage and is called a blastula or blastocyst, and possesses a cavity filled with fluid (blastocyst cavity). The free blastocyst has now reached the uterus and usually implants into the wall of the uterus during the 6th or 7th day (Figures 6–7 through 6–18). In abnormal situations, implantation of the blastocyst may occur in the ovary, in the fallopian tube, or in the wall of the body cavity or intestine. These embryos may abort spontaneously.

The outer layer of the blastocyst is called the trophoblast. In one region of the blastocyst a cluster of cells formed below the trophoblast is called the inner cell mass. The inner cell mass forms the embryo proper,

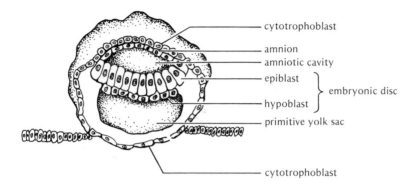

cytotrophoblast

amnion

amniotic cavity

epiblast

hypoblast } embryonic disc

primitive yolk sac

cytotrophoblast

Figure 6–17. Human blastocyst at the eighth day.

while the trophoblast forms the extraembryonic membranes (the membranes that surround the embryo). Implantation occurs by trophoblast cell invasion into the uterine tissue, in a way that some researchers believe may resemble the invasion of tumors into surrounding tissue. The blastocyst cavity is below the embryo proper rather than in the middle of it, so it is not exactly like the blastocoel of other embryos.

A space forms above the portion of the inner cell mass that becomes the embryo proper. This space becomes the amniotic cavity that helps protect the embryo. The embryo forms from the inner cell mass. During the second week the blastocyst, epiblast and hypoblast of the blastocyst differentiate in the inner cell mass. This marks the beginning of gastrulation. The upper layer, as in the chick, is the epiblast. The lower layer is the hypoblast. The embryo at this time continues to bury itself in the uterine lining. By the end of the second week the primitive streak

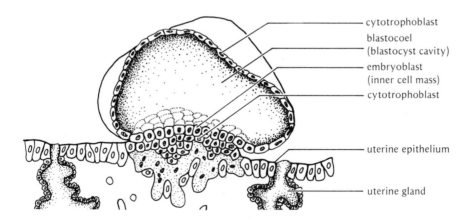

cytotrophoblast

blastocoel
(blastocyst cavity)

embryoblast
(inner cell mass)

cytotrophoblast

uterine epithelium

uterine gland

Figure 6–18. Human blastocyst implanted in the uterine wall.

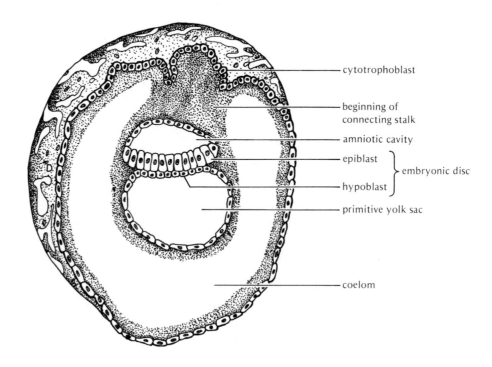

Figure 6–19. Human blastocyst at the end of the second week.

Figure 6–20. Human blastodisc at the end of the second week.

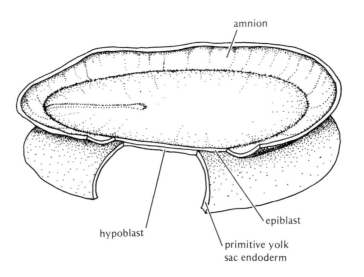

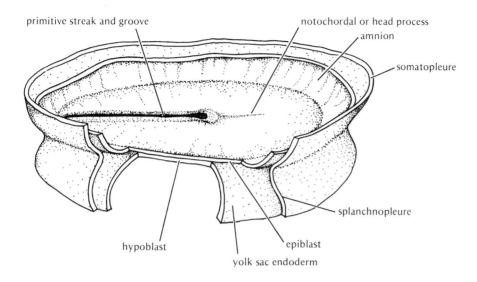

primitive streak and groove

notochordal or head process

amnion

somatopleure

hypoblast

epiblast

yolk sac endoderm

splanchnopleure

Figure 6–21. Human blastodisc at fifteen to sixteen days.

appears on the surface of the embryonic disc. This marks the start of gastrulation. A primitive groove forms down the middle of the primitive streak. Cells from the top layer (epiblast) move into the primitive groove to form the third embryonic layer, the mesoderm.

Human gastrulation is therefore somewhat similar to that of birds. A major difference between the two is the presence of a large amount of yolk in the bird, and none in the higher mammals. Thus, although the mammalian blastodisc sits atop a fluid-filled space, gastrulation occurs in much the same way as in birds and reptiles. This strongly suggests that mammalian ancestors had large yolky eggs, the yolk disappearing with time as development inside the mother evolved.

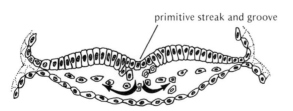

primitive streak and groove

Figure 6–22. Movement of cells through the primitive groove.

Placenta

The placenta brings the blood of the mother and fetus in close contact, so that nutrients and oxygen can be transferred to the developing fetus and wastes can be removed from fetal circulation. By three months of gestation in man, the placenta is well defined and continues development as the fetus matures. The human placenta forms from both the maternal and fetal tissues.

Before the placenta develops, the early human embryo absorbs nutrients from uterine fluid through its surface epithelium (surface layer). After implantation in the uterine wall, the blastocyst trophoblast layer begins to grow and spread out, sending outgrowths deeper into the uterine lining. Inside the trophoblast, many spaces develop. The invading trophoblast reaches blood capillaries of the uterine lining and breaks down the walls of these vessels. Blood therefore flows into the trophoblast spaces that serve to nourish the fetus. The surface of the trophoblast forms finger-like projections (villi). The placenta is a combination of the embryonic trophoblast villi and the uterine wall. The villi become bathed in the maternal blood as a result of breakdown of the uterine blood vessels. Thus, gas exchange and nutrient diffusion is facilitated between mother and fetus. The fetus can also aquire immunity to

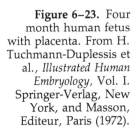

Figure 6–23. Four month human fetus with placenta. From H. Tuchmann-Duplessis et al., *Illustrated Human Embryology*, Vol. I. Springer-Verlag, New York, and Masson, Editeur, Paris (1972).

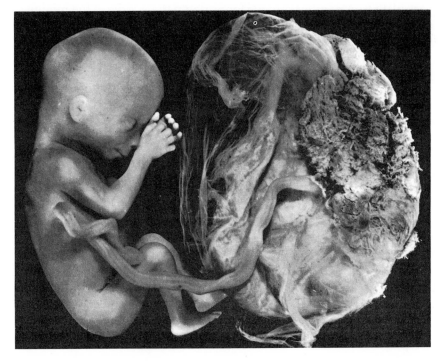

certain diseases as the result of maternal antibodies that can be passed from the mother's blood to the fetus via the placenta.

Many drugs, chemicals, and alcohol are passed to the fetus through the placenta. Forming embryonic organs are in their most sensitive states during early embryonic development and are very likely to be adversely affected by drugs, chemicals, alcohol, and pathogens. Mothers who take drugs, smoke, or drink alcohol are exposing their developing children to agents known to cause damage to embryonic tissues. The sedative thalidomide is one example of a drug that causes extensive damage to fetal limb, heart, and gut development when taken by mothers during the first two months of pregnancy, during the time these organs are developing.

Human Birth Defects

About two or three percent of human births result in malformed babies. There are all sorts of malformations that can occur, and the mechanisms that cause them are varied. Ten percent of women who took the drug thalidomide gave birth to babies with severe limb anomalies, which are otherwise rather rare. Some babies are born with a defect in the closure of the abdominal wall. In these cases the viscera protrude externally. Other babies are born with an open neural tube. Some twins are born unseparated. A single median eye is present in some babies. Some individuals are born with holes between their heart chambers, persistence of gill clefts, blindness, or lack of hearing.

The factors that cause developmental abnormalities include infective agents such as viruses, X rays, certain drugs and food additives, improper nutrition, mutations and chromosomal abnormalities arising from biological error, and possibly others.

The many factors that cause defects do so by a variety of mechanisms. These include improper gene dose (as we saw with Klinefelter's syndrome and Trisomy 21) arising from abnormal chromosome number; effects on inducers or the responding tissues, effects on fusions of bilateral structures resulting in fusion failure, effects on the processes resulting in persistence of embryonic conditions, effects on completion of development of an organ or tissue and inhibition of cell proliferation. It is obvious that factors can affect almost any aspect of embryonic development. Exactly how these agents cause defects is not well understood, however.

A key to understanding the effects of the multitude of agents and factors in causing developmental defects is the concept of the sensitive periods. Certain structures are sensitive to specific agents at given times and only at given times. The German measles virus only causes blindness in the baby if the mother has this disease between about the fifth and eighth weeks in her pregnancy. This is so because the virus must be present at the time of the closure of the lens cup. Once the lens cup closes, virus can no longer enter the lens and destroy it. Similar sensitive periods for the effects of German measles virus on other structures are also known. Ear destruction by the virus can only occur during the seventh to twelfth weeks of pregnancy. Damage to the heart occurs from the fourth to the ninth weeks.

Some drugs, such as those that inhibit RNA synthesis, cause defects only if they are present when the embryonic cells are synthesizing RNA. For example, RNA synthesis occurs during cleavage and drugs such as actinomycin D that inhibit RNA synthesis also inhibit development. Also, new RNA is synthesized during organogenesis. At this time RNA synthesis inhibitors have major effects on development. Once organs are formed and little new RNA is being synthesized, RNA synthesis inhibitors do not cause major defects. So it is clear that an agent can cause defects only if that agent is present at a time when processes are occurring in the embryo that are sensitive to that specific agent. We should also keep in mind that some agents produce severe defects in some species but not in others. Thalidomide, for example, causes severe limb defects in humans and rabbits, but not in rats at comparable doses. Thus, to conclude that a given drug is safe for man because it does not harm a given animal species is, at best, risky. Drugs must be tested on several different mammalian species, and even then their safety with respect to humans is seldom known for sure.

Experiments with Mammal Embryos: Are Blastomere Fates Fixed?

Beatrice Mintz and others have performed a set of exciting experiments that shed light upon the ability of mammalian blastomeres to form tissues other than those they usually form.

Separate cleaving mouse embryos or cells from such embryos were aggregated together and implanted into the uterus of foster mother mice. These embryos developed and resulted in the birth of normal mice. Specific cells could be marked with radioactive labels or with genetic markers such as specific enzyme characteristics. By combining parts of embryos and following the fate of labeled cells, it was found that each cell of the 4-cell embryo can form trophoblast or inner cell mass. In an experiment with rabbit embryos, it was also found that if seven of the eight cells of the 8-cell stage were destroyed, the remaining cell could form an entire embryo that gives rise to a fully formed adult. In addition, if one or several labeled cells from the inner cell mass of a mouse blastocyst are injected into another mouse blastocyst, a large portion of the developing host embryo tissues contain descendents of the labeled cell. These results suggest that the fates of cells in the early mouse embryo are not fixed. Such studies suggest that a great deal of damage could be done to early mammalian embryos without affecting normal development, and this should help the embryo survive traumas that may occur early in development, before implantation occurs.

By the blastocyst stage, however, although the inner cell mass cell fates may not be fixed with respect to the tissue they can form in the embryo proper, the fates of many blastomeres appear to be fixed, at least in their ability to form embryo proper versus trophoblast. It appears that the position of a cell in the early embryo is important in fixing the fate of that cell. Hillman and Graham found that fifteen whole mouse morulas could be aggregated around a central labeled morula. A giant composite embryo formed with the central morula forming inner cell mass only and not trophoblast. If the surrounded morula is on the inside of such a composite for more than eight hours, it loses its ability to form trophoblast. It has also been found that by the blastocyst stage, trophoblast cells of one embryo can not form part of the embryo proper if injected into another blastocyst.

These and other experiments suggest that it is the position of cells in the early embryo that helps fix the fate of those cells. Early in development, prior to the blastocyst stage, the fates of cells in forming trophoblast or embryo proper are not fixed. By the blastocyst stage, however, cells appear to become limited to forming either trophoblast or embryo proper, as a result of factors that influence these cells because of their position in the embryo. These questions will be discussed in greater detail in Chapters 7 and 8. We should keep in mind, however, that although the fate of a cell to form trophoblast or embryo proper may be fixed by the blastocyst stage, due to positional factors, the fates of cells within the inner cell mass to form specific parts of the embryo proper are not as yet fixed.

Summary

In summary, embryos of higher mammals develop in the uterus and are nourished through the placenta; thus there is no need for a great deal of stored yolk. Early development in mammals such as man is similar to that of reptiles and birds, the major difference being the lack of yolk in the eggs and embryos of higher mammals. That yolk once did exist in the ancestral eggs is likely, because the development of mammals conserves many of the mechanisms that apparently evolved with the yolky eggs of reptiles and birds.

Early development in the human embryo can be studied in culture. Steptoe and his group were able to remove human eggs from the ovary, fertilize them outside the body, culture them outside of the mother for several days, and then implant the blastocysts into the mother's uterus. These embryos can develop to term, resulting in the birth of normal babies. Successful culture of early human embryos outside the mother will lead to a better understanding of early development in man. Intimate knowledge of human development after the blastocyst stage, however, will probably lag behind knowledge of such development in related mammals such as the mouse. As our understanding of artificial culture conditions improves, it is likely that mammalian embryos will develop in culture for longer and longer periods, allowing careful study at all times. In fact, mouse embryos presently can develop in culture for nearly half of their normal gestation period (to about 8½ days). These embryos can therefore be studied with the intensity that only continuous monitoring would allow. It is likely that society would not allow similar long-term experiments with human embryos because it would be inhumane to subject them to conditions that could cause pain, defects, or death. Our understanding of human development will still progress because experimentation with mouse and rabbit embryos, for example, will lead to developments that are likely to substantially increase our understanding of mammalian development. Such understanding will lead to medical advances in the area of birth defects, genetic diseases, and other maladies that affect human infants.

PRIMARY GERM LAYER DERIVATIVES SECTION 2

IN THIS BRIEF SECTION we will outline the major structures derived from the ectoderm, mesoderm and endoderm. This outline will provide an introduction to the derivatives of these three primary germ layer in vertebrates in general. Some differences from the general plan exist in specific organisms and these will be dealt with in later chapters. In the next chapter on organogenesis, we will examine the development of many of these derivatives in detail, look at exceptions to the rule, and discuss mechanisms involved in organ formation. This outline serves the useful purpose of providing an overview of the primary germ layer derivatives as an introduction to the more detailed study of organogenesis.

Ectodermal Derivatives

The outer embryonic layer, the ectoderm, is composed of three regions: prospective neural tube, prospective neural crest, and prospective epidermis. The major derivatives formed from each of these regions are outlined below.

Neural Tube Derivatives

The neural plate, as described in Chapter 5, forms the neural tube as a result of interactions with the archenteron roof that are discussed further in the chapter on induction. The neural tube, influenced by these inductive interactions, differentiates into the brain, posterior pituitary gland, optic vesicles (that give rise to the retina of the eye as described in Chap-

Figure 6–24. Divisions of the vertebrate brain.

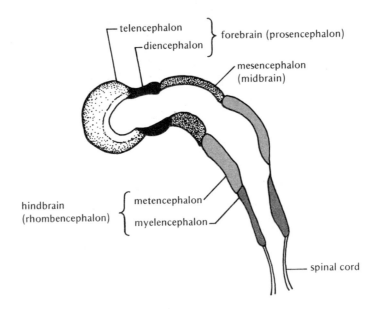

ter 7), spinal cord, and motor nerves that originate in the ventral portion of the neural tube and innervate muscles.

The brain differentiates into three major regions: forebrain (prosencephalon), midbrain (mesencephalon) and hindbrain (rhombencephalon) (Figure 6–24). The anterior portion of the forebrain is the telencephalon that forms the cerebral hemispheres and olfactory centers. The posterior portion of the forebrain is the diencephalon and forms the thirst center (thalamus), the hunger center (hypothalamus), the posterior pituitary, and optic vesicles. The midbrain or mesencephalon forms the visual interpretation centers (optic tecta). The anterior portion of the hindbrain, the metencephalon, gives rise to the cerebellum, while the posterior portion of the hindbrain, the myelencephalon, forms the medulla of the brain (Figure 6–24).

Neural Crest Derivatives

In Chapter 5 we noted that as the neural tube separates from the rest of the ectoderm and drops below the surface of the embryo, a narrow region at the crests of the neural folds also moves within the embryo. This region, the neural crest, consists of cells that become migratory,

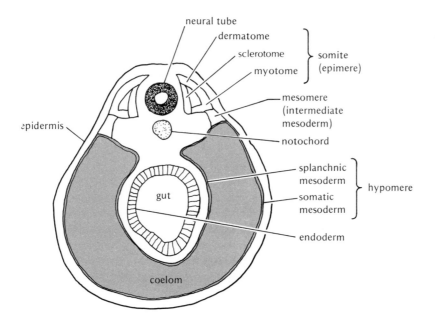

Figure 6–25. Mesoderm divisions.

traveling to distant parts of the body. These migratory neural crest cells form sensory nerves and ganglia (that receive impulses from sense organs), autonomic ganglia (that control involuntary activities), the adrenal medulla (inner part of the adrenal gland), all the pigment cells in the body with the exception of pigmented retina cells (derived from the neural tube), cartilages in the voice box and head, and some ectodermal muscles.

Epidermal Derivatives

After the neural tube and neural crest have dropped below the surface of the embryo (Chapter 5), the remaining surface ectoderm is the epidermis. Epidermal derivatives can be divided into two types: those derived from epidermal thickenings and those derived from the rest of the epidermis. Epidermal placode derivatives are: some head nerves, the lens of the eye, olfactory structures, the inner ear, and the taste buds. The remainder of the epidermis forms the outer layer of the skin, hair, horns, nails, mouth and anal linings, and the anterior pituitary.

Mesodermal Derivatives

The middle embryonic layer, the mesoderm, became situated between the ectoderm and endoderm as a result of the gastrulation process. Figure 6–25 shows the major subdivisions of the mesoderm. These are: the dorsal or upper division (on the back of the animal), called the epimere or somite; the mesomere or intermediate mesoderm; and the hypomere, the remainder of the mesoderm. Let us simply outline the major derivatives arising from these regions.

Epimere (Somite) and Mesomere (Intermediate Mesoderm) Derivatives

The somite or epimere is divided into three regions: myotome, dermatome, and sclerotome. The outer region of the somite is called the dermatome and gives rise to the dorsal portion of the dermis of the skin (dermis on the back of the animal). The inner part of the somite consists of the myotome that forms the back muscles and the sclerotome that gives rise to the vertebral column.

The mesomere or intermediate mesoderm gives rise to the kidneys, gonads, and associated structures.

Hypomere Derivatives

The hypomere is composed of two parts, somatic mesoderm and splanchnic mesoderm. Somatic mesoderm is the hypomere material closely associated with the epidermis. It is the outer portion of the hypomere in the lateral (side) and ventral (belly) regions of the embryo. The lateral and ventral parts of the dermis of the skin, the limb buds, and the outer portion of the peritoneum (the coelom lining) are derived from somatic mesoderm.

Splanchnic mesoderm is the hypomere material closely associated with the endoderm of the embryo. The inner part of peritoneum (the lining of the body cavity), smooth (gut) muscle, the heart and blood vessels, and embryonic blood cells are derived from the splanchnic mesoderm.

Thus, the three regions of the mesoderm form a variety of structures. The development of many of them will be examined in Chapter 7. In this brief outline of primary germ layer derivatives we have looked at the derivatives of the outer and middle embryonic layers. We now turn our attention to the inner germ layer, the endoderm.

Endodermal Derivatives

The inner embryonic layer is the endoderm. This germ layer surrounds the gut cavity. The structures derived from the endoderm are either part of the gut tube proper or result from outpocketings of the tube. These relationships are diagrammed in Figure 6–26. Let us examine the major endodermal derivatives of the embryo.

The gut tube proper is divided into three major regions: foregut, midgut, and hindgut. The anterior portion of the foregut is the pharynx.

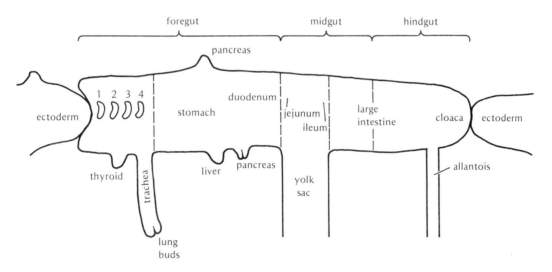

Figure 6–26. Endodermal derivatives. Diagram of the gut tube and its outpocketings.

Figure 6–27. Diagram of human embryo showing endodermal derivatives. From H. Tuchmann-Duplessis et al., *Illustrated Human Embryology*, Vol. I. Springer-Verlag, New York, and Masson, Editeur, Paris (1972).

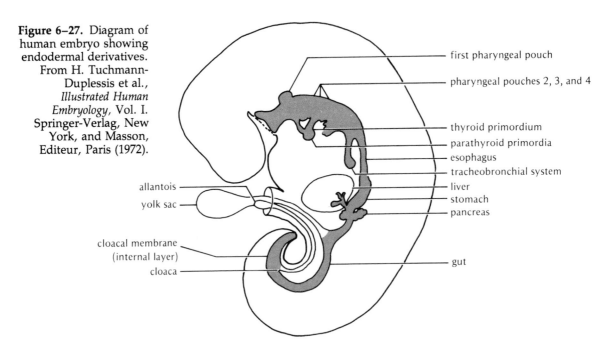

first pharyngeal pouch

pharyngeal pouches 2, 3, and 4

thyroid primordium

parathyroid primordia

esophagus

tracheobronchial system

liver

stomach

pancreas

allantois

yolk sac

cloacal membrane (internal layer)

cloaca

gut

The middle region forms the stomach, while the posterior region of the foregut becomes the duodenum. The small intestine consists of three regions: duodenum, jejunum, and ileum. The anterior segment of the midgut is the jejunum and the posterior region forms the ileum and part of the large intestine. The hindgut is divided into the remainder of the large intestine and the cloaca. Some authors consider the entire large intestine to be midgut, but this is not an important distinction.

Let's briefly examine the endodermal derivatives that arise from the gut tube proper. As can be seen in Figure 6–26, four pairs of pouches develop from the lateral walls of the pharynx. These are the pharyngeal pouches. Pouch one forms parts of the middle ear. Tonsil tissue is derived from pouch two (and also from the pharyngeal walls). The thymus (involved in formation of antibody-forming cells) and parathyroid glands (involved in calcium and phosphorus metabolism) appear to be derived mostly from pharyngeal pouches three and four. Part of the thymus may be of ectodermal origin. The pharynx tube itself forms the esophagus that leads to the stomach, and ventrally, forms the trachea that ends in the lung buds. The thyroid gland is an anterior outpocketing from the ventral wall of the pharynx (Figure 6–27).

We indicated that the rest of the foregut tube becomes the stomach and duodenum. Outpocketings also arise from the posterior portion of the foregut tube. The pancreas arises from a dorsal outpocketing and ventral outpocketing, and the liver arises from the ventral wall, posterior

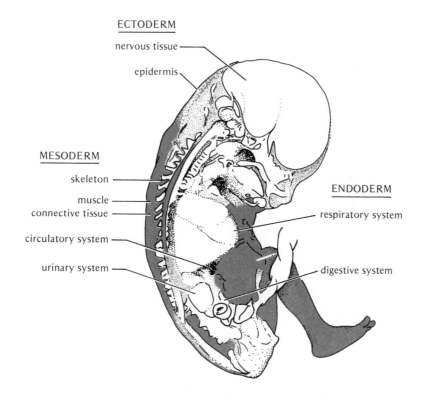

ECTODERM

nervous tissue

epidermis

MESODERM

skeleton

muscle

connective tissue

circulatory system

urinary system

ENDODERM

respiratory system

digestive system

Figure 6–28. Major germ layer derivatives. From Tuchmann-Duplessis et al., *Illustrated Human Embryology*, Vol. I. Springer-Verlag, New York, and Masson, Editeur, Paris (1972).

to the trachea invagination. In the bird and mammal embryo, where a yolk sac is present, the sac is directly connected ventrally with the midgut. If an allantois is present, it is derived from and connected, ventrally, with the hindgut. The anterior region of the hindgut forms part of the large intestine, and the posterior region forms the cloaca. The cloaca, in turn, gives rise to the urinary bladder, urethra, and rectum.

This concludes our examination of the major derivatives of the endoderm. It should be noted that there are three places in the body where the endoderm directly contacts the ectoderm without any mesoderm separating the two. These places of endoderm-ectoderm contact are at the stomodeum (mouth opening), the cloaca, and the pharyngeal pouches.

In this brief section we have outlined major derivatives of the three germ layers of the embryo. This outline should serve as a generalized study summary of the primary germ layer derivatives and their relationships with one another. It also sets the stage for a more detailed analysis of the development of many of these structures in Chapter 7. Figure 6–28 is a summary study guide of the major germ layer derivatives described in this section.

Readings

Early Human Embryo

Austin, C. R., and R. V. Short (eds.), *Reproduction in Mammals*, Book 2, *Embryonic and Fetal Development*, Cambridge University Press (1972).

Carr, D. H., Chromosomal Abnormalities in Clinical Medicine. *Prog. Med. Genet.*, 6:1 (1969).

Corner, G. W., *Hormones in Human Reproduction*, Atheneum, New York (1963).

Hillman, N., M. I. Sherman and C. F. Graham, The Effect of Spatial Arrangement on Cell Determination During Mouse Development. *J. Embryol. Exp. Morph.* 28:263–278 (1972).

Hsu, Y. C. Differentiation *in vitro* of Mouse Embryos to the Stage of Early Somite. *Develop. Biol.* 33:403–411 (1973).

Mintz, B., Clonal Basis of Mammalian Differentiation. *Symp. Soc. Exp. Biol.* 25:345–369 (1971).

Moustafa, L. A., and R. L. Brinster, The Fate of Transplanted Cells in Mouse Blastocysts *in vitro. J. Exp. Zool.* 181:181–202 (1972).

Redding, A., and K. Hirschhorn, *Guide to Human Chromosomal Defects.* Birth Defects Original Article Series, 4, National Foundation, New York (1968).

Rugh, R., *The Mouse: Its Reproduction and Development*, Burgess, Minneapolis (1968).

Steptoe, P. C., and R. G. Edwards, Reimplantation of a Human Embryo with Subsequent Tubal Pregnancy. *Lancet* 1–7695:880 (1976).

Tuchmann-Duplessis, H., G. David, and P. Haegel, *Illustrated Human Embryology.* Springer Verlag, New York (1971).

Primary Germ Layer Derivatives

Balinsky, B. I., *An Introduction to Embryology*, 4th ed. W. B. Saunders, Philadelphia (1975).

De Haan, P. L., and H. Ursprung (eds.), *Organogenesis*, Holt, Rinehart & Winston, New York (1965).

Hamilton, W. J., J. D. Boyd, and H. W. Mossman, *Human Embryology*. Blakiston Co., Philadelphia (1962).

Mossman, H. W., Comparative Morphogenesis of the Fetal Membranes and Accessory Uterine Structures. *Contrib. Embryol.* Carnegie Institute, Washington 26:129–296 (1937).

Nelson, O. E., *Comparative Embryology of the Vertebrates*, Blakiston Co., New York (1953).

Patten, B. M., *Human Embryology*. Blakiston Co., Philadelphia (1946).

Patten, B. M., *Foundations of Embryology*, McGraw-Hill, New York (1958).

Rugh, R., *Experimental Embryology*, Burgess, Minneapolis (1948).

Witschi, E., *Development of Vertebrates*, W. B. Saunders, Philadelphia (1956).

CHAPTER 7

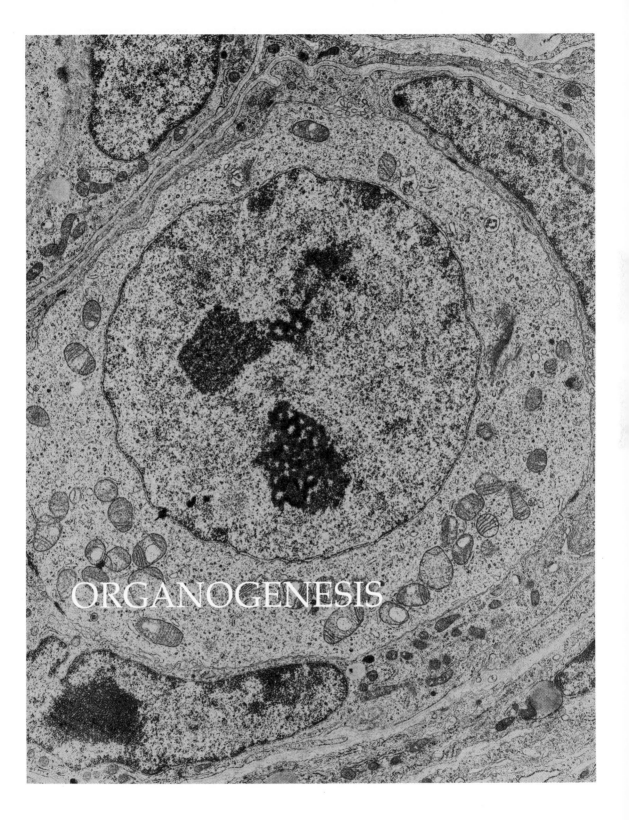

ORGANOGENESIS

EYE DEVELOPMENT

WE HAVE EXAMINED the early development of embryos and have outlined the many derivatives of ectoderm, mesoderm, and endoderm. Now, let us zero in on some of these derivatives in more depth, stressing the development of organs and organ systems that convey concepts.of general importance in developmental biology.

We will examine organogenesis—the development of organs and organ systems—from both an anatomical and an experimental viewpoint. In this way, we will be able to understand how our knowledge of the development of organs has grown from the research in the field. We will first consider the vertebrate eye, and then discuss the neurological system, the heart, the development of limbs, the urogenital system, and the immune system.

The Vertebrate Eye

The development of the eye has been a stimulating topic of investigation for many laboratories. Experiments on mechanisms of eye development often involve elegant operations on the eye-forming region. In this section, we will look at some of these experiments to try to understand the nature of the mechanisms that control eye embryogenesis. We will also describe some interesting experiments that deal with the area of eye regeneration. Before we deal with experiments let us first look at the structural aspects of eye development to gain an understanding of the basic system that provides the framework for the experiments on mechanisms of development and regeneration that will be described.

Eye Development: Summary of Events

The major events in eye development can be briefly summarized as follows.

1. Contact with the roof of the archenteron (the notochord) causes the eye cup rudiments that are part of the neural plate (the prospective neural tube) to develop into the optic vesicles.

optic vesicles

2. The optic vesicles form as lateral evaginations (outgrowths) of the posterior portion of the forebrain (diencephalon).

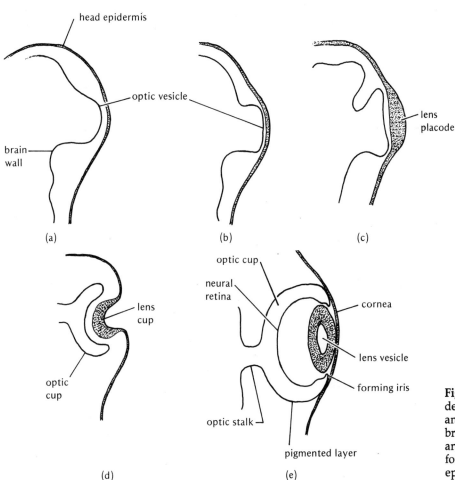

Figure 7–1. Steps in eye development. Retina and iris form from the brain wall while the lens and part of the cornea form from the epidermis.

eye cups

pigmented layer

neural (sensory) retina

lens forming ectoderm

3. The optic vesicles contact the prospective lens ectoderm. Contact with this area influences the development of the optic vesicles to form the eye cups. The eye cups consist of an outer pigmented layer and an inner layer of neural (sensory) retina.

4. The lens forming ectoderm (epidermis) has previously interacted with underlying foregut endoderm and portions of the prospective heart mesoderm. These interactions apparently maintain lens-forming competence in the prospective lens ectoderm. The final induction, however, that causes the lens to form arises from contact of the lens forming ectoderm with the tip of the optic vesicle. The lens then forms, first by thickening, and then by folding or cellular rearrangements of the lens forming epidermis.

cornea

5. The lens and optic cup, by touching the overlying epidermis and mesenchyme, induce formation of the transparent protective covering of the eye, the cornea.

choroid coat

6. The choroid coat and sclera, outer coats of the eye, develop from mesenchyme that accumulates around the eyeball.

sclera

7. The iris of the eye, the structure that regulates the size of the pupil, develops from the rim of the optic cup.

iris

Some of these events are summarized in Figure 7–1. A drawing of a fully developed eye is shown in Figure 7–2.

Mechanisms of Eye Development (Optic Vesicles)

Exactly what causes the optic vesicles to form from the brain wall is not known. We do know that the archenteron roof induces the neural tube to differentiate into many components, including the optic vesicles. Specific genes in the prospective optic vesicle cells probably become activated to form the specific messages that code for optic vesicle proteins. The actual process of evagination of the optic vesicles followed by cup formation of these vesicles may involve cytoplasmic microtubules and microfilaments that cause cells to change shape. Changes in the adhesiveness of cells in the optic vesicle area may also be involved in causing evagination and cup formation. These mechanisms are discussed in depth in Chapter 9. Recall that similar mechanisms were implicated in

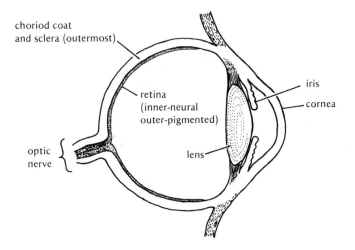

choriod coat
and sclera (outermost)

retina
(inner-neural
outer-pigmented)

iris

cornea

optic
nerve

lens

Figure 7–2. Drawing of sagittal section of human eyeball, showing major parts discussed in this section.

neural tube formation (Chapter 5) and in gastrulation (Chapter 4). The expansion of the optic vesicles appears to be aided by the accumulation of fluid in the brain ventricles. These ventricles are continuous with the optic vesicles. Coulombre and Coulombre inserted a glass tube through the wall of the eye to prevent the build-up of fluid pressure. This resulted in abnormal eye development. It is also likely that regional patterns of mitosis and cell growth may play roles in optic vesicle development.

Differentiation of the optic vesicles seems to be dependent upon its touching the surface ectoderm (prospective lens). If such contact fails to take place, the prospective neural retina tends to form pigmented retina instead of neural retina. In addition, formation of the optic cups from the optic vesicles normally appears to be dependent upon contact with the developing lens ectoderm. Mesenchyme (embryonic connective tissue) that accumulates around the eyeball also appears to be important for normal eyeball development. If this mesenchyme is removed, the optic vesicles do not develop normally.

Let's sum up what we have seen so far. The optic vesicles evaginate from the brain wall. After contact with the prospective lens, the single-layered optic vesicles become pushed in, or cup shaped. The inner layer of the cup becomes the neural retina (sensory retina) and the outer layer becomes the pigmented retina. Optic vesicle development is dependent upon archenteron roof induction, fluid build up, contact with mesenchyme, and contact with the prospective lens epidermis.

The eye rudiment (in the neural plate and optic vesicle stages) can be split in half. Each half can give rise to a complete eye. Thus, the eye rudiment can self regulate. That is, the cells that remain after a part is removed are not fixed in their fate, but instead some cells can change

self regulation

Figure 7–3. Layers of
the neural retina.
Arrows show impulse
path.

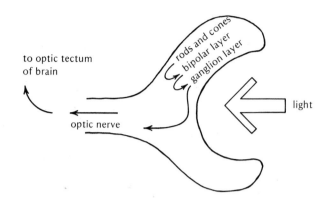

their prospective fates to form the complete structure. That the fates of
the early prospective parts of the eye are not fixed, can also be demonstrated by placing ear vesicle or nasal placode in contact with the prospective pigmented retina of the optic cup. These abnormal inducers can
cause neural retina to develop from the pigmented layer. Also, if the tip
of the optic vesicle fails to contact the prospective lens, the prospective
neural retina tends to form pigmented retina.

rods and cones   The rods and cones, the light-sensitive receptor cells of the neural
retina, develop in the outermost part of the retina. That is, they develop
in the layer closest to the choroid coat and farthest from the lens and
pupil. In order for light to reach the rods and cones, it must first pass
ganglion cells   through a layer of ganglion cells, followed by a layer of bipolar neurons
(Figure 7–3). The light excites the rods and cones. The impulse is passed
bipolar neurons   on to the bipolar cells and then on to the ganglion cells that transmit the
message to the brain via the optic nerve. Optic nerve fibers from an eye
optic chiasma   enter the side of the brain opposite from that eye. The optic chiasma is
the place at which the nerve fibers from the two eyes cross. In mammals,
not all optic nerve fibers cross. Some enter the brain on the same side as
the eye of origin. Differentiation of the neural retina begins with the
ganglion cells, followed by the bipolar neurons. The rods and cones are
last to differentiate.

A very intriguing question that we may ask is: what causes the nerve
processes from the retina of the eye to enter and connect to very specific
regions of the brain? In other words, when the nerve cells from the eye
enter the optic tectum, the visual interpretation center of the mesencephalon of the brain, what causes these cells to attach to specific regions of
the optic tectum? It is known that in chick embryos, for example, nerves
from the dorsal part of the retina "home" to the ventral part of the optic
tectum. Nerves from the ventral retina "home" to the dorsal portion of
the optic tectum. A fascinating approach to this problem has recently
been taken by Barbera, Marchase and Roth. The details of this work will
be given in Chapter 9, but let's summarize these experiments here in the

context of eye development. These workers hypothesized that adhesive recognition may play an important role in governing where the nerve cell endings from the eye will finally attach in the brain. That is, the cells will stay where they "stick" the best. This hypothesis was tested by rotating labeled single cells from the dorsal *or* ventral retina with pieces of dorsal *or* ventral optic tectum. The researchers measured the number of labeled dorsal or ventral retina cells that adhered to each tectal half. The results indicated that dorsal retina cells adhered best to ventral tectum and ventral retina cells adhered best to dorsal tectum. The results suggested that adhesive recognition may indeed play an important role in controlling "retino-tectal" nerve hook-ups. Dorsal retina nerves hook up to ventral tectum and also stick best to ventral tectum as shown by Roth's group. Ventral retina cells, likewise, hook up to and stick best to dorsal tectum. At this point, however, we can only suggest that adhesiveness does play the key role in controlling the right "hook-ups". The future should provide new insights in this intriguing area.

The molecular nature of the selective adhesion of dorsal retina cells to ventral optic tectum and ventral retina cells to dorsal optic tectum is being investigated by Roth's group. Preliminary evidence that is far from proving anything suggests that the nature of this specific adhesion may involve specific enzymes on one cell surface binding to sugar chains on the adjacent cell surface. These experiments will be explored further in Chapter 9, "Mechanisms of Morphogenesis."

Before moving on to the parts of the eye derived from epidermis, let us conclude this discussion by noting that the optic cup gives rise to other structures besides the neural retina. The outer wall of the optic cup becomes the pigmented coat posteriorly and thins out anteriorly to form the iris and the basal part of the ciliary body that helps control lens **ciliary body** shape. The pigmented coat cells give rise to cytoplasmic processes that interdigitate with the outer parts of the rods and cones. The neural retina and pigmented coat thus become locked together. The pigmented coat appears to help nourish the neural retina by transporting nutrients between the blood vessels of the choroid coat and the cells of the neural retina.

Mechanisms of Lens Development

Some of the mechanisms involved in lens development have been elucidated. We have already seen how contact with the lens forming ectoderm plays a key role in development of the optic cup. Let us begin our study of lens formation by examining the other part of the latter relationship. That is, when the optic vesicle touches the lens forming epidermis, does this contact also play a role in inducing lens formation? The answer to this question is yes. The final induction or direct cause of the lens forming ectoderm to form a lens comes from contact or proximity with the optic vesicle (Figure 7–4). Exactly how such contact causes lens formation is not well understood.

The ability of the prospective lens ectoderm to form lens is maintained by previous contact with the foregut endoderm and portions of the prospective heart or head mesoderm. The final induction of the lens is caused by proximity or contact with the optic vesicle. Molecules may pass from nearby tissue to the lens forming ectoderm, causing lens formation to occur. Abnormal inducers such as ear vesicle and guinea pig thymus can also cause lens formation in competent ectoderm. Environmental influences such as temperature also affect final lens induction. The nature of induction, at least what we know about it, is discussed in more detail in Chapter 8.

What is the nature of the hypothetical inducer that appears to pass from the optic vesicle to the lens forming ectoderm? The answer to this question is not known, but there is evidence that substances do pass from the optic vesicle to the lens forming ectoderm. At the time of induction, one can observe a decrease in cytoplasmic basophilia (acidity) and a decrease in the number of ribosomes in the tip of the optic vesicle (prospective neural retina) and a corresponding increase in cytoplasmic basophilia and an increase in the number of ribosomes in the lens forming ectoderm. These results, however, do not prove that something actually passes from the optic vesicle to the lens ectoderm. More direct evidence for passage of material comes from experiments in which the optic vesicles were labeled with radioactive amino acid (C^{14} phenylalanine). This label began to appear in the lens forming ectoderm. This result suggests that something is passed between the optic vesicle and lens forming ectoderm, but it does not indicate exactly what is passed or what the nature of the hypothetical inducer is.

There have been some additional experiments that illustrate the relationship between optic vesicle and lens forming ectoderm. If the optic vesicle is removed before it reaches the lens forming epidermis, no lens usually forms. A thin layer of cellophane placed between the optic vesicle and lens forming ectoderm blocks lens formation. A thin slice of agar, however, which may allow passage of molecules, inserted between optic vesicle and lens forming ectoderm did not block lens formation. It

Figure 7–4 (opposite). Seventy-two hour chick embryo showing eyecups and lenses on either side of the large diencephalon cavity. Courtesy of Peter Armstrong.

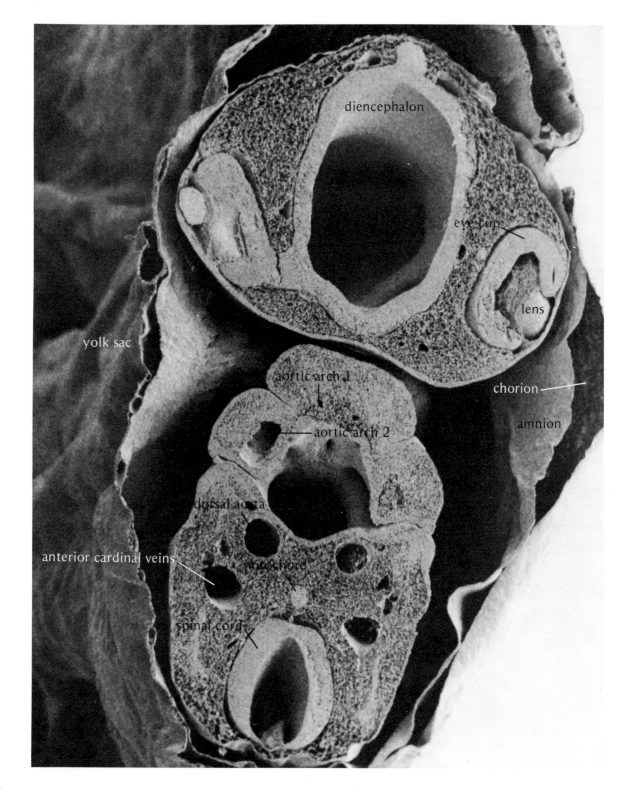

Figure 7–5. Lens fiber formation. Epithelial cells elongate into fiber cells (cortex fiber cells). Nucleus fiber cells are those that were formed during early lens development. From Papaconstantinou, *Science*, 156:338, Copyright 1967 by the American Association for the Advancement of Science.

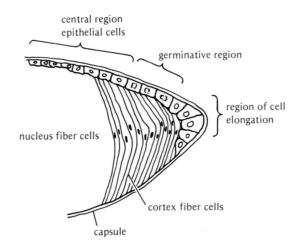

should be noted, however, that in certain species and under certain conditions lens formation can occur to some extent without final induction by the optic vesicle. This may occur as the result of some other factors that can influence the lens forming ectoderm, such as interaction with the foregut endoderm, or heart or head mesoderm, or environmental influences such as temperature. For example, low temperatures appear to favor lens induction by head mesoderm, making contact with the optic vesicles less necessary. Most lenses that form without induction by the optic vesicles are, however, not normal. Thus, we can conclude that although many factors may stimulate lens formation, contact or proximity with the optic vesicle provides the final induction, under normal conditions, for the development of the lens.

What factors actually cause the lens forming ectoderm to fold and form the lens vesicle? In Chapters 5 and 9 we examine some of the factors that may cause a flat sheet of cells to fold, forming a cup and then a vesicle. These factors include cytoplasmic microtubules and microfilaments that change the shape of cells, so that folding of cell sheets can occur. Also, changes in adhesiveness of cells with their neighbors may play a role in vesicle formation. Another way in which a sheet of cells can fold to form the lens vesicle may involve tight cell attachment and mitosis. If the lens forming ectoderm cells become firmly attached to each other so that no lateral cell movement can occur, and if mitosis increases the mass of these cells, the sheet of cells will buckle inward (See Fig. 9–11, Chapter 9). Microfilament bundles and extracellular materials have been observed at the outer surface of the lens forming ectoderm. These may play a role in limiting expansion of the cell layer so that folding will occur.

The lens begins to form as a thickening in the epidermis called the lens placode. The lens placode then invaginates to form the lens vesicle.

Fiber cells form from lens epithelial cells. Within the fiber cells, proteins called crystallins form. These proteins give the lens the ability to focus light upon the retina. Contact of the lens with the presumptive retina of the optic vesicle appears to induce lens fiber differentiation (Figure 7–5).

crystallins

Lens Regeneration. The lens can regenerate in vertebrates such as salamanders. In such animals, if the normal lens is removed, a lens regenerates from the dorsal rim of the iris of the eye (Figure 7–6). Thus, a lens, which is usually an epidermal derivative, can regenerate from the iris, which is a neural derivative. The events involved in lens regeneration from the iris include loss of pigment granules in the iris, and cell division. The cells then undergo typical lens formation, forming epithelial cells and lens fibers. If the lens is removed from the eye, but then inserted back into the same eye, no new lens is formed. The iris does not begin to undergo change. These results suggest that the lens itself inhibits the iris from forming another lens. Only mature lenses and not undifferentiated lenses can inhibit regeneration of a lens from the iris. When lens proteins from mature lenses are separated by gel electrophoresis, it has been found that two of the lens proteins, when placed in the eye, can inhibit lens regeneration from the iris. Thus, lens protein from the mature, differentiated lens may be released into the eye fluid and may block formation of additional lenses from the iris. These experiments on lens regeneration suggest that the salamander eye has evolved an elegant means of regenerating a lens from cells that normally do not form the lens. Along with this development, a means of preventing formation of additional lenses in a normal eye also has evolved.

Eye Development and Viral Damage. Before leaving the lens, let's mention how studies on lens development have led to an understanding of an important virus-caused birth defect in human beings, namely blindness caused by German measles (rubella) virus. It has been known for a long time that if a pregnant woman is infected with German measles virus during the first three months of pregnancy, her newborn child may

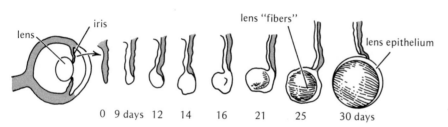

Figure 7–6. Lens regeneration from iris of salamander eye. From N. K. Wessells, *Tissue Interactions and Development*, p. 170, based on experiments in R. W. Reyer, *Quart. Rev. Biol.* 29 (1954).

Figure 7–7.
Susceptibility of the lens
to virus infection.
Studies by Saxen's
group indicate that the
embryonic lens is only
susceptible to infection
by German measles
virus before closure of
the lens cup.

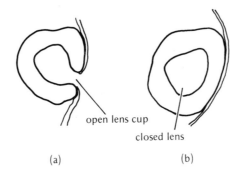

open lens cup

closed lens

(a)   (b)

be blind. If she contracts the disease after the first trimester of pregnancy, blindness rarely occurs in the newborn child. It is known that in the human embryo, eye development occurs during the first six weeks. At six weeks, the lens vesicle has closed and lens fiber differentiation takes place. It has been suggested by Robertson, Blattner, and Williamson that the lens and other organs that arise from ectodermal invaginations are only susceptible to the harmful effects of certain viruses as long as they are open to the exterior (e.g., cup stage or earlier). Once they close, viruses can no longer enter the structures (Figure 7–7). Saxen's group in Finland showed that this was indeed true for the German measles virus induced lens cataract in human embryos. Using chick embryos and human embryos obtained from therapeutic abortions, these workers showed that the virus-sensitive period was limited to the open stage of the lens vesicle. Closed lenses could only be infected if virus was introduced surgically directly into the lens vesicle. Thus, a careful study of the developing lens has led to important information regarding the nature of a major congenital abnormality.

Other Eye Structures

The cornea, the transparent covering of the front of the eye, is derived from epidermal ectoderm that lies over the anterior portion of the eye and from mesoderm (mesenchyme) cells that lie under the ectoderm. The optic cup, and especially the lens, induce and maintain the cornea. The mesenchyme of the cornea is continuous with the sclera (the protective coat that is seen as the white of the eye). The epidermal portion of the cornea is continuous with the skin or eyelid epithelium. The development of corneal transparency involves the dissolution of pigment granules in the prospective cornea. The lens and eye cup stimulate corneal development. If the eye cup is removed, the cornea does not develop. The lens alone can be transplanted under other epidermis. This epidermis then loses its pigment and differentiates into cornea. The eyeball also seems to be needed to maintain corneal transparency. If the eyeball is removed, chromatophores invade the cornea and transparency is lost. The cornea becomes similar to normal skin.

The shape of the cornea is important in allowing light waves to bend in the proper way. The curvature of the cornea is caused by pressure exerted by the eye fluid against the cornea. The coats around the eye include the outermost protective white of the eye (the sclera), and the layer between the sclera and retina (the choroid coat). The choroid coat is pigmented and contains blood vessels that nourish the eye. The choroid coat and sclera develop from mesenchyme that accumulates around the eyeball. Factors that may cause these coats to form may come from the pigmented retina coat or from the neural retina.

Summary

In summary, the eye develops from a complex series of reciprocal interactions that occur among the optic vesicles, epidermis, and mesenchyme. The nature of the so-called "inducers" is unknown. Some of the mechanisms that may be involved in optic vesicle, lens, and cornea development have been described. Mechanisms that cause formation of the optic cup and lens vesicle may include cell shape changes caused by cytoplasmic microtubules and microfilaments, differential cell adhesiveness, and mitosis. The experiments by Roth's group suggest that selective adhesive recognition may be involved in the proper hook-up of optic nerve fibers with the brain. Our understanding of the molecular basis of eye morphogenesis is still in its infancy. These molecular mechanisms will eventually be elucidated. In this age of biochemistry that time may not be very distant.

SECTION 2 # DEVELOPMENT OF NERVOUS INTEGRATION AND BEHAVIOR

HOW DO NERVES get to and innervate specific end organs? How do nerves that originate in the central nervous system get to where they are going to innervate the multitude of sense organs and muscles in our body? How does behavior develop? These questions are among the most absorbing and also, as yet, the most poorly understood problems in developmental biology today. Hypotheses that attempt to deal with these questions have been proposed, and some concrete experimental evidence is slowly improving our understanding of the problems of nerve integration and behavior development. Here, we will briefly survey these problems and potential solutions.

Hypotheses of Nerve–End Organ Hookup

A variety of hypotheses have been proposed to explain how nerves get to and hook up with specific end organs. A few of these will be described here. We can categorize these hypotheses as chemical, electrical, mechanical, and adhesive recognition models of nerve growth and end organ innervation. Before we look at these models, it should be stressed that the complete story that eventually will explain the complex interactions that occur during nerve–end organ hookup will probably be a combination of more than one of the models that will be described. Therefore we should keep in mind that it is unlikely that any one of the hypotheses will turn out to be the whole answer. We will see why this is the case as we examine the experimental evidence for each model.

Chemical Model

The chemical model suggests that the direction of nerve outgrowth is in response to a gradient of specific chemicals. This is a chemotaxis hypothesis, which suggests that nerves grow towards specific chemicals. Recall that we confronted the question of chemotaxis once before, in Chapter 2, where we saw that known chemicals given off by female gametes of certain plant forms specifically attract male gametes of those species. There is little concrete experimental evidence in support of a chemical model of directional nerve growth except in one major area, that of a substance called nerve growth factor.

nerve growth factor

A protein has been purified from mouse submaxillary glands. This protein stimulates massive nerve outgrowth from ganglia (Figure 7–8). This protein is called nerve growth factor and appears to be present in many sources, such as mouse sarcoma tumor, snake venom, and mouse salivary glands. The factor was discovered when Bueker, in order to see if nerves innervated rapidly growing tissue, implanted mouse sarcoma 180 tumor tissue into the body wall of two to three day chick embryos. This tumor tissue stimulated massive nerve outgrowth from adjacent ganglia. Rita Levi-Montalcini suggested that the tumor contains a nerve growth factor. In early attempts to determine if the factor was nucleic acid or protein, it was subjected to snake venom that contains the enzyme phosphodiesterase, which degrades nucleic acid. In control experiments, the snake venom itself was used to see if any nerve outgrowth occurred. To the surprise of the investigators, the venom was about one thousand times more potent than the tumor homogenate in promoting nerve outgrowth.

Salivary glands of mice, similar to glands that produce snake venom, were found to have nerve growth factor activity up to ten thousand times higher than the tumor homogenate and ten times higher than snake venom. Tiny quantities of nerve growth factor (such as 0.5 micrograms per gram of body weight) cause a four to six fold increase in developing sympathetic ganglia.

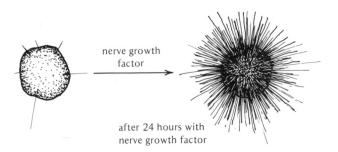

nerve growth
factor

after 24 hours with
nerve growth factor

Figure 7–8. Effect of nerve growth factor on chick embryo sensory ganglion. A seven day embryo sensory ganglion is cultured *in vitro* in a medium containing nerve growth factor. From experiments by Rita Levi-Montalcini and B. Booker.

Antibodies were prepared against nerve growth factor by Cohen and Levi-Montalcini by injecting purified factor into rabbits. When this antibody was injected into newborn rats, mice, rabbits, and kittens, it prevented the growth of sympathetic chain ganglia (nerve cell clusters of part of the nervous system in the spinal cord area). These experiments suggest that a substance similar to nerve growth factor exists in the embryo and may be responsible for the growth and maintenance of sympathetic and sensory nerve cells.

How does nerve growth factor work? Does nerve growth factor influence the direction of nerve growth or is it just a general growth stimulating agent? We can not as yet provide complete answers to these questions. What does nerve growth factor do to cells that respond to it? First of all, if one examines nerve cells exposed to nerve growth factor, large quantities of microtubules and microfilaments (see Chapter 12) appear in the cytoplasm. These fibrous elements may play a role in nerve growth and differentiation, and the cytoplasmic flow in nerve axons. In addition to the formation of such elements in nerve cells exposed to factor, RNA synthesis and protein synthesis are stimulated in such cells. Lipid synthesis and glucose oxidation are also enhanced in the presence of nerve growth factor.

Nerve growth factor, like insulin, appears to bind to specific surface receptors on nerve cell membranes. This can be observed by examining the rate of binding of radioactively labeled nerve growth factor to the cell surface. It may be that nerve growth factor, like many hormones, has a primary effect on the cell membrane. By binding to the surface membrane receptor, nerve growth factor may set off a series of reactions that are transported to the cytoplasm and nucleus by some sort of chemical messenger such as cyclic AMP (adenosine monophosphate). Some recent work suggests that nerve growth factor may be a protein-degrading enzyme (a protease). There is no conclusive evidence as yet, however, regarding the sequence of events by which nerve growth factor acts to stimulate nerve growth. There is also no solid evidence that supports the contention that nerve growth factor is responsible for the highly specific sort of directional nerve growth required for hooking up specific nerves with specific end organs. It may be that nerve growth factor is responsible for nerve cell growth and development, but not for guiding nerves to their specific end organs.

Electrical Model

Might it be that nerves grow toward an end organ because of electrical currents or gradients between the outgrowing nerve and the end organ?

There is no conclusive evidence to support this model of nerve growth. Some experiments, however, suggest that nerves do respond to electricity. If direct electric current is applied to amputated forelimb stumps of laboratory rats, some regeneration occurs. Cartilage, bone marrow, bone, blood vessels, muscles, and nerves develop in this tissue. It may be that electric currents stimulate nerve growth.

When nerve explants were placed in a culture dish and an electric current was directed through the culture medium, nerve outgrowth was always found to be in the direction of the current. It seems, therefore, that nerves do grow in response to electric current. A more careful repeat of this experiment was performed by Paul Weiss. He found that the nerves would grow in the direction of such current even after the current was turned off! An examination of the culture plates by Weiss revealed that the current oriented the culture gel, causing the formation of fine grooves in the culture medium. These grooves formed in the direction of the electric current. The nerves did not follow the electric current itself, but instead followed the grooves formed by the electricity in the substratum. These experiments, therefore, do not provide any evidence for an electrical model of nerve growth. They do not, however, rule out the possibility that, in the body, small electric currents may play some role in guiding nerve growth, but evidence for this contention is still lacking.

Mechanical Model

The results of Paul Weiss's experiments, that nerves followed grooves in the culture medium, support a mechanical model of nerve guidance. These experiments were extended by Weiss by preparing little hollow frames filled with blood plasma. The frames were of different shapes. As the plasma clotted, lines of tension were set up in different directions corresponding to the shape of the frames. Weiss found that nerves always grew out along these lines or grooves. He termed such nerve growth in response to mechanical components or oriented components of the substratum contact guidance. The nerves, therefore, appear to grow according to mechanical guidelines. They use these guidelines as road maps to get to where they are going.

contact guidance

In the animal, an extension of this model suggests that the types of road maps used by nerves to get to where they are going include blood vessels. There is little doubt that nerves, *in vivo*, do use mechanical guidelines such as blood vessels to get to where they are going. Nerves can be observed to follow blood vessels and oriented tissues in the body.

We can conclude that the mechanical model is probably an important part of explaining how nerves get to where they are going.

It is unlikely, however, that such a model is the whole story or that contact guidance explains all of the aspects of nerve–end organ hookup. There appear to be too many highly specific nerve hookups that are found in end organs to be explained by contact guidance alone. For example, in the section on eye development, we note that nerves from the dorsal part of the retina hook up to very specific regions on the ventral part of the optic tectum of the brain, and nerves from the ventral part of the retina home to very specific regions on the dorsal portion of the optic tectum. The home bases on the optic tectum are extremely close together. The neurons appear to be able to locate, very precisely, tiny regions on the tectum. Several models have been proposed to explain highly specific nerve hookups. We'll mention one of these here. Some of the others will more appropriately be developed in the behavior section.

Adhesive Recognition Model

Roth and his colleagues proposed that nerves may find specific regions in their end organs by adhesive recognition. This model proposes that a nerve gets to its final resting place by a chemical recognition between the surfaces of the growing nerve cells and the surfaces of the cells in the region of the end organ where the nerve is destined to connect. This chemical recognition is proposed to be one that causes adhesion of the nerve to the end organ. It is this adhesion that keeps the nerve in its proper place and prevents the nerve from moving on to another area. The specific molecular groups responsible for such adhesive recognition may turn out to be sugar-binding enzymes on one cell surface that attach to sugar chains on the other surface. Some evidence in support of such an enzyme-substrate recognition has been obtained by Roth and co-workers. This matter of adhesive recognition being responsible for morphogenetic events such as nerve–end organ hookups is dealt with in detail in Chapter 9.

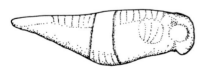

Figure 7–9. Reversal of skin on tadpole flank. From experiments by Sperry and Miner, 1951.

Development of Behavior

It may be that of all the functions that develop in the embryo, fetus, and newborn, behavior is the most complex. Understanding of the development of behavior is still in its infancy and far from the molecular level that has been reached in other areas. We are still at the experimental embryology level of investigation in this area. There is little doubt, however, that in the not too distant future we will understand the development of behavior at the molecular level as we now are beginning to understand areas such as early egg activation, cell differentiation, and others.

Basic questions in the development of behavior are: Do nerves that innervate end organs have knowledge of the region before they innervate the area? Or are nerves indifferent and only learn where they are or what they must do after they innervate the end organs? These questions are important to the story of behavior because they deal with the problem of what influences what in the setting up of nerve pathways essential for behavior.

Sperry and Miner performed experiments that shed light on these problems. They rotated a large strip of skin on the flank of a frog tadpole 180°. This skin was oriented such that what was once dorsal skin was now in a ventral position and ventral skin in a dorsal position (Figure 7–9). The skin was allowed to heal into place with this new orientation. In the newly metamorphosed frog, irritation in the dorsal part of the original skin (now lying ventrally) resulted in the frog scratching its back. Irritation of the ventral part of the original skin (now lying dorsally) resulted in the frog scratching its belly. Thus the frog responded according to the old orientation of the skin, not the new. This behavioral response could result, for example, from two possibilities:

1. The skin attracted the appropriate nerve. For example, belly (ventral) skin (even if it was in a dorsal position) attracted a belly nerve.

2. The skin altered the way in which the nerve reacted, or, in other words, instructed the nerve about what kind of skin it really was.

An examination of the innervation of the reversed skin revealed that the nerves in the specific region innervated the skin in that region. In other words, a normal back (dorsal) nerve innervated the belly skin that was now on the back and a normal belly nerve innervated the back skin that was now on the belly. So originally ventral axons that now innervate the back skin that has been moved to the stomach area appear to aquire dorsal properties and may realign their interconnections with other neurons in the central nervous system (spinal cord) so that motor impulses go to nerves that innervate dorsal muscles. Thus possibility number two is likely, namely that the skin (end organ) can instruct nerves as to what they should do. Steinberg, however, suggests that an additional alternative may exist. It may be that an anatomically dorsal nerve has some physiologically ventral fibers and an anatomically ventral nerve may have some physiologically dorsal fibers. Even if this is the case, it appears that the skin (end organs) has some sort of instructive role in causing the right nerve fiber to become activated.

We should stress that in many cases of nerve regeneration, the regenerating nerves appear to know exactly where to go as if they are part of a printed circuit. Behavioral evidence that supports the contention that nerves know exactly where to go with respect to end organ hookup comes from experiments with amphibian eyes. If the eyes of frogs or salamanders are rotated 180° and the optic nerves cut, vision is restored following nerve regeneration from the eyes to the visual centers in the brain, the optic tecta. The behavior of the animal, however, is such that it responds as though every part of its visual field was upside down. Thus it appears that the regenerating nerve fibers reach their original end stations in the optic tecta. This has been confirmed by direct anatomical observations with regenerating optic nerves in fish. Nerves, therefore, seem to know exactly where to go (Figure 7–10).

The concept that early behavior develops as a result of nerve–end organ hookup and not from learning is supported by a variety of additional studies. Many muscles in the embryonic chick become active only after nerves hook up to these muscles. Thus, the nerves appear to influence the muscle in enabling it to function. Also in the chick, the intricate hatching movements appear precise and integrated right from the beginning without any "practice" at earlier stages.

If salamander embryos are reared in an anesthetic drug that prevents muscle activity, they develop but are paralyzed. When the larvae are returned to water, without the drug, they immediately behave like normal swimming larvae that were totally raised without the drug. Thus all of the early behavioral activities were not learned and began immediately as soon as the muscles were allowed to act. So certain behavior appears to be controlled totally by inherent qualities of the nerve-muscle hookup. Likewise, in the chick embryo, some muscular movements occur before any functional sensory nerve connections exist. These movements appear totally controlled by motor nerve–muscle hookups.

Interesting experiments have been performed with kittens that suggest that there is a time in development in which the vision nerves are plastic and able to be specified by environmental influences. After this period of plasticity, the nervous pathways are fixed and learning can

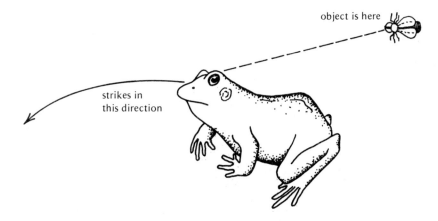

object is here

strikes in
this direction

Figure 7–10. Behavior of frog after 180° rotation of eyes. After Sperry, R. W., in *Handbook of Experimental Physiology*, S. S. Stevens, ed., John Wiley & Sons, pp. 236–280, 1951.

not undo the specification that occurred during the plastic period. The plastic period appears to exist in the one to two month old kitten. A similar plastic period may appear in the two to four year old human child. If kittens are reared during the plastic period in environments consisting only of visually vertical stripes or horizontal stripes, they do not cope well with the real world in terms of visual detection. The cats raised with horizontal stripes walk into vertical objects such as chair legs without apparently detecting them. The cats raised with vertical stripes do not respond to horizontal visual cues such as the chair seat. The animals lacked the ability to detect anything made of lines that were not present during the earlier plastic period.

The importance of such studies to human development is clear. For example, if a two to four year old child has severe squint or astigmatism, permanent visual problems can occur because brain pathways are fixed at that age. Attempts at correcting such a problem after the plastic period may not be possible. It is likely that many types of behavior in human beings are initiated by pathways that develop during the plastic period. Exposure of the child to a variety of learning experiences during the two to four year old period may be the best way to assure that the child will learn well in later years.

HEART DEVELOPMENT

THE HEART PUMPS BLOOD through the vessel system and thereby supplies the body tissues with food and oxygen and removes wastes that accumulate in these tissues. Obviously it is an organ we can not do without. The development of the heart involves a series of cellular migrations, fusions, and specific differentiation. It is a system that reflects a multitude of morphogenetic mechanisms described in Chapter 9. Let us examine experiments that have led to our present understanding of heart embryogenesis. First, however, it would be useful to summarize the events that occur in heart development. In this way we will have a general outline of these events in our mind before we examine the experiments that have led to our current understanding of heart

Heart Embryogenesis: Summary of Events

The major events that occur during heart development in the chick embryo can be summarized as follows.

1. The heart begins to differentiate as two vesicles originating in the hypomere mesoderm on either side of the developing foregut. It will be shown later how these vesicles originate.

2. As the ectoderm and endoderm fold to form the head fold of the embryo, to lift the anterior end of the embryo off the yolk, the two heart rudiments are brought together in the ventral midline. The two heart rudiments fuse. Each of the rudiments consists of an inner lining, the endocardium, and an outer muscle layer, the myocardium (Figure 7–11). These events occur between the 25th and 30th hour of incubation in the chick embryo or between the 7 and 20 somite stage. In human beings, similar events occur during the third week, or by about the 8-somite stage.

endocardium

myocardium

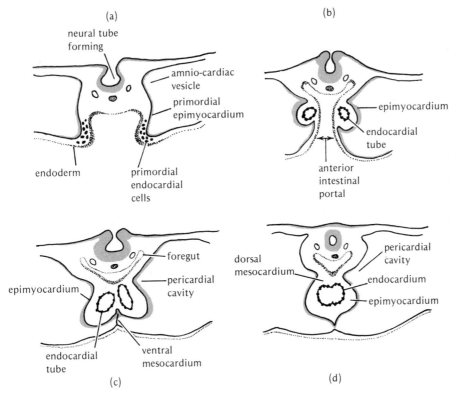

(a)

neural tube forming

amnio-cardiac vesicle

primordial epimyocardium

endoderm

primordial endocardial cells

(b)

epimyocardium

endocardial tube

anterior intestinal portal

(c)

foregut

pericardial cavity

epimyocardium

endocardial tube

ventral mesocardium

(d)

dorsal mesocardium

pericardial cavity

endocardium

epimyocardium

Figure 7–11. Cross sections of heart development in 25 to 30 hour chick embryos. From T. W. Torrey, *Morphogenesis of the Vertebrates*, John Wiley & Sons, Inc.

3. Fusion of the paired heart vesicles occurs first at the anterior, head end, and proceeds posteriorly. The first region formed after fusion is the most anterior portion of the heart, the truncus or conus arteriosis that leads to the ventral aorta blood vessels and the ventricle, the thick walled muscular pumping chamber. Next to form by fusion of the paired rudiments is the atrium, the heart chamber that delivers blood to the ventricle. The last part of the heart to form is the sinus venosus, the heart chamber that receives venous blood.

truncus or conus arteriosis

ventricle

atrium

sinus venosus

4. The heart tube bends to form an S-shape. The heart begins to beat just after the beginning of fusion of the paired heart rudiments, just before the conus arteriosus forms.

Heart formation in amphibians is in many ways similar to that described for the chick. There are no body foldings in the amphibian embryo, to bring the heart areas together. Instead, the prospective heart mesoderm cells on each side of the embryo converge ventrally (Figure 7–12). Mesenchyme-like loose cells appear to come from the right and left regions of the ventrally converging hypomere splanchnic mesoderm.

Figure 7–12. Development of amphibian heart.

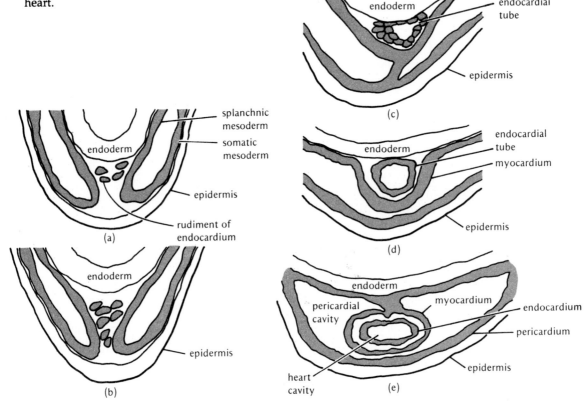

These cells form the endocardium, the inner heart lining as a thin-walled tube. The heart endocardial tube divides at both ends. At the anterior end are the two ventral aortae and at the posterior end the two vitelline veins that bring the first venous blood to the heart join with the heart. As the hypomere regions from each side converge and fuse ventrally, the splanchnic mesoderm envelops the endocardial tube, forming the muscular myocardium. As in the chick, the heart tube in the amphibian twists, forming an S-shape. The vitelline veins, and two common cardinal veins (formed by the union of veins from the head and posterior regions of the body), enter the most posterior chamber, the sinus venosus. This chamber leads into the thin-walled atrium, which in turn joins with the thick-walled ventricle and conus arteriosus as in the chick (Figure 7–13). This general S-shape is the basic condition found in all vertebrates. Later in embryogenesis, however, hearts of different vertebrate groups develop differently, sometimes forming multiple chambers that adapt the hearts to the physiological needs of specific animal groups.

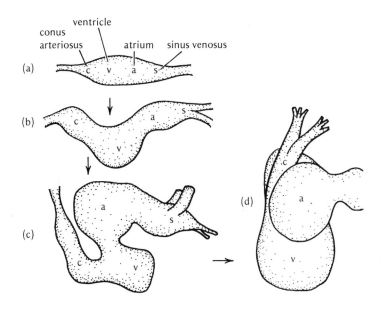

Figure 7–13. Twisting of frog heart rudiment. From Joy B. Phillips, *Development of the Vertebrate Anatomy* (1975). The C. V. Mosby Co., St. Louis.

Experimental Analysis of Mechanisms of Heart Development

The analysis of chick heart development has been accomplished by a number of elegant experiments by Mary Rawles, Robert DeHaan, and others. These experiments illustrate how radioactive labeling, microsurgery, and transplantation techniques can be used to elucidate the nature of the development of a major organ. By isolating portions from any part of the pre-streak blastoderm of chick embryos in culture, it was found that beating tissue developed. Thus, heart forming potential is not restricted to any localized region of the early epiblast. As the embryo progresses to the primitive streak stage, only cells from the posterior half of the blastoderm develop into beating tissue when isolated in culture— the heart forming region is becoming more localized as development proceeds. Mary Rawles showed that certain fragments from the blastoderm of a head-process–stage embryo would give rise to heart tissue when cultured on the vascular chorioallantoic membrane of an older embryo. She determined that in the head-process stage embryo—the stage when notochordal cells move forward from Henson's node—the heart-forming regions appeared to be localized in the epiblast at either side of the midline. These cells seemed to move through the primitive

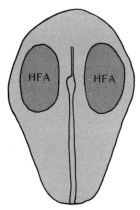

Figure 7–14. Heart forming area (HFA) in head process stage of chick blastoderm.

groove and take up residence on either side of the head process (Figure 7–14).

DeHaan and his colleagues Rosenquist and Stalsberg performed a beautiful series of experiments designed to determine more accurately the fates of very specific regions of the blastoderm in forming the heart. These workers transplanted fragments of blastoderm from a radioactively labeled donor embryo to an unlabeled recipient. The fate of many such specific fragments can be followed by autoradiography. By this method, and by carbon marking (using specks of carbon to mark specific areas) and microcinematography, it was determined that in the mid-streak stage, when the primitive streak is well developed, heart-forming cells lie in paired regions about midway down the length of the streak, extending from the midline about halfway to the edge of the embryo. The cells enter through the primitive groove and by the end of the head-process stage the heart forming mesoderm is organized in two separate regions. The most anterior parts of the heart-forming regions fuse to give rise to the most anterior part of the heart, the conus arteriosus. The middle parts of the heart-forming regions fuse to give rise to the middle part of the heart, the ventricle, and the posterior portion of the heart forming regions fuse to give rise to the most posterior parts of the heart, the atrium and sinus venosus (Figure 7–15). This brings us back to the summary description of events given previously. The experimental analysis has helped us to understand the events that lead up to the formation of the two mesodermal vesicles that fuse, with which we began the summary section.

How does the heart-forming mesoderm get to where it is going? DeHaan and colleagues, using the labeling technique just described, determined that the heart forming mesoderm of the head process embryo moved forward relative to the endoderm. This was accomplished by transplanting a radioactively "hot" square of endoderm-mesoderm into a "cold" host embryo, as done previously. After a few hours the embryo was sectioned and examined by autoradiography. The results showed that the radioactive mesoderm moved forward relative to the endoderm (Figure 7–16). As will be seen shortly, the heart forming mesoderm appears to use the associated endoderm as a road map to help it get where it is going.

We can now ask the questions: Do any inductive interactions of the type found in other systems occur during heart development? What is the relationship between the heart forming mesoderm and closely associated endoderm? In the amphibian system, some evidence for inductive interactions in heart formation can be found. If the endoderm is removed from newt embryos, the heart never develops. Thus, the intimate contact between endoderm and heart-forming mesoderm in the amphibian embryo seems to be required for heart development.

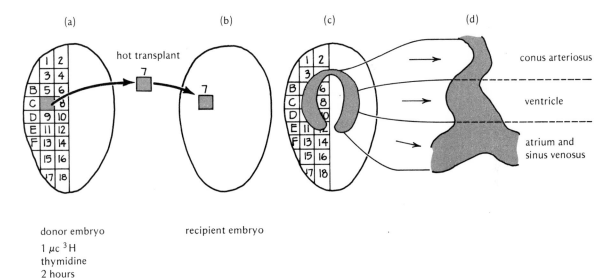

donor embryo
1 μc ³H
thymidine
2 hours

recipient embryo

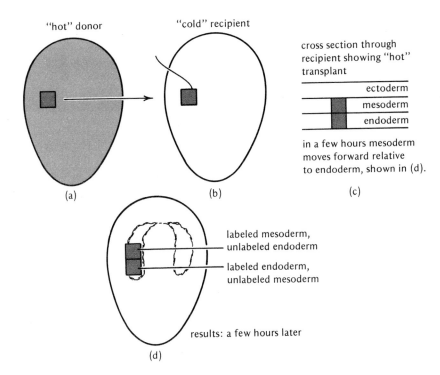

Figure 7–15 (above).
Experimental
determination of chick
heart fate-map. "Hot"
explant from (a)
transplanted to cold
recipient (b). (c) and (d)
show the fate-map of
the chick heart forming
regions projected upon
a chick heart, developed
from the experiments
shown in (a) and (b).
After R. L. DeHaan,
Ann. N.Y. Acad. Sci.,
1965, and R. L. DeHaan
in *The Emergence of Order
in Developing Systems,*
M. Locke, ed. (1978)
Academic Press.

Figure 7–16 (left).
Forward migration of
heart forming
mesoderm relative to
endoderm in chick
embryo. After
experiments by DeHaan
and colleagues.

Certain agents can be used to separate the pre-heart mesoderm and endoderm in the chick embryo. These substances include sodium citrate and ethylene diaminetetraacetic acid (EDTA) that bind divalent cations. If head-process–stage chick embryos are treated with sodium citrate, the endodermal layer can be nearly completely removed, leaving the ectoderm and mesoderm relatively intact. Many small twitching heart vesicles are formed in these embryos, but they do not migrate together to form a heart. These experiments suggest that the endoderm is the specific substratum (cell layer) needed for prospective heart cells to migrate upon to get to the heart region. If a piece of endoderm with the precardiac mesoderm attached is removed from the embryo and cultured *in vitro*, the mesoderm migrates along the endoderm surface and differentiates into a twitching heart vesicle. It appears that the endodermal cells in the precardiac area elongate, furnishing an oriented substratum for cardiac tissue migration. Twitching vesicles will form without endoderm, but they do not form a heart unless the endoderm "guides" the mesoderm cells to the right place. Adhesive gradients may exist in the endoderm that may assist the prospective heart cells in moving in a specific direction. For example, the pre-heart cells may move in the direction of increasing "stickiness" of the associated endoderm. It is clear that the nature of the interactions that occur between the heart-forming mesoderm and endoderm are still not well understood.

A final question that can be asked concerning the nature of heart development is: What causes and controls heart beating? It has been shown that differentiation in heart muscle cells occurs after the cells have stopped dividing. At the 10-somite stage in the chick embryo, heart beating begins. Beating begins in the ventricular region and spreads to the rest of the heart as the other regions differentiate. The heart beat begins irregularly, then follows a rhythmic rate of about 35 beats per minute, increasing to about 85 beats per minute as the ventricle has looped, and after sixty or so hours of incubation, the chick heart rate reaches the typical embryonic rate of about 115 beats per minute.

If a heart that is contracting at about 115 beats per minute is cut into three sections, the sinoauricular part (sinus venosus plus atrium or auricle), the ventricular part, and the conus part, only the sinoauricular portion beats at 115 contractions per minute. The other two fragments revert back to the slower early embryo rate. This experiment suggests that the sinoauricular region acts as a "pacemaker" that spreads the faster beating rate to the other parts of the heart. The factors controlling heart rate seem, however, to be very complex. Experiments with chick cells have shown that, in culture, the pulsation rate is inversely related to the size of the cell cluster (with seven day heart cells). A larger cell cluster tends to beat more slowly than a smaller cell cluster. It is therefore unlikely that pacemaker cells alone govern the absolute rate of whole heart beating.

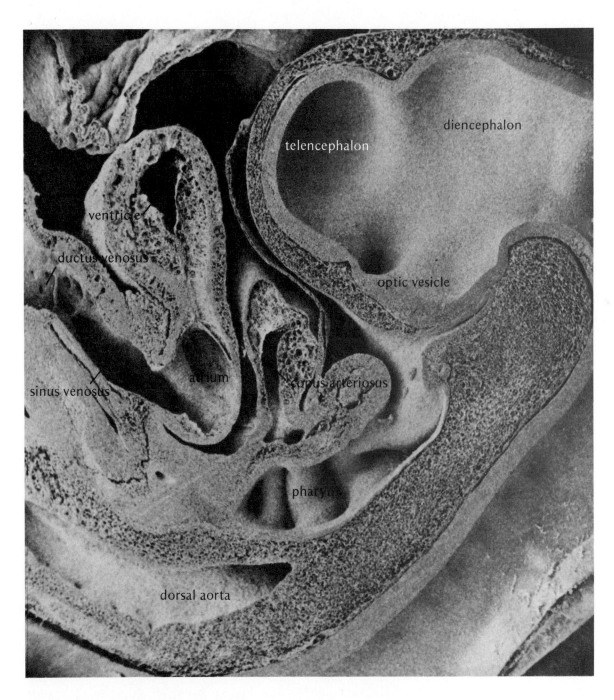

Figure 7–17. Seventy-two hour chick embryo showing heart chambers. Courtesy of Peter Armstrong.

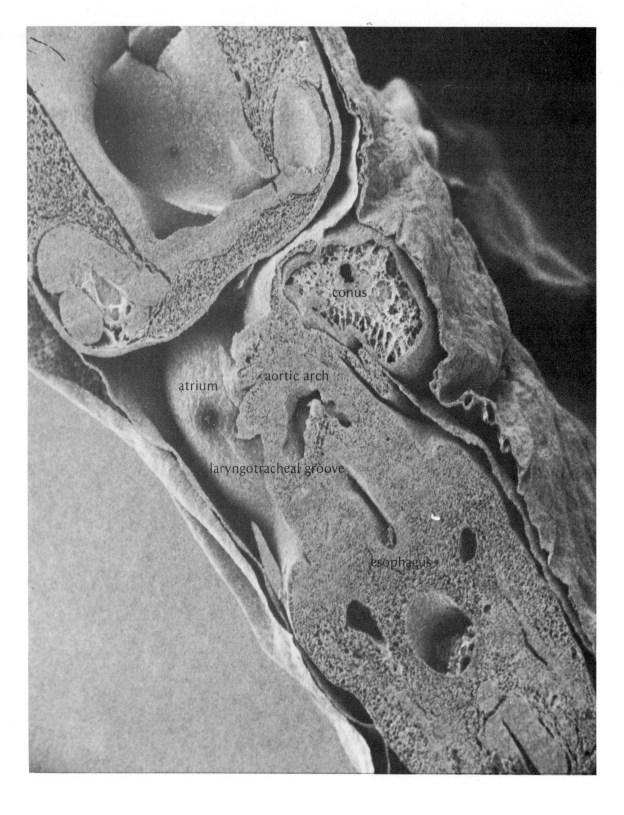

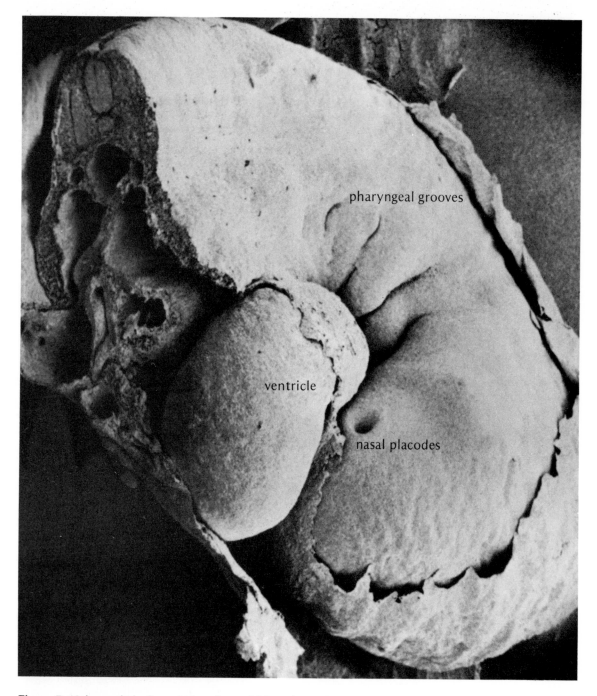

Figure 7–18 (opposite). Seventy-two hour chick embryo showing heart chambers. Courtesy of Peter Armstrong.

Figure 7–19 (above). Seventy-two hour chick embryo showing heart chambers. Courtesy of Peter Armstrong.

Instead, other factors such as volume of heart tissue present also play a role in governing heart rate. In addition, DeHaan has shown that potassium ion concentration plays a key role in controlling heart cell beating. A lower potassium ion concentration appears to increase the percentage of cells that beat.

In summary, a variety of mechanisms seem to be involved in heart development. Migrations of prospective heart-forming mesoderm cells in the chick embryo occur on an endoderm substratum, eventually forming two mesodermal vesicles on either side of the developing foregut. Fusion of these heart rudiments in an anterior to posterior direction occurs to form the early heart. The newly formed heart tube twists into a looped structure with specific chambers. This twisting may be caused by migrations of sheets of cells and changes in cell shape (see Chapter 9). Finally, after cell division stops, heart muscle cell differentiation takes place and spontaneous beating occurs. Much remains to be learned about heart development but we have come a long way in our understanding of this area of embryology, thanks to the work of Rawles, DeHaan, and others.

LIMB DEVELOPMENT AND REGENERATION

Limb Development

THE DEVELOPMENT OF LIMBS has been extensively studied during this century, and a great deal of interesting information has surfaced regarding limb development. Much of the understanding we now have about limb development has resulted from experiments that involve operations on developing limbs of vertebrates. We will stress some of these experiments here. In this section we will examine basic limb development and some of the concepts that have developed concerning the mechanisms that control limb development. In addition, after investigating the components of normal limb development in vertebrates, we will take a brief look at the intriguing process of limb regeneration. Some vertebrates can regenerate entire limbs after the normal limbs are amputated. An understanding of regeneration in these organisms may lead to methods of stimulating limb regeneration in man. As will be seen, we have come a long way towards an answer to the regeneration problem. We are on the threshold of major, medically useful developments in this area.

A Limb Fate Map

Before we examine the component processes involved in limb development, let's briefly consider the origin of the limb buds and the nature of a limb bud fate map. Recall from Chapter 6 that limbs are derived from the somatic mesoderm of the hypomere. Cells from the somatic mesoderm accumulate in the limb-forming regions and migrate to a position right below the epidermis. The epidermis right above this accumulation of mesoderm cells thickens in many vertebrates, and is termed the apical ectodermal ridge (Figure 7–20). Thus the limb bud consists of a core of somatic mesoderm covered by epidermis. The epidermis is often differ-

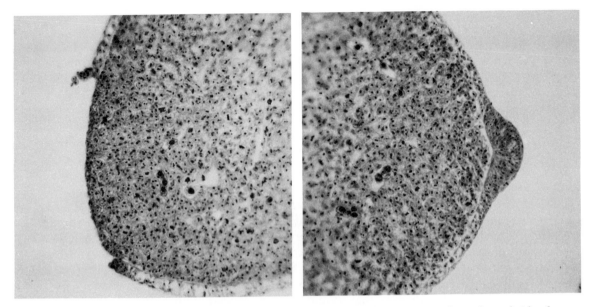

Figure 7–20. Tip of chick embryo wingbud. (a) Apical ectodermal ridge has been removed. (b) Apical ectodermal ridge is present. From Saunders, *J. Exp. Zool.* 108:363-403 (1948). Courtesy of J. Saunders.

apical ectodermal ridge

entiated into an apical ectodermal ridge. A summary of the formation of the limb bud is shown in Figure 7–21.

Figure 7–22 shows a fate map of a chick forelimb developed by Saunders. Fate maps such as this one are worked out by labeling different portions of the limb bud with carbon particles or radioactive tracers, and following the fate of these marked regions in the adult limb. Recall from Chapter 4 that a similar procedure was used by Vogt to construct whole embryo fate maps. Note in the fate map that all the bones of the limb are derived from the mesoderm portion of the limb bud. We will return to a discussion of the role of the mesoderm versus the ectoderm in limb formation later in the story.

The Limb-Forming Field

embryonic field

The limb bud and surrounding area is an example of an embryonic field, an area that simply possesses a set of specific properties. The subject has

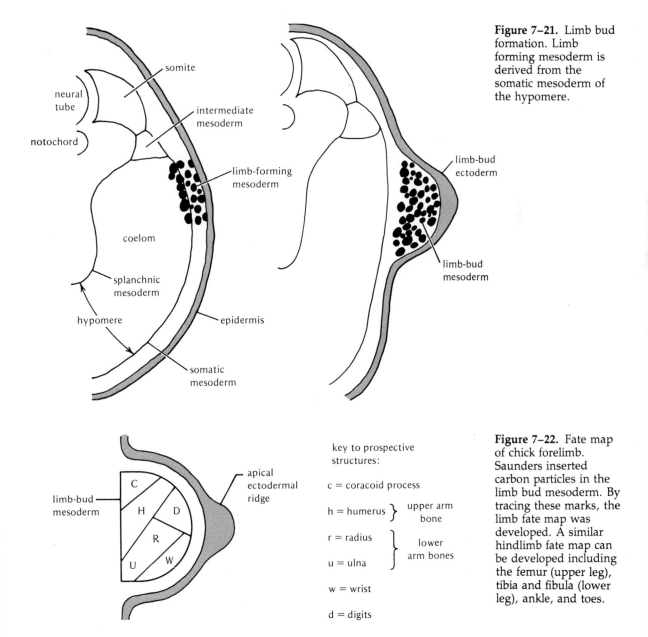

Figure 7–21. Limb bud formation. Limb forming mesoderm is derived from the somatic mesoderm of the hypomere.

Figure 7–22. Fate map of chick forelimb. Saunders inserted carbon particles in the limb bud mesoderm. By tracing these marks, the limb fate map was developed. A similar hindlimb fate map can be developed including the femur (upper leg), tibia and fibula (lower leg), ankle, and toes.

key to prospective structures:

c = coracoid process

h = humerus } upper arm bone

r = radius ⎫
           ⎬ lower arm bones
u = ulna ⎭

w = wrist

d = digits

been most thoroughly examined in relation to the limb system, so before examining the details of limb development, let us define what is meant by an embryonic field and why the limb bud area is called an embryonic field. By doing this we will begin to understand the nature of the limb bud and its properties as a developmental system.

What are the properties of an embryonic field? Embryonic fields possess the following four properties:

1. The potency of tissue to form a given structure is more widespread in the embryo than only the prospective structure.

2. The expression of the potency of tissue to form the structure decreases with distance from the prospective structure.

3. Parts of the primordium of the structure (the embryonic tissue bud that gives rise to a structure) can yield the complete structure.

4. Augmentations of the primordium can give rise to the complete structure.

Developing amphibian limb buds, eye, ear, and heart primordia are examples of embryonic fields. That is, these primordia possess the set of four properties given above. This has been determined to be so as a result of numerous experiments that indicate that a given primordium area has each specific field property. A brief examination of the four properties given above indicates that an embryonic field is an area that has the ability to regulate itself. In other words, if some pieces of the area are removed, a complete structure can still be formed, so cells in an embryonic field must be able to recognize their own position and their own fate. These cells also must be able to change their fates in response to disruption of the normal primordium. Let us begin to understand the limb-forming area by examining interesting experiments that indicate that the limb-forming area is indeed an embryonic field. After we get a "feel" for the nature of the limb-forming area, we will turn to an examination of experiments that elucidate the roles of the different portions of the limb bud in controlling limb formation.

The first field property is that the potency to form a given structure is more widespread than the prospective structure. The amphibian embryo is choice material for experiments on the nature of the limb-forming area because operations on embryos are easily accomplished, embryos are readily available, and they develop in plain water. Byrnes, Braus, and Balinsky performed experiments on frog embryos that support the first field property. If a normal forelimb bud is removed from the embryo, a normal limb still develops from the tissue surrounding the normal limb bud. Thus for the amphibian limb-forming area, the potency to form the limb is indeed more widespread than the limb bud itself. Field property one is therefore satisfied.

The second field property is that the expression of the potency to form the structure decreases with distance from the prospective structure. Balinsky used abnormal inducers (such as ear vesicle or nose rudi-

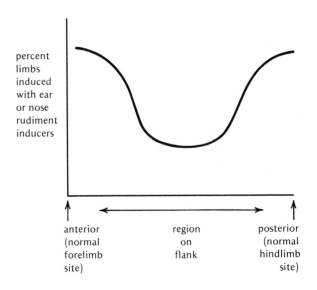

percent
limbs
induced
with ear
or nose
rudiment
inducers

anterior   region   posterior
(normal   on   (normal
forelimb   flank   hindlimb
site)   site)

Figure 7–23. Limb induction in newts with abnormal inducers. Balinsky found that it was easier to induce limbs closer to the normal limb forming site. Limbs formed nearest the normal forelimb site resembled forelimbs. Limbs formed nearest the normal hindlimb site resembled hindlimbs. Limbs formed between the two sites possessed hindlimb and forelimb characteristics.

Figure 7–24 (below). Induction of limbs with artificial inducers. (a)-(c) Limb induction in newt with nose rudiment, (a) and (b) show operation. Position of normal limb buds shown in (a). Induced limb shown in (c). From: I. B. Balinsky, *An Introduction to Embryology*, p. 387, The W. B. Saunders Co., (1975).

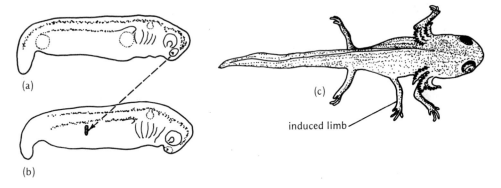

(a)

(b)

(c)

induced limb

ment) to induce limb formation in the area of the flank between the normal forelimb and hindlimb sites in the amphibian embryo. These abnormal inducers cause the underlying mesoderm cells to accumulate under the epidermis to form a limb bud, and an additional limb developed. Balinsky found that the further away from the normal limb-forming site one went, the more difficult it was to induce a limb with the abnormal inducers. The area on the flank equidistant from the normal forelimb site and normal hindlimb site was least able to form a limb (Figures 7–23, 7–24.). In addition, limbs produced closer to the normal forelimb site were forelimb-like (four digits and thin). Limbs formed closer to the normal hindlimb site resembled hindlimbs (five digits and thick). In between the sites, the limbs that did form possessed characteristics of forelimbs and of hindlimbs. Thus, for the amphibian limb-forming area, the expression of the potency to form the limb decreases

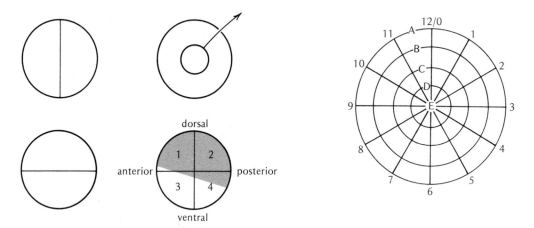

Figure 7–25 (left). Parts of limb bud can give rise to complete limb. Harrison and Swett found that halves (a) and (b), the periphery (c) or single quadrants (d) could give rise to complete limbs. The shading in (d) indicates best limb forming ability.

Figure 7–26 (right). Polar coordinates in a positional information field. Each cell is assumed to have information with respect to its position on a radius (the line from A through E) and its position around the circle (0 through 12). Positions 12 and 0 are identical, so that the sequence is continuous. (From S. J. Bryant and L. E. Iten, *Devel. Biol.* 50:212 (1976).

with distance from the prospective limb. So, field property two is also satisfied.

Recall again that field property three is that parts of the primordium can yield the complete structure. Experiments that support this property of amphibian limb forming area come from the work of Harrison, Swett, and others. These workers removed portions of the limb bud as shown in Figure 7–25. If half or even three-quarters of the limb bud was removed, the remaining portion could form a complete, normal limb. If the limb bud was divided into quadrants, certain quadrants formed limbs at greater frequency than did other quadrants. Parts of the amphibian limb bud can yield a complete limb. Thus, field property three is satisfied.

Recall that the fourth field property is that augmentations of the primordium (increasing the amount of tissue) can give rise to the complete structure. If an extra piece of limb mesoderm is stuffed under the ectoderm together with the normal mesoderm in an amphibian embryo, a single complete limb is still formed. Field property four is thus satisfied.

The limb-forming area is therefore truly an embryonic field because experimental evidence indicates that the limb-forming region possesses all four field properties. How can cells in the limb-forming area adjust to all of the operations performed to jointly form a normal limb? This problem has not yet been fully solved. French, Bryant, and Bryant, however, have recently developed a theory to explain how cells in fields such as limb buds may be able to assess their positions and therefore regulate development of the structure. They suggest that each cell has some sort of molecular information that gives it knowledge of its position on the radius of a circle and around the circle. Thus, it is believed that the cells in fields recognize position in terms of a circle and their place on the circle—in terms of a pattern like polar coordinates (see Figure 7–26). Experiments that will eventually prove or disprove the "circle" theory of positional information are being performed. In addition, the molecular nature of the positional information is also under investigation. Many experiments have already provided some support of the "circle" rule. For example, if an X-irradiated amphibian limb stump is provided with a complete piece of nonirradiated skin (epidermis and dermis) containing a full circle of cells from the area, regeneration occurs. If the skin does not contain many cells from the circumference of the circle, regeneration fails to occur.

Mechanisms of Limb Development

The limb bud consists of two major parts: the limb-bud mesoderm, and the ectodermal covering, the apical ectodermal ridge. Let us begin our examination of some of the mechanisms of limb development by investigating the following questions. What is the role of the limb-bud mesoderm in limb development? What is the role of the apical ectodermal ridge?

Some investigators believe that the apical ectodermal ridge induces the underlying limb-bud mesoderm to differentiate into limb bones (Saunders-Zwilling Model of Limb Development). Others believe that the apical ectodermal ridge does not induce mesodermal differentiation but instead serves as a specific protective boundary that covers the mesoderm (Amprino Hypothesis of Limb Development). As we will see, there is no solid proof for an inductive function of the apical ectodermal ridge. Some indirect evidence suggests that induction may occur, but at this point in time the question of whether or not the apical ectodermal ridge does actually induce mesodermal differentiation remains unresolved. We will come back to the apical ectodermal ridge shortly, but let us now turn to an interesting related question.

What determines whether a hindlimb or forelimb is formed from a limb bud? In other words, what determines the property of "limbness"? The answer to this question is unambiguous. It is the limb mesoderm that determines if the limb formed will be fore or hind. The following experiment indicates that this is so. Saunders and Zwilling transplanted mesoderm from a hindlimb (leg) bud under the apical ectodermal ridge of the forelimb (wing) bud in the chick embryo. They also did the reciprocal experiment of transplanting mesoderm from the wing bud under the apical ectodermal ridge of a leg bud. In the first case (leg mesoderm under wing bud) a leg (with scales and claws) developed. In the second case (wing mesoderm under leg bud) a wing developed (with feathers, etc.). Thus limbs are mesodermally specific.

In chick embryos the apical ectodermal ridge is also essential for limb development. Nonlimb ectoderm cannot replace the apical ectodermal ridge. If the apical ectodermal ridge is removed from a limb bud, or if the ridge is removed and replaced by non-limb ectoderm, all further limb outgrowth is stopped. Additional evidence for the importance of the apical ectodermal ridge comes from Zwilling experiments with the chick "wingless" mutation. This mutant forms wing-buds but wings fail to develop. If the mutant wingless mesoderm is replaced with mesoderm from a normal nonmutant wing bud, a wing fails to develop. It was noticed that the mutant "wingless" epidermis does not possess an apical ectodermal ridge. Thus, even if normal mesoderm is present under "wingless" mutant epidermis, no apical ridge develops and no wing develops.

Other experiments also show that the apical ectodermal ridge plays an important role in limb development. If a second apical ectodermal ridge is grafted onto a limb bud beside the original ridge, the distal portions of the limb that develop are doubled (two bones are formed side-by-side). The apical ectodermal ridge therefore appears to be of importance in controlling limb outgrowth. Also, prospective thigh mesoderm develops into toes when grafted right below the apical ectodermal ridge of a wing. Thus toes appear at the end of the wing. Although the "legness" of the grafted mesoderm still remains, the wing apical ectodermal ridge appears to dramatically influence the thigh mesoderm, causing it to form more distal structures, toes.

Evidence has just been presented that indicates that the apical ectodermal ridge plays an important role in influencing outgrowth of the limb-forming mesoderm. It should also be noted that the mesoderm also appears to exert an influence on the apical ectodermal ridge. Recall the "wingless" chick mutant. If "wingless" mutant mesoderm is stuffed under the apical ectodermal ridge of a normal chick wing bud, the ridge degenerates. If a thin sheet of mica is placed between the ridge and mesoderm of a normal chick limb-bud, the ridge flattens. These experiments indicate that the limb mesoderm appears to "maintain" the apical ectodermal ridge. Perhaps a factor is released by the mesoderm that acts to maintain the apical ectodermal ridge.

Apical ectodermal ridges can be placed in a small porous basket in a culture dish that contains limb bud mesoderm outside of the basket. The ridges appear to be able to prevent death of mesoderm cells. In the absence of the ridges, the mesoderm cells begin to die. The factor supplied by the ridge that prevents mesoderm cell death may be the same material that influences limb mesoderm outgrowth in the intact limb bud. Similar culture experiments also indicate that a factor seems to be produced by the limb bud mesoderm that prevents flattening of the apical ectodermal ridge. We are now in the age of biochemistry and it is reasonable to expect that in the next decade or so the molecular nature of the factors that influence limb mesoderm outgrowth and apical ectodermal ridge maintenance may be elucidated.

What other factors are involved in limb development? In Chapter 9, "Mechanisms of Morphogenesis," we will examine some of the interesting mechanisms that appear to control the development of form in embryos. It is appropriate to mention here that cell death clearly plays an important role in limb development. The digits on your hands and feet, that is, your fingers and toes, develop as the result of cell death occurring in the regions between the digits. Cell death also helps shape the upper arm, forearm, and elbow region. Let us just mention one experiment here that gives us a bit of insight into the nature of cell death in limb development. If cells are taken from that portion of a stage 17 chick embryo wing bud that is destined to die at stage 24 and grafted to the somite area of the embryo or placed in culture, death of these grafted cells occurs when the embryo reaches stage 24 (or when the cells would have reached stage 24 in the culture experiments). Thus it seems that although stage 17 embryo cells look normal, they are "programmed to die" and will die at the right time. The cells, however, can be prevented from dying if they are grafted to the dorsal side of the limb bud. If the cells programmed to die are taken from a stage 22 embryo (instead of stage 17), however, they can not be saved from dying even if transplanted to the dorsal portion of the limb bud. Thus a "death clock" may be set by stage 17 but it can be turned off by certain environmental influences. Once the embryo reaches stage 22, however, the "death clock" counts down and can not be turned off, no matter what environment the cells are placed in. In cases in which the "death clock" can be turned off, it seems that limb mesoderm provides some sort of factor that can change the program. The nature of the "death clock" and the factors influencing it are not well understood. We will, however, get additional insight into the nature of cell death as a morphogenetic mechanism in Chapter 9.

Limbs develop in a proximo-distal sequence. In other words, the limb regions closest to the body differentiate first and the digits are last to differentiate. It appears that cell density is greatest in the proximal regions while cell density is lower near the growing tip because cells can move away from each other more easily at the tip. It is known that when cells are very closely packed, contacting each other on all sides, growth and movement of such cells may cease. This is termed **contact inhibition** of growth and movement. When growth (cell division) ceases, differentiation often begins to occur. Thus, it may be that the closeness of cell contact in the different regions of the forming limb plays an important role in controlling the program of limb differentiation. The means by which cell contact inhibits continued growth and movement are not well understood. It is, however, the topic of a great deal of current investigation. It might be mentioned, as an aside, that one of the reasons why cells in tumors continue to grow and spread is that tumor cells, compared to nontumor cells, appear less inhibited by contact with each other. Thus, the mechanism of contact inhibition of growth and movement that appears to be important in developmental systems seems to be defective in tumor cells. We will return to this topic in Chapter 13.

contact inhibition

Limb Axes

How does limb orientation come about? An area of limb-bud mesoderm that exists near the posterior junction of the limb bud with the body appears to play a key role in determining limb orientation. This area is called a **zone of polarizing activity (ZPA)** (Figure 7–27). If such a block of tissue (a ZPA) is transplanted beneath the apical ectodermal ridge of a chick wing bud, an additional wing develops in this area. The posterior side of the new wing is always facing the implanted ZPA. The ZPA appears to determine the anterior-posterior axis of the limb. It also appears to stimulate limb outgrowth.

zone of polarizing activity (ZPA)

The dorsal-ventral axis of the limb seems to be at least, in part, determined by limb bud ectoderm. The limb bud mesoderm can be removed, dissociated into single cells, and then repacked and placed under the ectoderm. The limb forming from such an implant has a dorsal and ventral side but does not seem to have any anterior-posterior differences. ZPA cells would have been totally dispersed in the implant and thus no anterior-posterior axis could develop. The dorsal and ventral (top-bottom) surfaces that develop in such an implant are located next to the original dorsal and ventral sides of the ectodermal cover that surrounds the mesodermal implant. Thus, the ectoderm appears to play a role in determining the dorsal-ventral axis of the developing limb. The means by which the ZPA and ectoderm act in determining limb orientation are as yet not well understood.

Limb Regeneration

Why can adult newts regenerate legs while adult human beings cannot? The answer to this question would go a long way toward increasing the possibility of developing a technique that may promote regeneration of amputated human limbs. The answer to this question is not yet in hand. A great deal of exciting information has, however, recently become available that begins to solve part of the regeneration puzzle. Before we look into this work, let us briefly summarize the regeneration abilities of various vertebrates.

Newts and salamanders (urodele amphibians) have the most remarkable powers of regeneration. These organisms can regenerate entire limbs, jaw parts, gill parts, eye parts, and tails. Frogs (anuran amphibians) can only regenerate normal limbs in the tadpole stage. As adults, frogs can not regenerate limbs. Lizards can regenerate tails; fish can regenerate fins but not tails; birds can regenerate parts of the beak.

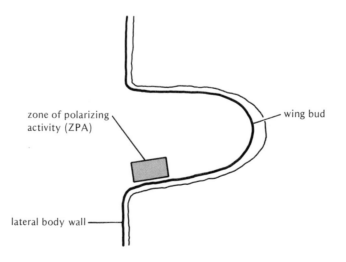

Figure 7–27. Location of zone of polarizing activity in chick wing bud. Based on findings by J. W. Saunders.

zone of polarizing activity (ZPA)

wing bud

lateral body wall

Thus the only adult vertebrates that normally regenerate limbs are the newts and salamanders.

Mammals are also unable to regenerate limbs after simple amputation but can regenerate many internal tissues. For example, a large portion of the liver can be removed. Regeneration will restore the liver tissue. It is of interest that in newly born opossums, regeneration of limbs can occur if prior to amputation a piece of brain tissue is implanted into the limb. It should be noted that the infant opossum limb is not fully differentiated and is in the same state as the frog tadpole leg when it begins to lose regenerative ability. Three important points should be kept in mind before we move on. The first is that the presence of additional nervous tissue can stimulate limb regeneration. This is not only true of infant opossums, but also of other organisms such as frogs. The second important point is that less differentiated tissue may be more capable of regeneration than more differentiated tissue. The third point is that limb regeneration can be stimulated in organisms that do not normally regenerate limbs. This raises the hope that someday it might be possible to stimulate limb regeneration in man. We will return to the topic of stimulating limb regeneration after we examine what actually occurs during limb regeneration in vertebrates.

Sequence of Events in Limb Regeneration

The sequence of post-amputation events that occurs in the newt limb can be clearly outlined as follows.

1. The wound is covered by local epidermal cells that spread over the cut surface by active ameboid movement.

2. The muscle and cartilage just below the wound covering dedifferentiate. Increased activity of degradative enzymes appears in this tissue. The bone and cartilage matrix disintegrate. Muscle cells and cartilage cells lose their differentiated appearance and become transformed into dedifferentiated embryo-like cells.

3. Undifferentiated, embryo-like cells accumulate under the epidermal covering of the wound. These undifferentiated cells, together with the epidermal covering, are called the **regeneration blastema** or **regeneration bud**. The undifferentiated cells that make up this blastema appear to come from surrounding tissue and not from distant parts of the body. This was shown to be the case by labeling exper-

regeneration blastema

regeneration bud

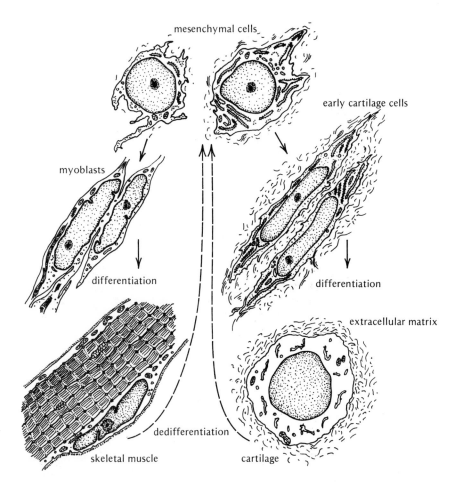

mesenchymal cells

early cartilage cells

myoblasts

differentiation

differentiation

extracellular matrix

skeletal muscle

dedifferentiation

cartilage

Figure 7–28. Cytology of limb regeneration. Cartilage and muscle apparently dedifferentiate into mesenchymal cells early in regeneration. The mesenchymal cells then form the muscle and cartilage of the newly regenerated limb. From E. D. Hay in D. Rudnick, ed., *Regeneration* (1962) The Ronald Press Co.

iments. Radioactively labeled limb cells (muscle, cartilage and bone, epidermis, and nerve Schwann cells) all dedifferentiated and took part in forming the regeneration blastema. Also, only local limb X-ray irradiation blocked regeneration, while whole body irradiation (where the limb was shielded from the X-rays) did not. These results suggest that local limb cells play the major role in forming the regeneration blastema.

4. The blastema cells begin active division and growth.

5. Division and growth decrease and differentiation of the cells begins to take place. This differentiation includes synthesis of muscle proteins and cartilage matrix by the newly forming muscle and cartilage cells respectively. As mentioned earlier with regard to normal limb development, when cell division and growth cease, conditions favorable for differentiation develop. Thus, new limb bones and muscle are formed and regeneration is complete. A summary of these events is given in Figure 7–28.

Before we turn to the interesting problem of stimulation of limb regeneration, let us briefly look at experiments that will help us to understand some of the basic mechanisms that operate in regenerating limbs. One important principle of limb regeneration is that if a limb is amputated, regeneration at the cut surface involves structures distal to the cut. Thus if the cut is through the lower arm, parts of the lower arm distal to the cut, the wrist and digits, will regenerate. If the cut is through the upper arm then the parts of the upper arm distal to the cut, the lower arm, the wrist, and the digits regenerate. The cells in the area of the cut surface must have information that allows regeneration of only the distal (missing) structures. In our earlier discussion on normal limb development, it was noted that French, Bryant, and Bryant have formulated a theory of how cells can determine position. This theory involves a cell's ability to recognize its position on a circle. It may be that regeneration of

Figure 7–29. Regeneration of the newt *Triturus cristatus*. (a)-(f) Consecutive stages of regeneration of a forelimb amputated above the elbow. (g)-(m) Consecutive stages of regeneration of a hindlimb amputated above the knee. From G. Schwidefsky, *Roux. Arch.* 132:57–114 (1934).

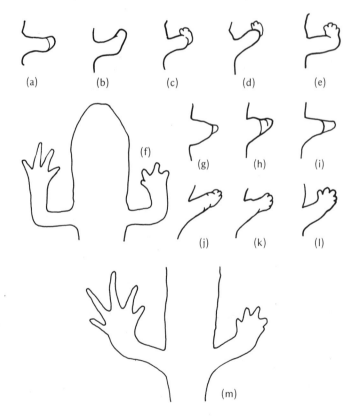

structures distal to the cut occurs because cells in the region of the cut recognize what needs to be made by their ability to recognize their place on a circle around the cut area. Future work will determine whether this concept does truly play an important role in limb development and regeneration.

Another important set of experiments that helps us to understand limb regeneration is that of Zwilling's and Searls's groups. Zwilling dissociated wing bud mesoderm into single cells, mixed the cells up, and grafted this group of cells under the ectoderm on the body of a chick embryo. A normal limb formed with the bone and muscle in the right places. Did the mesoderm cells "sort out," the pre-cartilage cells moving to the center of the limb bud and the pre-muscle cells moving to the periphery of the bud? Or did the positions that cells found themselves in determine what the cells would form? Searls and Janners labeled chick embryos with tritiated thymidine. They then removed either the central core of the limb bud (prospective cartilage) or the peripheral portion of the limb bud (prospective muscle). Each labeled region was implanted into the limb bud of a cold (unlabeled) chick embryo, but at right angles to the long axis of the developing bones. Thus each "hot" implant spanned both prospective muscle and prospective cartilage in the cold limb bud (Figure 7–30). What happened? The results clearly showed that the tissue that formed from the hot implant became cartilage in the

TRITIATED THYMIDINE-LABELED EMBRYO

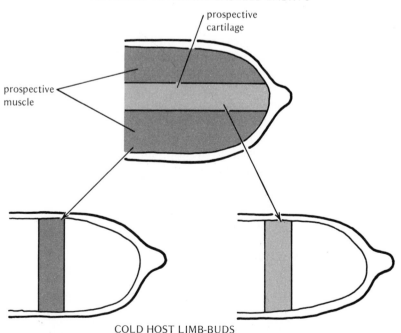

COLD HOST LIMB-BUDS

Figure 7–30. Experiment showing effects of cell position on tissue differentiation in limb buds. In both (a) and (b) the implant forms muscle in the peripheral regions and cartilage in the central core region of the limb buds. Modified and redrawn from R. L. Searls and M. Y. Janners, *J. Exptl. Zool. 170*, 365 (1969).

central cartilage-forming region of the host bud, and became muscle in the peripheral muscle-forming region of the host bud. Thus, in the developing limb-bud, the fate of prospective muscle and cartilage cells is not irreversibly fixed but can be changed by the environment around the cells. In the developing limb bud, cell position is a key factor in controlling differentiation.

But what about regenerating adult limbs? When the cartilage and muscle dedifferentiate, do cells originally derived from cartilage only redifferentiate into cartilage or can these cells redifferentiate into muscle? Do cells originally derived from muscle only redifferentiate into muscle or can these cells form cartilage? The answer to this question in regenerating salamander limbs is still a bit unclear. An experiment, however, suggests that although cartilage can dedifferentiate, the resulting cells can redifferentiate mainly into cartilage. In the case of muscle, however, cells that dedifferentiate from muscle tissue may be able to form both muscle and cartilage. Before amputation, radioactively labeled salamander cartilage, cartilage plus perichondrium, muscle, or epidermis was grafted into irradiated host salamander limbs. Host limbs were irradiated to discourage participation of host cells in the regeneration process. The results of the experiment were quite interesting. Only muscle seemed capable of giving rise to all of the component tissues in the regenerated limb (cartilage, perichondrium—the connective tissue around cartilage, connective tissue of joints, fibroblasts, and muscle). Cartilage only appeared to give rise to cartilage, perichondrium, and connective tissue, but not muscle. Epidermis did not appear to give rise to any tissue in the regenerated limb. What do these results mean? The results suggest that muscle may be able to dedifferentiate and redifferentiate into many tissues, while cartilage can only redifferentiate into cartilage and connective tissue. It should be noted, however, that muscle contains several cell types, including myoblasts, connective tissue, and blood elements, so it is still unclear if any cells in the adult limb can really redifferentiate into totally different cell types. Thus, in the early limb bud, the fates of cells are clearly not irreversibly fixed. In the adult limb, however, it remains to be seen just how fixed are the fates of the component cell types.

Stimulation of Regeneration

Let's now return to the problem of stimulating limb regeneration. We have already alluded to one important factor that stimulates limb regeneration, namely, nervous tissue. Recall that limb regeneration could occur in the infant opossum if nervous tissue was implanted into the limb. In salamanders, nerves also appear to be important in regeneration because, normally, limbs will not regenerate after the nerves have been removed from the limbs. The salamander limb, however, can develop in the embryo even if nerves are kept out of the developing limb. These limbs, therefore, do not contain nerves. If these limbs are amputated and nerves are kept out of the area, the limbs still regenerate. Thus, a limb that had had no nerves could regenerate without nerves. A limb that has had nerves prior to amputation is dependent upon nerves for regeneration. So limbs appear to develop an "addiction" to nerves if they were originally present.

Exactly how nerves stimulate regeneration is not yet well understood. An important factor seems to be the quantity of nervous tissue present, not the type of nerves. Regeneration is stimulated by most types of nervous tissue and does not appear to be dependent on whether the nerves are sensory or motor. In fact, if a limb nerve is deviated from its normal position in the newt and placed under the skin in a region not too far removed from the normal limb, a new limb can form in this area. Nerves, therefore, appear to exert a profound influence on regeneration.

What do nerves do to stimulate regeneration? If nerves are removed from a regenerating limb, RNA, DNA, protein synthesis, and synthesis of extracellular glycosaminoglycan (a kind of glycoprotein molecule) decrease. Singer isolated and partially purified a basic protein from whole brains or brain nerve endings that stimulates protein synthesis in regenerating limbs. This material may turn out to be the nerve "factor" that stimulates regeneration. Nerves may also stimulate regeneration by stimulating blood vessel growth to the area. Increased blood supply to the regenerating limb may directly promote development by supplying nutrients, oxygen, hormones, and other substances to the regeneration blastema.

Schwann sheath cells surrounding nerves appear to be able to give rise to the entire regeneration blastema and to the entire regenerated limb. Salamander limbs can be irradiated with X-rays so that regeneration does not occur. Regeneration will occur, however, in such an irradiated stump if unirradiated nerve is implanted into the amputation site. By taking nerves from a pigmented salamander and implanting these nerves into the irradiated stump of a non-pigmented salamander, it was shown by Wallace that the entire regenerated limb arises from the donor nerve Schwann sheath cells and fibroblasts, connective tissue cells. Thus, not only can nerves stimulate regeneration but nerve sheath cells and fibroblasts apparently can give rise to all of the bone and muscle cells of the regenerated limb!

Cell damage at the wound surface appears to play an important role in stimulating regeneration. A newly regenerated limb can form in newts even if no amputation has occurred. A tight ligature around a limb or introduction of cartilage breakdown products under the skin can cause partial or complete limb regeneration to occur. Products from damaged cells, therefore, may stimulate regeneration. This is supported by experiments in which beryllium nitrate is applied to the amputation surface. This salt suppresses regeneration, possibly by binding substances that are released from damaged cells that stimulate regeneration. The beryllium treatment must be carried out immediately after amputation, or it will not inhibit regeneration. This is interpreted to mean that the salt must bind the products from damaged cells immediately or these products will set into motion the regenerative process. Another amputation of the beryllium treated stump allows regeneration to occur unless beryllium is applied again. These experiments suggest that products of damaged cells stimulate limb regeneration.

Becker has reported that if direct electric current is applied to amputated forelimb stumps of laboratory rats some regeneration occurs. Cartilage, bone marrow, bone, blood vessels, nerves, and muscles all begin to develop. The limb formed, however, is incomplete and only grows to the first joint. Electric fields may play an important role in regeneration, and application of small currents may become an important means of stimulating limb regeneration in organisms that usually cannot regenerate. In adult frogs, regeneration can be stimulated by implantation of electrogenic couples that enhance the distal negativity of the stump. The role of an electrical potential difference in limb regeneration is not well understood. Studies from many systems, however, suggest that conditions favorable for regeneration include electrical gradients, with the tip of the structure being more negatively charged than the proximal portion of the structure. The future will tell if electrical fields in organisms are of importance in controlling many developmentally significant processes.

To sum up, the limb bud has remarkable properties of self-regulation that enable it to form a limb even under conditions of massive mutilation. Adult limbs can regenerate in newts and salamanders but not in most other vertebrates. The sequence of events in limb regeneration include processes of cell migration, dedifferentiation of limb tissues, and redifferentiation of newly regenerated limb tissue. Nerve tissue, products of cell damage, and application of electric current can stimulate limb regeneration in systems that do not usually regenerate limbs. Work in this field offers hope that eventually it will be possible to stimulate limb regeneration in human beings.

UROGENITAL ORGANOGENESIS

WE NOW TURN to the development of the urogenital system. In addition to describing the embryology of this system, we will examine the types of mechanisms that appear to play important roles in its development.

What is the urogenital system? Aren't we talking about studying two systems—the urinary and reproductive systems? The urogenital system is really two systems—the kidney system and the reproductive system. We will examine the development of these systems as a single system because the embryogenesis of the structures of these systems is interrelated. In fact, as you will see, the structures that belong to one system at one stage in development sometimes become a part of the other system at a later embryonic stage. We will therefore describe the development of their structures together.

Before we begin, let us define what we mean by the urogenital system a little more carefully. The urogenital system consists of the kidneys and accessory structures involved in the formation of urine, and the gonads and accessory structures utilized for reproduction. We should mention that the kidneys are by no means the only organs involved in excretion. For example, some fish excrete salt through their gills. Salt is excreted by the nasal glands of some marine birds and reptiles, and by the rectal glands of sharks and rays. Skin glands (sweat and mucous glands) excrete salt, nitrogenous wastes and water. Other structures that are involved in regulation of body fluids and excretion include the liver, lungs, and salivary glands. Thus although we will consider the kidney here, we must keep in mind that the kidney is not the only organ involved in waste elimination and maintenance of the body fluids.

Kidney

We will begin our story of urogenital development by examining the embryology of the kidney in the more primitive vertebrates. Then we will work our way up to the advanced vertebrate systems. Discussion of the development of the genital structures will follow the discussion on kidney development. The interrelationships between the structures used for excretion and reproduction will be shown.

In Chapter 6 we noted that the kidneys and gonads are derived from the intermediate mesoderm or mesomere. This is the region of mesoderm between the somite and hypomere. The kidneys and gonads are therefore mesodermal in origin. Let us now examine the development of the functional kidney of aquatic embryos and larvae: namely the kidney of embryonic fish and of tadpoles. This kidney is called the pronephros. Some of the structures of this kidney become part of the urogenital system of higher vertebrates, as we will see later.

pronephros

Pronephros

The pronephros, the functional kidney of fish and amphibian embryos, develops from the intermediate mesoderm (mesomere). The specific area of the mesomere that gives rise to the pronephros is a region consisting of segmented portions of intermediate mesoderm called nephrotomes. The cells in the nephrotomes separate to form an internal cavity, the nephrocoel. In this way pronephric tubules form from the nephrotomes and contain internal nephrocoels. The mechanisms of tubule formation are unclear but appear to involve the types of morphogenetic cellular rearrangements discussed in Chapter 9. As shown in Figure 7–31, the pronephric tubules are continuous with the coelom by ciliated funnels (nephrostomes) and fuse at the other end to form the pronephric duct. This duct elongates in a posterior direction and fuses with the cloaca. The pronephric duct therefore directly transfers wastes into the cloaca for elimination.

nephrotomes

nephrocoel

pronephric tubules

nephrostomes

pronephric duct

glomus

A network of fine blood vessels (glomus) is associated with the ciliated funnels of the pronephric tubules. Wastes leave through the lining of these blood vessels into the coelomic area right beside the openings of the pronephric funnels (nephrostomes). The wastes are picked up by these funnels, thus entering the pronephric tubules. The wastes are transported to the cloaca via the pronephric duct.

The pronephros is functional in fish embryos and amphibian larvae. We should also note that a pronephros does form in reptiles, birds, and mammals but, as will be seen in the next section, never becomes functional.

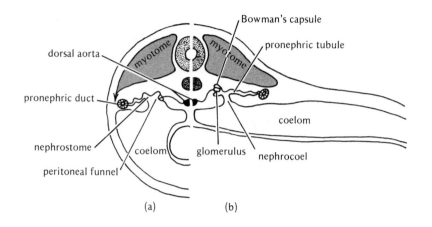

(a)   (b)

Figure 7–31. Diagram of cross section through embryos. (a) Anamniote (fish or frog embryo). Glomerulus is considered external since branch of aorta pushes into side of coelom. (b) Amniote (bird, reptile, or mammal embryo). Glomerulus is surrounded by Bowman's capsule and is referred to as internal. Most amniotes and anamniotes possess internal glomeruli. From J. B. Phillips, *Development of Vertebrate Anatomy*, p. 434 (1975) The C. V. Mosby Co.

Mesonephros

As the larval fish and amphibians become adults, the pronephric tubules disintegrate. Thus, an important mechanism involved in formation of the adult kidney is differential death of the embryonic pronephric tubules. The pronephric duct, however, remains intact. A second set of tubules develops from the nephrotomes in the intermediate mesoderm by aggregations of these mesoderm cells. This second set of tubules forms posterior to the pronephric tubules and connect to the pronephric duct after the first set of tubules disintegrate. The new tubules are the **mesonephric tubules**. The old pronephric duct is now called the mesonephric duct or the Wolffian duct (Figure 7–32), since the mesonephric tubules have attached to it. Waddington and O'Connor have found that the formation of mesonephric tubules appears to be induced by the pronephric duct. If the pronephric duct is blocked from growing posteriorly and does not reach the area that forms the mesonephric tubules (mesonephrogenic mesoderm), well-developed tubules never form. Thus one important mechanism that seems to be involved in mesonephric tubule morphogenesis is induction by the pronephric duct. The nature of induction is explored in Chapter 8.

The dorsal aorta branches into fine vessel clusters called glomeruli that contact the mesonephric tubules. The portion of each tubule that contacts the glomerulus expands and invaginates to form Bowman's capsule. Wastes are filtered from the blood in the glomeruli through Bowman's capsules of the mesonephric tubules and into the tubules proper. Each Bowman's capsule plus its glomerulus is a functional renal unit called a Malpighian body. Wastes are therefore filtered from the blood directly into the mesonephric tubules. This is more complex than

mesonephric tubules

mesonephric duct

Wolffian duct

glomeruli

Bowman's capsule

Malpighian body

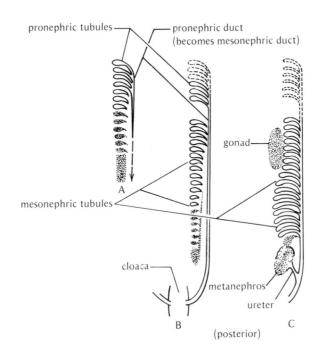

Figure 7–32.
Development of kidney
in vertebrates. (a)
Pronephric duct
becomes mesonephric
duct. (b) and (c)
Relationship of
pronephric duct,
mesonephric duct and
ureter to kidney
development. Modified
from Burns, "Urogenital
System," in *Analysis of
Development*, Willier,
Weiss, and Hamburger,
eds., 1955 by the W. B.
Saunders Co.,
Philadelphia, Pa.

the situation in the pronephros, in which wastes could be picked up from the coelom area by the nephrostomes of the pronephric tubules. Kidney function in the mesonephros (and also in the more advanced metanephros) is not simply a filtration of wastes from the glomeruli into Bowman's capsules. Instead, many processes occur in the more advanced kidneys, including reabsorption of nutrients, hormones, and other substances from the kidney tubules back into the blood.

The mesonephros is the functional kidney of adult fish and amphibians. Some individuals term the mesonephric kidney of many adult fish

opistonephros

and amphibians an opistonephros because it resembles both the mesonephros and metanephros (a more advanced kidney form). It resembles the metanephros in terms of lack of nephrostomes and connection of mesonephric tubules to large collecting ducts.

It should be stressed that the mesonephric kidney is not only present and functional in adult fish and amphibians. It is also the functional kidney of embryonic reptiles and birds and some mammals in which wastes are not adequately removed by the mother (such as the pig). In other mammalian embryos in which the mother (through the placenta) effectively removes embryonic wastes, the mesonephros does not function as an excretory organ. Let us now turn to an examination of

metanephros

the development of the metanephros, the functional kidney of adult reptiles, birds, and mammals.

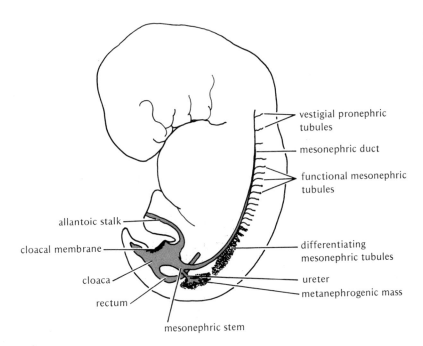

vestigial pronephric tubules

mesonephric duct

functional mesonephric tubules

allantoic stalk

cloacal membrane

cloaca

rectum

mesonephric stem

differentiating mesonephric tubules

ureter

metanephrogenic mass

Figure 7–33.
Diagrammatic sagital section through mammaliam embryo showing location of three types of kidneys. From J. B. Phillips, *Development of Vertebrate Anatomy*, p. 439, (1975). The C. V. Mosby Co., St. Louis.

Metanephros

A new duct, the ureter, arises as an outgrowth at the point at which the mesonephric duct joins the cloaca. The outgrowth begins as an evagination, the ureteric bud. The developing ureteric bud grows and branches into the metanephrogenic mesoderm, the posterior portion of the nephrotome of the mesomere that gives rise to the metanephric kidney. In a way similar to the development of the mesonephros, the metanephric tubules differentiate from the metanephrogenic mesoderm as a result of induction by the ureter (Figures 7–32, 7–33). That the ureteric bud induces formation of metanephric tubules is shown by the fact that metanephric tubules fail to develop if the ureteric bud does not enter the metanephrogenic mesoderm. The metanephros is complex, and includes Bowman's capsules and associated glomeruli as does the mesonephros (Figure 7–31).

In summary, the pronephros, mesonephros, and metanephros develop in the intermediate mesoderm or mesomere. Each consists of tubules connected to a main excretory duct. We will now turn to a discussion of the genital system. It will become apparent that some structures that are part of the embryonic kidney system may switch roles and become incorporated into the genital system. Some of these mechanisms will be reexamined in the chapter on mechanisms of morphogenesis, Chapter 9.

ureter

ureteric bud

Genital System

The gonads function in reproduction allowing perpetuation of the species. It is the germ cells formed in the gonads that directly combine to form the new organism. In Chapter 6 we learned that the gonads are of mesodermal origin. But what about the germ cells? What is their origin? We will examine this intriguing question first and then move on to a discussion of gonad development.

Origin and Migration of the Primordial Germ Cells

At first thought one would assume that the primordial germ cells, the cells that appear to give rise to the reproductive cells that will form functional gametes, arise in the developing gonad. This is not the case. Instead these cells appear to arise in different areas in different organisms. They then migrate to the gonad area or are transported to this area via the bloodstream. It will be seen that the primordial germ cells are not necessarily of mesodermal origin.

A simple way of summarizing the origin of the primordial germ cells in some groups of vertebrates follows.

Amphibians—Primordial germ cells appear to originate in the vegetal endoderm of the early frog embryo. In salamaders the cells appear to originate in the hypomere of the mesoderm.

Birds and Reptiles—Primordial germ cells appear to originate in the extraembryonic endoderm.

Human Beings—Primordial germ cells appear to originate in the yolk sac endoderm.

In some invertebrates, however, primordial germ cells appear to be distinguishable before germ layer formation, and it may be that these cells should not be grouped as being derived from a specific germ layer.

How does one go about identifying the origin of the primordial germ cells? Several types of experiments led to the conclusions above. The area suspected of giving rise to the germ cells could be irradiated with ultraviolet light or with X-rays. The area could be excised from the embryo, or killed with a hot needle. In some systems, primary germ cells have specific staining properties and may be distinguishable in the early embryo by using special stains. In the former experiments, when the germ cells are killed, the gonad develops devoid of germ cells. In this manner one can identify the region of the early embryo that gives rise to the germ cells. It is clear that the primordial germ cells do not originate in the gonad, but arrive there from another region of the embryo.

A second question that should be asked is: How do the primordial germ cells get to the gonad? The mechanisms by which germ cells get to the gonads are not fully understood. In birds, primordial germ cells appear to enter the blood vessels for transport to the gonad area. In frogs, primordial germ cells appear to migrate in the dorsal endoderm to the gonad area. In mice and men, the primordial germ cells that appear to originate in the endoderm can be observed (with specific stain for alkaline phosphatease, an enzyme localized in primordial germ cells) to migrate to the gonad area. Thus primordial germ cells get to their destination by migration or transport by the blood. This is reminiscent of the ways in which tumor cells spread in the body—by invasion (migration) or metastasis (blood transport). Interesting comparisons of this sort between tumor and embryonic cells will be discussed in Chapter 13.

To say that primordial germ cells get to their destination by mechanisms such as migration or blood transport is only part of the explanation. How does a migrating primordial germ cell know where to stop and stay? It is possible that specific adhesive recognition between the primordial germ cell and the region of destination may occur. That is, primordial germ cells may stop and stay because they specifically stick to the gonad area. This mechanism of specific adhesive recognition will be described in more detail in Chapter 9.

In summary, primordial germ cells in vertebrates appear to originate in the endoderm or mesoderm. It may be, however, that these cells should not be classified as germ layer derivatives and should instead be considered as special cell types apart from the germ layers. The mechanisms by which germ cells arrive at their destination in the genital region include migration and blood transport. Specific adhesive recognition may play a role in determining where the germ cells stop and stay. Let us now turn to a discussion of gonad development.

Gonad Development

In the previous section we noted that the primordial germ cells migrate to the gonad area. The germ cells, therefore, originate elsewhere in the embryo and by mechanisms that appear to include migration and selective cell adhesion, "home" to the gonad region. Where is this gonad region? How does the gonad develop? In the following discussion, keep in mind that the gonads and associated structures usually develop in pairs just as the kidneys do.

As indicated in Chapter 6, in vertebrates the gonads arise in the posterior intermediate mesoderm—the region right above the posterior splanchnic mesoderm of the hypomere. The region of mesoderm that **germinal ridge** forms the gonad begins to thicken and is called the germinal ridge. The primordial germ cells have previously migrated into this germinal ridge. Some researchers feel that there is no conclusive evidence that the primordial germ cells that have migrated into the germinal ridge give rise to the gametes. Instead, some feel that the primordial germ cells simply stimulate gonad development. The gametes may arise from the gonad mesoderm. The types of experiments discussed in the last section, however, suggest that the primordial germ cells are indeed gamete precursors. No proof that they are not has, as yet, materialized.

Indifferent Gonad

There is some controversy about the details of the events that occur in the development of the gonads. Evidence supports the contention that **germinal epithelium** the surface of the germinal ridge, called the germinal epithelium, that contains some of the primordial germ cells buds off to form the primary sex cords. Others believe that the primary sex cords develop from **primary sex cords** strands of mesoderm from neighboring areas. In any case, as seen in Figure 7–34, the primitive gonad now consists of an outer germinal

Figure 7–34. Differentiation of the indifferent amphibian gonad into an ovary or testis. Relative predominance of cortex or medulla is shown. From R. K. Burns (after Witschi) in *Survey of Biological Progress I,* Academic Press.

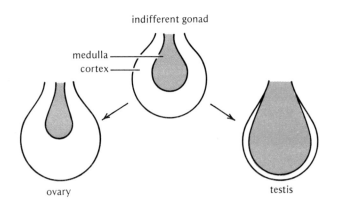

epithelium or cortex and an inner region made up of the primary sex cords called the medulla. It should be noted that at this stage in development the gonad is neither an ovary nor a testis. Instead, it is called an indifferent gonad. At this point in development we are at the indifferent stage of sexual development or, in other words, at a stage before a testis or ovary develops.

cortex

medulla

indifferent gonad

indifferent stage of sexual development

It is not only the indifferent gonad that characterizes the indifferent stage. This stage is also characterized by the presence of all the ducts and tubules that are usually present only in either adult males or adult females. In genetic males, certain parts of the gonad and certain ducts degenerate before adulthood. In genetic females, different portions of the gonad and certain other ducts and tubules degenerate before adulthood. Thus, the genital system of the male versus the female develops as a result of death of some structures and growth of others. Let us examine this absorbing story in a little more detail.

Genetic Females

In genetic females, the primary sex cords of the gonad degenerate. Thus the inner medulla of the gonad becomes reduced. The outer region, or cortex of the gonad, however, develops and the primordial germ cells contained in the cortical region become clustered in groups and become surrounded by follicle cells that protect and nourish the developing oocytes. These clusters of cells are called secondary sex cords or nests of oogonia. The female gametes or eggs develop from oogonia that undergo maturation as described in Chapter 1.

secondary sex cords

nests of oogonia

The urogenital ducts that are present at the indifferent stage of sexual development include the mesonephric duct with its tubules and Bowman's capsules (as described earlier in the chapter) and the Müllerian ducts. The Müllerian ducts develop in the intermediate mesoderm, grow posteriorly, and fuse with the cloaca. In genetic female reptiles, birds, and mammals, the mesonephric ducts and associated tubules degenerate. The Müllerian ducts possess a funnel-shaped anterior opening called the ostium tubae. This opening is close to the ovary. As eggs are ovulated, they fall into the ostium tubae and into the Müllerian ducts. Since the Müllerian ducts function to transport eggs, they are the oviducts of the female (Figures 7–35, 7–36).

Müllerian ducts

ostium tubae

oviducts

Thus, in genetic female vertebrates, the primary sex cords of the gonad degenerate and the cortex of the gonad becomes the important

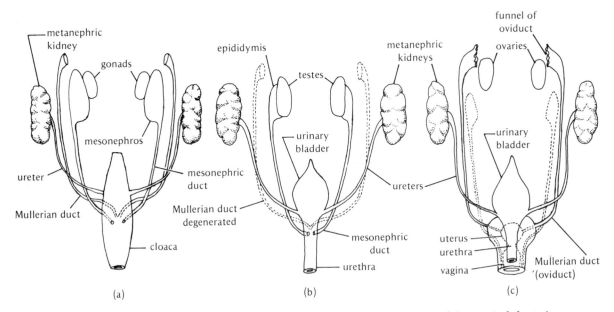

Figure 7–35. Diagram showing transformations of the genital ducts in mammalian embryos in transition from an indifferent stage (a) to the male (b) and female (c) conditions. From B. I. Balinsky, *An Introduction to Embryology,* 4th ed., W. B. Saunders Co.

gamete-producing region. Secondary sex cords, or nests of oögonia, develop in the cortex of the ovary. Mesonephric ducts and tubules degenerate (in reptiles, birds, and mammals) while the Müllerian ducts persist, forming the functional oviducts. It is clear that the adult gonad with its associated structures develops as a result of degeneration of some embryonic regions and persistence and development of others. After considering the development of the male gonad and accessory structures, we will briefly return to an examination of some of the mechanisms that appear to control genital development.

Genetic Males

In genetic males, the inner region of the gonad, the medulla, that includes the primary sex cords, develops. The primordial germ cells present in the primary sex cords develop into spermatogonia. Spermatogonia, as described in Chapter 1, form the male gametes, the spermato-

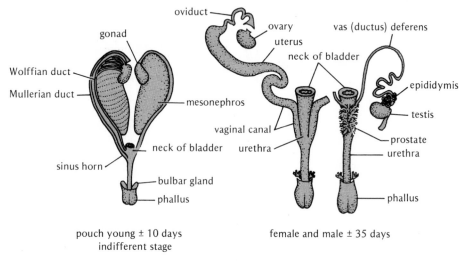

Figure 7–36. Sexual differentiation in young opossums. (a) Indifferent stage (about 10 days) (b) Female and male embryos (about 35 days). From R. K. Burns, *Survey of Biological Processes I*, Academic Press.

zoa. Development of the primary sex cords involves hollowing out. As a result of this process, the primary sex cords become hollow structures containing the primordial germ cells. These hollow structures are the seminiferous tubules. These tubules connect to adjoining groups of cells (rete cord cells) which hollow out forming a network of tubules called the rete testis. The tubules of the rete testis connect to the mesonephric tubules. The mesonephric tubules join the mesonephric duct. Sperm therefore are released from the seminiferous tubules into the ducts of the testis and finally into the mesonephric duct that is connected with the cloaca (Figures 7–35, 7–36). Here is a perfect example of how some ducts and structures originally only associated with the urinary system become incorporated into the reproductive system. We should also note that in genetic males, the Müllerian ducts degenerate after the indifferent stage of sexual development. In addition, remember that the mesonephric duct still maintains a urinary function in fish and amphibians (along with a reproductive function). In reptiles, birds, and mammals, however, the mesonephric duct carries only sperm and is called the sperm duct or ductus deferens. In these organisms, you will recall that the urinary function is taken over by the metanephros with its associated ureter.

In summary, in genetic male vertebrates, the medulla of the gonad develops and seminiferous tubules form from the primary sex cords. The mesonephric ducts persist and act as sperm ducts while the Müllerian ducts degenerate.

seminiferous tubules

rete cord cells

rete testis

ductus deferens

Figure 7–37. Effect of sex hormones on sex duct development in chick embryos. (a) Normal female embryo, no added hormones, eighteen day incubation. (b) Male embryo treated with female hormone. Both oviducts present and enlarged. (c) Normal male embryo, no added hormones, seventeen day incubation, no oviducts. (d) Female embryo treated with male hormone. Oviducts mostly absent, mesonephric ducts enlarged. From: Willier, Weiss, and Hamburger, *Analysis of Development,* © 1955 by the W. B. Saunders Co., Philadelphia, Pa.

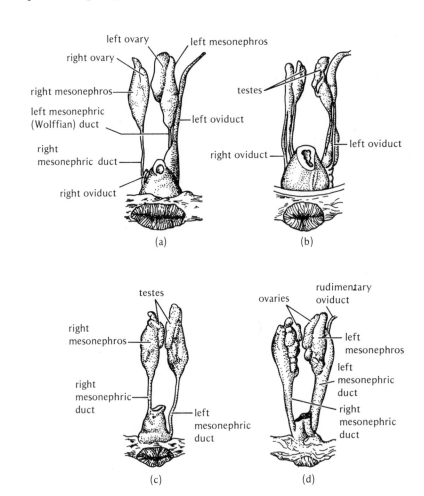

Mechanisms of Gonad Development

In Chapter 9 we will examine examples of mechanisms that appear of general importance in controlling morphogenesis. Here we will look into mechanisms that appear to control some of the changes that occur during genital development in each specific sex. We have noted that in genetic males and genetic females, at the indifferent stage of sexual development, structures of both male and female reproductive systems are present. We observed that after the indifferent stage certain of these structures degenerate while others develop, depending upon the specific sex of the individual. What causes these specific genital changes in each sex? The answer to this question is not fully understood.

It is likely that sex hormones play an important role in controlling some of these changes. As will be seen in Chapter 9, such hormones appear to be able to control the level of lysozomal enzymes in cells. Lysozomal enzymes are degradative enzymes in cell digestive organelles called lysozomes. Levels of these enzymes such as cathepsins, acid phosphatase, and ribonuclease are observed to increase in the sexual ducts that degenerate. For example, before breakdown of the Müllerian ducts in genetic males, the levels of these enzymes in these structures increase. These enzymes may directly cause disintegration of the structures.

It was suggested above that sex hormones may influence the levels of these enzymes in the structures. Injections of male hormones (androgens) into genetic females causes breakdown of the Müllerian ducts. If female hormones (estrogens) are inoculated into genetic males, on the other hand, the Müllerian ducts persist (Figure 7–37). It is as yet unclear, however, if these hormones play a key role in early sexual differentiation in the embryo. Some evidence suggests that they do. For example, Frank Lillie observed that in twin cattle where their fetal blood supplies intermingle, if a genetic male and a genetic female embryo are developing together, the genetic female becomes masculinized. It is thought that the male twin releases male sex hormones that modify the female. It can also be shown that if a piece of a gonad of one sex is transplanted near the developing gonad of an embryo of the other sex, genital structures of the recipient are modified in the direction of the sex of the donor gonad. These experiments, however, do not prove that sex hormones are playing the key role in controlling genital differentiation in the early embryo. It is likely that future work will definitively indicate exactly what factors control the differentiation of genital structures in each sex in the embryo.

DEVELOPMENT OF THE IMMUNE SYSTEM

WE FIGHT OFF infectious diseases and probably even diseases such as cancer with our immune system. It is this system that contains the cells that travel throughout our tissues to seek out and destroy foreign cells and substances. It is this system that produces antibodies that specifically combine with foreign invaders of all sorts. Let us first examine the development of this system and then briefly discuss how the system works.

Development of Tissues and Cells of the Immune System

Chick Embryo

thymus glands

bursa of Fabricius

In the chick embryo, two glands are associated with the early development of cells that become competent to respond to foreign substances (antigens). These organs are the thymus gland and the bursa of Fabricius. The thymus gland, as mentioned in Chapter 6, develops from the pharyngeal pouches (Figure 7–38). The bursa of Fabricius develops as an out-pocketing of the hindgut (Figure 7–39).

The stem cells that give rise to the cells of the immune system appear to be derived from the yolk sac wall in the chick embryo. These cells can be observed as possessing darkly staining basophilic cytoplasm.

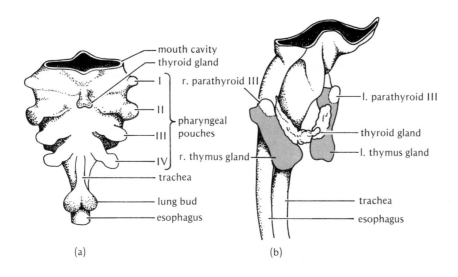

mouth cavity
thyroid gland
I
r. parathyroid III
II
pharyngeal pouches
III
IV
r. thymus gland
trachea
lung bud
esophagus

l. parathyroid III
thyroid gland
l. thymus gland
trachea
esophagus

(a)   (b)

Figure 7–38. Derivation of the thymus. (a) Parts of the third and fourth pharyngeal pouches (endoderm) that migrate down to the chest cavity (b) form the thymic lobe. From G. L. Weller, *Contrib. Embryol. Carneg, Inst.* 24:93 (1933).

The stem cells enter the rudiment of the thymus at six to seven days of incubation in the chick embryo. These stem cells appear to give rise to white blood cells called lymphocytes. Large numbers of medium and small lymphocytes can be observed in the thymus gland by 12 days of incubation.

Stem cells originating in the yolk sac enter the bursa of Fabricius at twelve to thirteen days of incubation. These stem cells appear similar to those that entered the thymus at six to seven days. By seventeen days, many lymphocytes are present in the bursa, as was observed earlier in the thymus.

By fourteen days of incubation in the chick embryo, the production of antibody by lymphocytes can be detected. The type of antibody that is produced at this time is termed IgM or immunoglobulin M. This antibody has a molecular weight of about 900,000 daltons. Another major type of antibody, IgG or immunoglobulin G (molecular weight, 150,000 daltons), however, is not found until hatching. We will return to a discussion of these and other antibodies shortly.

IgM

immunoglobulin M

IgG

immunoglobulin G

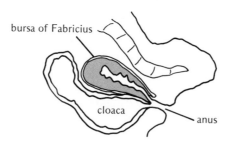

bursa of Fabricius
cloaca
anus

Figure 7–39. Bursa of Fabricius in chicken. From Burnet, *Self and Not Self* (1969) Cambridge University Press.

Mouse Embryo

In the mouse embryo, stem cells that "seed" the organs of the immune system also appear to originate in the yolk sac wall. These stem cells seem to travel from the yolk sac to the fetal liver and finally to the bone marrow. As in the chick embryo, the lymphoid stem cells of the mouse embryo can be identified by their large size and darkly staining cytoplasm. The stem cells enter the thymus at eleven days of gestation. By sixteen days, lymphocytes can be observed in the fetal liver.

Origin of T and B Lymphocytes

We saw that in the chick two major glands become seeded with lymphoid stem cells, the thymus and bursa of Fabricius. Similar events occur in the mouse in that the thymus is seeded by lymphoid stem cells. A bursa of Fabricius has not been identified in the mouse, but tissue with

T cells

similar function is present. The lymphocytes that are produced by the thymus and bursa in the chick embryo have different functions. The

cellular immunity

thymus lymphocytes (termed T cells) play a major role in cellular immunity. Cellular immunity involves a direct attack by a T cell on a foreign

B cells

cell. The bursa lymphocytes (termed B cells) are active in the production of humoral immunity. Humoral immunity involves the synthesis of free

humoral immunity

soluble antibodies that combine with foreign antigens.

T cells mature in the thymus from the stem cells and become able to respond to foreign cells (become competent). These T cells then enter the circulation. B cells (in the chick embryo) mature in the bursa from the stem cells that originally seeded it. Around the time of hatching, B cells from the bursa enter the circulatory system and seed the bone marrow, spleen, and lymph nodes. The bone marrow becomes the major source of proliferating B cells in the adult. The origin of the B cells that confer humoral immunity in the mouse embryo is not well understood. It is assumed that the B cell population developed from stem cells somewhere in the body (possibly the fetal liver) and then seeded the bone marrow. The likely source of B cells in adult mammals is, therefore, the bone marrow.

The role of the bursa and thymus in conferring immunity has been shown by surgical removal experiments. In the chick embryo, removal of the bursa, before B cells have been released for seeding other organs, results in failure of the development of an immune response. After B cells have been released from the bursa, however, removal of this gland does not affect the development of immunity. Similarly, removal of the thymus (in, for example, the mouse embryo) before T cells have been released (at about birth) results in decreased immunity. Removal of the thymus after birth, when T cells have already been released into the circulation, has little affect on immunity. What do we mean by immunity? How do T and B cells function?

Function of T and B Cells in the Immune Response

We have seen that T and B lymphocytes mature in the thymus and bursa (birds) respectively. These cells are released into the circulatory system, seeding other lymphoid tissues. We have also mentioned the T cells play a major role in cellular immunity, while B cells form circulating antibodies (humoral immunity). How do the T and B cells respond to foreign antigens during the immune response? A very large variety of substances exist that might elicit an immune response. How, then, can cells of the immune system respond to these substances in a specific manner? Let us begin to examine this question with the B cell system responsible for humoral immunity.

Humoral Immunity

B Cell Clones

B cells synthesize antibodies in response to antigenic stimulation. Mature antibody-secreting B-cells are called plasma cells. Substances that are antigenic (able to elicit the production of antibodies) include proteins, glycoproteins, and polysaccharides that are folded in complex molecular patterns (see Chapters 11 and 12). The B cells recognize some of these patterns as foreign. They synthesize an antibody that specifically combines with the antigen.

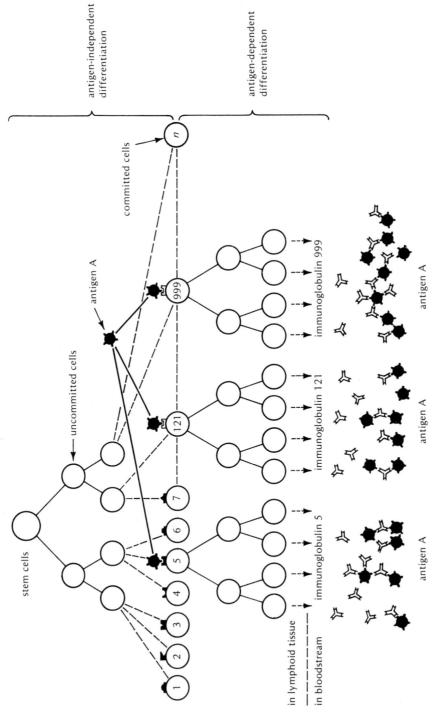

Figure 7–40. Clonal selection theory of antibody formation. From G. Edelman, The Structure and Function of Antibodies, *Scientific American* 223:34. Copyright © 1970 by Scientific American, Inc. Used with permission.

Several hypotheses have been developed to explain the nature of the specific response of B cells to antigenic stimulation. One of the most widely accepted hypotheses is that developed by Sir Macfarlane Burnet and others, and called the clonal selection hypothesis (Figure 7–40). This hypothesis is not restricted to B cells, as we will discuss later. The hypothesis assumes that each B cell can only produce one kind of antibody, but that each B cell has all the genes needed for all types of antibody synthesis. It is proposed that a B cell can only synthesize one type of antibody because it becomes restricted or determined for such a single response. In other words, only one set of genes that regulate the production of one kind of antibody can be activated. Other genes for other antibodies remain inactive throughout the life of the cell.

clonal selection hypothesis

B lymphocytes are thus determined: their fates are fixed so that they can produce only one type of antibody. When the appropriate antigen comes along that can elicit a response in a given determined lymphocyte, this antigen interacts with the B cell surface. Such interaction occurs because the committed lymphocyte has already synthesized a small amount of specific antibody. This antibody is incorporated into the surface membrane of the B cell. The specific antigen interacts with the surface antibody. This interaction, it is proposed, stimulates the lymphocyte to divide, resulting in a clone (cells derived from one parent cell) of cells that possess the specific surface antibody and are committed to the synthesis of the one specific antibody (Figure 7–41). In this way a large population of specific antibody-forming cells can be produced to respond to a given antigen. This allows the production of large quantities of antibody and permits an effective immune response upon antigenic stimulation.

There is good evidence that the clonal selection hypothesis is correct. For example, if lymphocytes are exposed to two different antigens, and then single B cells are placed in microdrops, each microdrop contained antibody against antigen 1 or 2 but not both. Thus it seems that one B cell can only respond to one antigen.

It should be noted that humoral immunity does not appear to be totally dependent upon B cells and their descendents. Removal of the thymus at an early stage was found to result in less effective production of circulating antibodies in response to some antigens. In addition, inoculation of B cells alone into animals that have been irradiated to destroy their immune systems, did not result in effective reaction to antigens. Inoculation of B plus T cells permitted effective antibody production by the B cells. Thus it seems that T cells and perhaps other white blood cells such as macrophages assist the B cells in their role of producing specific antibodies. At present it is believed that a strong humoral immune response involves several cell types. Macrophages receive antigenic stimulation and in some way process it. A signal from the macrophages

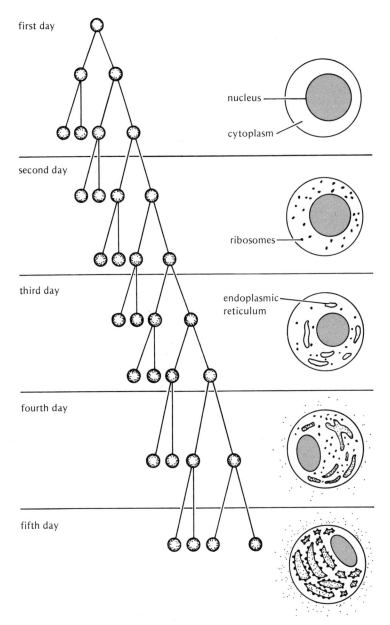

is then sent to T cells. The T cells then are able to interact with B cells, stimulating the B cells to produce large quantities of specific antibody. The basis of these interactions is not well understood.

A brief examination of the nature of antibodies is appropriate now that we have surveyed the cell types involved in antibody production. We'll then move on to a look at T cell mediated immunity.

Antibodies

We briefly mentioned two types of circulating antibody called IgM and IgG. In addition to these, another type of antibody that can be found in the blood is IgA. Other classes of antibody (IgD and IgE) are found as minor components in the serum and will not be described here. IgM has a molecular weight of about 900,000 daltons, while the molecular weight of IgG and IgA is about 150,000 daltons.

IgG is the major immunoglobulin in the serum of immunized adult mammals such as man. As shown in Figure 7–42, each IgG molecule is composed of four polypeptide chains held together by disulfide bonds and noncovalent associations. Two of these chains are called light chains, each having a molecular weight of 23,000 daltons. The other two are also identical and are called heavy chains. The heavy chains with a molecular weight of 53,000 daltons are a few times the size of the light chains.

The important functional sites on a specific IgG molecule, the antigen combining sites, possess unique amino acid sequences not found on an IgG molecule made against a different antigen. These sites consist of a part of the light and heavy chains (the variable regions). The remaining portion (constant region) of the light and heavy chains of all IgG molecules have generally identical amino acid sequences.

It is the variable regions of the antibody molecule that account for the specificity of antibodies in combining with antigens. How do B cells make antibodies that possess polypeptide chains that consist of a con-

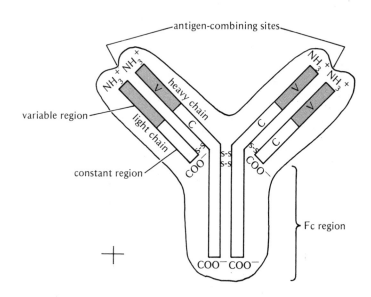

Figure 7–42. Immunoglobulin Molecule. The antigen combining sites are located between the V regions of the light and heavy chains. From Hood, Weissman & Wood, *Immunology* (1978) p. 5. The Benjamin/Cummings Publishing Company, Menlo Park, California. Used with permission.

stant plus a variable region? This is an important puzzle that is not yet fully understood. It is important because one would expect that a single gene codes for a single polypeptide chain. But how can a single gene code for a constant amino acid sequence plus a variable amino acid sequence in one polypeptide chain?

Two genes appear to code for each polypeptide chain in immunoglobulin molecules. The experimental evidence that supports this contention includes molecular hybridization work. These experiments (discussed in Chapter 10) suggest that, for example, light chain messenger RNA has complementary base sequences with two different DNA fragments of embryonic cells. Also, by sequencing DNA it has been shown that variable region genes are not adjacent to constant region genes.

How are polypeptide chains with products from two separate genes constructed in antibody-forming cells? This could occur by combining genes or gene products at the DNA, RNA, or polypeptide level. Recent work suggests that some of the combining occurs at the RNA level. It appears that high molecular weight nuclear RNA undergoes processing events that remove RNA that is intervening between the variable and constant immunoglobulin message sequences. These processing events bring the variable and constant region RNA message sequences together. This combined message is then transported to the cytoplasm for translation into immunoglobulin polypeptide.

Variable and constant region genes may also be combined at the DNA level. If this occurs, it could account for the problem of the mechanism by which a given B cell will synthesize only one type of specific immunoglobulin molecule in response to only one type of antigen. This is the case because the rearrangement of the light and heavy variable and constant genes in the DNA could activate transcription of the combined variable-constant gene sequence. This may be a critical step in the differentiation of antibody-forming cells. An excellent discussion of the problem of antibody synthesis and antibody-forming cell differentiation is found in Hood, Weissman, and Wood (1978).

Cellular Immunity

As indicated earlier, two systems of immunity protect vertebrates against foreign invaders: the humoral and the cellular response. The humoral immune response was summarized in the previous section and is based upon antibodies formed by B plasma cells, that circulate around the body in the blood serum. The cellular immune response is mediated by the T cells.

T cells, like B cells, appear to have antigen receptors on their surfaces that may or may not be antibodies. Unlike B cells, where it is known that the surface antigen receptor molecules are specific antibodies, the molecular nature of the T cell receptor is not known. Like B cells, each T cell carries only one kind of surface antigen receptor that will respond to only a few closely related antigen determinants. Antigen stimulated T cells give rise to killer T cells that directly kill foreign cells, and a variety of types of T cells that have other roles such as helping B cells to differentiate and proliferate.

The clonal selection hypothesis accounts for large scale response to specific antigens in the T cell system just as it does for the B cell system. The initial encounter with antigen causes T cell clones to proliferate and differentiate. As seen for B cells, it is likely that the specific antigen interacts with the surface of those T cells that have receptors for that antigen. This antigen-receptor interaction in some way causes that cell to proliferate, forming a clone of T cells that respond to the specific antigen. The B cell response is in the form of specific antibody synthesis. The T cell response (killer T cells) is in the form of direct assault upon the invading cell. It should be noted that in the case of both T and B cells, some cells, the memory cells, retain the capacity for continued proliferation upon subsequent encounter with the given antigen. Other cells are short lived and die in a few days after functioning to form an antibody (B plasma cells), or as killers (Killer T cells).

Maturation of T cell lines involves the establishment of a microenvironment in the embryonic thymus that stimulates maturation, seeding of the thymus by T cell precursors that arise in the yolk sac, proliferation and differentiation of T cell precursors into mature T cells in the thymus, and migration of T cells to the peripheral lymphoid tissues (Figure 7–44).

Before we conclude this discussion of the development of immunity with a look at how T cells act, it should be mentioned that determination of whether a cell is a T cell or a B cell is made possible by detection of specific cell surface markers. For example, in mice, antigens Ly-4 and Pc-1 are specific for the B cell line. Ly-4 is found on young B cells, while Pc-1 is found only on differentiated B cells (plasma cells). Thy-1 is a membrane glycoprotein antigen found on T line cells, but not on B cells. Many T cells (but not B cells) carry the antigen Ly-1 or Ly-2 and Ly-3.

Antigens that enter the body through the upper respiratory and gastrointestinal tract are filtered through local lymph nodes and specialized lymphoid organs such as the tonsils, adenoids, appendix, and Peyers patches (Figure 7–43). Antigens that enter the bloodstream are

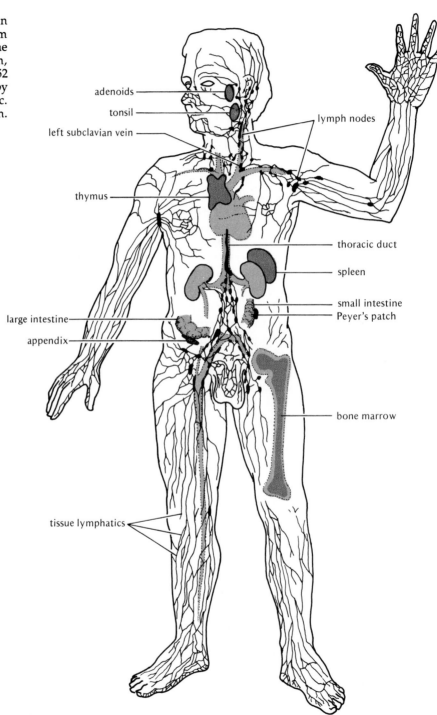

Figure 7–43. Human lymphoid system. From N. K. Jerne, The Immune System, *Scientific American* 229:52 Copyright © 1973 by Scientific American, Inc. Used with permission.

adenoids

tonsil

left subclavian vein

thymus

lymph nodes

thoracic duct

spleen

small intestine

Peyer's patch

large intestine

appendix

bone marrow

tissue lymphatics

filtered out by macrophages, the large white blood cells that line the blood vessels in the spleen, lungs, and liver. Lymphocytes (T and B cells) pass through the lymphoid organs and back into the bloodstream. In the lymphoid organs the T cells and B cells may contact antigens and macrophage-processed antigens. T cells reside in the diffuse surface cortex of the lymph nodes, while B cells migrate to discrete regions in the lymph nodes called follicles. B cells and T cells thus appear able to recognize specific regions of the lymph nodes. Contact with antigen causes the T and B cell clones to proliferate and differentiate.

When stimulation by specific antigens causes maturation, large numbers of killer T cells and antibodies produced by B-plasma cells are distributed by the lymph vessels to the bloodstream and body tissues. The killer T cells can be observed to accumulate in large numbers on the blood vessel walls near sites where the specific antigens have invaded the tissues. The killer T cells crawl through the blood vessel walls and enter the tissues, where they attack the foreign antigen-containing entities. While most B-plasma cells remain in the lymph node, secreting antibody which enters the circulation, most killer T cells, themselves, enter the circulation and travel to distant sites where they attack the foreign antigens (Figure 7–44).

In summary, the development of the immune system involves a complex series of steps that lead to the formation of active cells such as B-plasma cells and killer T cells that respond to specific antigens. These final cell types are derived from B and T cells that arise from stem cells in specific microenvironments such as those in the thymus, bursa of Fabricius, bone marrow, and fetal liver. Clones of B-plasma cells, killer T cells, and other associated cell types form in response to antigenic stimulation. Antigen interacts with specific surface receptors in the B and T cell lines. Each B cell responds to only one type of antigen and can only

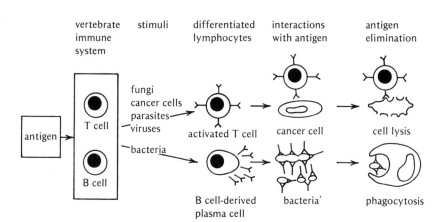

Figure 7–44. Reactions that occur upon stimulating T and B cells. From Hood, Weissman and Wood, *Immunology* p. 2. The Benjamin/Cummings Publishing Company, Menlo Park, California.

(a) immunoglobulin or lectin receptors with random lateral mobility

(b) patching with divalent or multivalent antigen or lectin

(c) movement of patches toward the cell pole

(d) capping at the cell pole

Figure 7–45. Movement of antigen receptors (such as immunoglobulins) in the cell surface of a lymphocyte caused by binding of multivalent antigens. From Hood, Weissman and Wood, *Immunology* (1978) p. 34. Benjamin/Cummings Publishing Company, Menlo Park, California.

synthesize antibody specific to this antigen. The mechanism by which a given B cell is programmed to synthesize only one kind of antibody may involve translocation of variable and constant region genes at the DNA level so that they are close to each other. This could commit the cell to exclusive production of a particular immunoglobulin molecule by activating transcription of the complete antibody genes that become adjacent to each other.

In Chapter 9 we will see that cell surface receptors can often move laterally in the plane of the cell membrane. Antigens that bind to T and B cell surface receptors could move these receptors together, in some way triggering events in the nucleus that lead to clone proliferation and specific antibody synthesis (in B cells). This idea is supported by the finding that multivalent molecules called lectins (see Chapter 9) can stimulate differentiation of T and B cells. These lectins bind to cell surface molecules, drawing them together (Figure 7–45).

Readings

Eye Development

Coulombre, A. J., The Eye. In *Organogenesis*, R. L. DeHaan and H. Ursprung, eds. Holt, Rinehart and Winston, New York, pp. 219–251 (1965).

Jacobson, M., Development of Neuronal Specificity in Retinal Ganglion Cells. *Develop. Biol.* 17:202–218 (1968).

Karkinem-Jaaskelainen, M., and L. Saxén, Advantages of Organ Culture Techniques in Teratology. In *Tests of Teratogenicity in Vitro*, J. D. Ebert and M. Marois, eds., Elsevier North-Holland, New York, pp. 275–284 (1976).

Lopashov, G. V., and O. G. Stroeva, Morphogenesis of the Vertebrate Eye. *Adv. Morphogen.* 1:331–378.

Reyer, R. W., Regeneration in the Amphibian Eye. In *Regeneration*, D. Rudnick ed., Ronald Press, New York, pp. 211–265 (1962).

Robertson, G. G., A. P. Blattner, and R. J. Williamson, Origin and Development of Lens Cataracts in Mumps Infected Chick Embryos. *Am. J. Anat.* 115:473–486 (1964).

Stone, L. S., Regeneration of the Retina, Iris and Lens. In *Regeneration in Vertebrates*, C. S. Thornton, ed., University of Chicago Press, pp. 3–14 (1959).

Wessells, N. K., *Tissue Interactions in Development*. W. A. Benjamin, Menlo Park, California (1977).

Development of Nervous Integration and Behavior

Barbera, A. J., R. B. Marchase, and S. Roth, Adhesive Recognition and Retinotectal Specificity. *Proc. Nat. Acad. Sci. U.S.* 70:2482–2486 (1973).

Cohen, S., Purification of a Nerve Growth Promoting Protein from the Mouse Salivary Gland and Its Neruocytotoxic Antiserum. *Proc. Natl. Acad. Sci. U.S.* 46:302–311 (1960).

Gaze, R. M., *The Formation of Nerve Connections*. Academic Press, New York (1970).

Hamburger, V., Emergence of Nervous Coordination. Origins of Integrated Behavior. *The Emergence of Order in Developing Systems*, M. Locke (ed.) Academic Press, New York, p. 251 (1968).

Hamburger, V., Specificity in Neurogenesis. *J. Cell. Comp. Physiol.* 60:31 (suppl. 1) (1962).

Jacobson, M., *Development Neurobiology*. Holt, Rinehart & Winston, New York (1970).

Jacobson, M., Development of Specific Neuronal Connections. *Science,* 163:543 (1969).

Jacobson, M., Retinal Ganglion Cells: Specification of Larval Connections in *Xenopus laevis. Science* 155:1106 (1967).

Levi-Montalcini, R., Growth Control of Nerve Cells by a Protein Factor and its Antiserum. *Science* 143:105 (1964).

Sperry, R. W., Optic Nerve Regeneration and Return of Vision in Anuran. *J. Neurophysiol.* 7:57 (1944).

Sperry, R. W., Embryogenesis of Behavioral Nerve Nets. In *Organogenesis,* R. L. DeHaan and H. Ursprung (eds.), Holt, Rinehart & Winston, New York (1965).

Weiss, P., *In vitro* Experiments on Factors Determining the Course of the Outgrowing Nerve Fiber. *J. Exp. Zool.* 68:393–448 (1934).

Heart Development

DeHaan, R. L., Cell Coupling and Electrophysiological Differentiation of Embryonic Heart Cells. In *Tests of Teratogenicity in Vitro,* J. D. Ebert and M. Marois, eds. Elsevier North-Holland, New York, pp. 225–232 (1976).

DeHaan, R. L., Emergence of Form and Function in the Embryonic Heart. In *The Emergence of Order in Developing Systems,* M. Locke, ed., Academic Press, New York, (1968).

DeHaan, R. L., Morphogenesis of the Vertebrate Heart. In *Organogenesis,* R. L. DeHaan and H. Ursprung, eds., Holt, Rinehart & Winston, New York, pp. 377–420 (1965).

DeHaan, R. L. and H. G. Sachs, Cell Coupling in Developing Systems: The Heart Cell Paradigm. *Current Topics in Developmental Biology,* A. A. Moscona and A. Monroy, eds., Academic Press, New York, Vol. 7:193–228 (1972).

Patten, B. M. *Foundations of Embryology,* McGraw-Hill, New York (1958).

Rawles, M. E., The Heart-Forming Areas of the Early Chick Blastoderm. *Psysiol. Zool.* 16:22–42 (1943).

Wilens, S., The Migration of Heart Mesoderm and Associated Areas in *Amblystoma punctatum. J. Exp. Zool.* 129:579–606 (1955).

Limb Development

Amprino, R., Aspects of Limb Morphogenesis in the Chicken. In *Organogenesis*, R. L. DeHaan and H. Ursprung, eds., Holt, Rinehart & Winston, New York, pp. 255–284 (1965).

Balinsky, B. I., Supernumerary Limb induction in the Anura. *J. Exp. Zool.* 188:195–202 (1974).

Becker, R. D., The Significance of Bioelectric Potentials. *Bioelectrochemistry and Bioenergetics* 1:187–199 (1974).

French, V., P. J. Bryant, and S. V. Bryant, Pattern Regulation in Epimorphic Fields. *Science* 193:969–981 (1976).

Harrison, R., Experiments on the Development of the Forelimb of Amblystoma, a Self-Differentiating Equipostential System. *J. Exp. Zool.* 25:413–461 (1918).

Hay, E. D., Metabolic Patterns in Limb Development and Regeneration. In *Organogenesis*, R. L. DeHaan and H. Ursprung, eds., Holt, Rinehart & Winston, New York, pp. 315–336 (1965).

Iten, L. E., and S. V. Bryant, The Interaction Between the Blastema and Stump in the Establishment of the Anterior-Posterior and Proximo-Distal Organization of the Limb Regenerate. *Develop. Biol.* 14:119–147 (1975).

Jabaily, J. A., and M. Singer, Neurotrophic Stimulation of DNA Synthesis in the Regenerating Forelimb of the Newt, *Triturus. J. Exp. Zool.* 199:251–256 (1977).

Milaire, J., Aspects of Limb Morphogenesis in Mammals. In *Organogenesis*, R. L. DeHaan and H. Ursprung, eds., Holt, Rinehart & Winston, New York, pp. 283–300 (1965).

Mizell, M., and J. J. Isaaco, Induced Regeneration of Hindlimbs in the Newborn Opossum. *Amer. Zool.* 10:141–155 (1970).

Rubin, L., and J. W. Saunders, Ectodermal-Mesodermal Interactions in the Growth of Limbs in Chick Embryo. *Develop. Biol.* 28:94–112 (1972).

Saunders, J. W., Jr., and J. J. Fallon, Cell Death in Morphogenesis. In *Major Problems in Developmental Biology*, M. Locke, ed., Academic Press, New York, pp. 289–316 (1966).

Searls, R. L. and M. Y. Janners, *J. Exp. Zool.* 170:365– (1969).

Singer, M., Neurotrophic Control of Limb Regeneration in the Newt. *Ann. N.Y. Acad. Sci.* 228:308–321 (1974).

Swett, F. H. Determination of Limb Axes. *Quart. Rev. Biol.* 12:322–339 (1937)

Thorton, C. S., Amphibian Limb Regeneration and Its Relation to Nerves. *Amer. Zool.* 10:113–118

Wallace, H., The Components of Regrowing Nerves Which Support the Regeneration of Irradiated Salamander Limb. *J. Embryol. Exp. Morph.* 28:419–435 (1972).

Wessells, N. K., *Tissue Interactions in Development*, W. A. Benjamin, Menlo Park, California (1977).

Zwilling, E., Limb Morphogenesis. *Adv. Morphog.* 1:329 (1916).

Urogenital Morphogenesis

Blackler, A. W., Transfer of Primordial Germ Cells Between Two Subspecies of *Xenopus laervis. Jour. Embryol. Exptl. Morph.* 10:641 (1962).

Burns, R. K., Role of Hormones in the Differentiation of Sex. In *Sex and Internal Secretions*, W. C. Young, ed., William and Wilkins, Baltimore, pp. 76–160 (1961).

Fraser, E. A. The Development of the Vertebrate Excretory System. *Biol. Rev.* 25:159–187 (1950).

Gruenwald, P., Development of the Excretory System. *Ann. N.Y. Acad. Sci.* 55:142–146 (1952).

Jost, A., Gonadal Hormones in the Sex Differentiation of the Mammalian Fetus. In *Organogenesis*, R. L. DeHaan and H. Ursprung, eds. Holt, Rinehart & Winston, New York, pp. 611–628 (1965).

Lillie, F. R., The Theory of the Free-Martin. *Science* 43:611 (1916).

Phillips, J. B., *Development of Vertebrate Anatomy.* C. V. Mosby, St. Louis, pp. 434–439 (1975).

Saxen, L., Inductive Interactions in Kidney Development. In *Control Mechanisms of Growth and Differentiation*, 25th Symposium, Society of Experimental Biology. Academic Press, New York (1971).

Swift, C. H., Origin and Early History of the Primordial Germ-Cells in the Chick. *Amer. J. Anat.* 15:483–516.

Willier, B. H., P. A. Weiss, and V. Hamburger, eds., *Analysis of Development.* W. B. Saunders, Philadelphia (1955).

Witschi, E., Migration of the Germ Cells of Human Embryos from the Yolk Sac to the Primitive Gonadal Folds. Carnegie Institution of Washington, publication 575, *Contributions to Embryology* 32:67 (1948).

Development of the Immune System

Auerbach, Development of Immunity. In *Concepts in Development*, J. Lash and U. R. Whittaker (eds.), Sinauer, Sunderland, Mass. (1974).

Burnert, S. M., *Self and Not-Self.* Cambridge University Press, London (1969).

Burnet, M., *The Clonal Selection Theory of Acquired Immunity.* Cambridge University Press, London (1969).

Edelman, G. M., Antibody Structure and Molecular Immunology. *Science* 180:830 (1975).

Edelman, G. M., The Structure and Function of Antibodies. *Sci. Amer.* 223:34 (1970).

Greaves, M. F., J. T. Owen, and M. C. Raff, *T and B Lymphoctes: Origins, Properties and Roles in Immune Responses.* Exerpta Medica, Amsterdam (1975).

Jerne, N. K. , The Immune System. *Sci. Amer.* 229:52 (1973).

Hood, L., Two Genes: One Polypeptide Chain—Fact or Fiction? *Fed. Proc.* 31:179 (1972).

Hood, L., and J. Prahl, The Immune System: A Model for Differentiation in Higher Organisms. *Adv. Immunol.* 14:291–351 (1971).

Hood, L. E., I. L. Weissman and W. B. Wood, *Immunology.* Benjamin/Cummings, Menlo Park, Calif. (1978).

Katz, D. H., *Lymphocyte Differentiation, Recognition and Regulation.* Academic Press, New York (1977).

Miller, J. F. A. P., Immunological Function of the Thymus. *Lancet* 2:748 (1961).

Nossal, G. J. V., and G. L. Ada, *Antigens, Lymphoid Cells and the Immune Response,* Academic Press, New York (1971).

Porter, R. and J. Knight (eds.), *Ontogeny of Acquired Immunity*, CIBA Foundation Symposium, Elsvier North-Holland, (1972).

Raff, M. C., Development and Differentiation of Lymphocytes. *Developmental Aspects of Carcinogenesis and Immunity*, T. J. King ed., Academic Press, New York, pp. 161–172 (1974).

Secarz, E., L. A. Herzenberg, C. R. Fox, eds., *The Immune System II: Regulatory Genetics.* Academic Press, New York (1977).

Szenberg, A., and N. L. Warner, Dissociation of Immunological Responsiveness in Fowls with a Hormonally Arrested Development of Lymphoid Tissues. *Nature* 194:146 (1962).

Weissman, I. L., Thymus Cell Migration. *J. Exp. Med.* 126:291 (1967).

Weissman, I. L., G. A. Gutman, and S. H. Friedberg, Tissue Localization of Lymphoid Cells. *Ser. Haematol.* 8:482 (1974).

CHAPTER 8

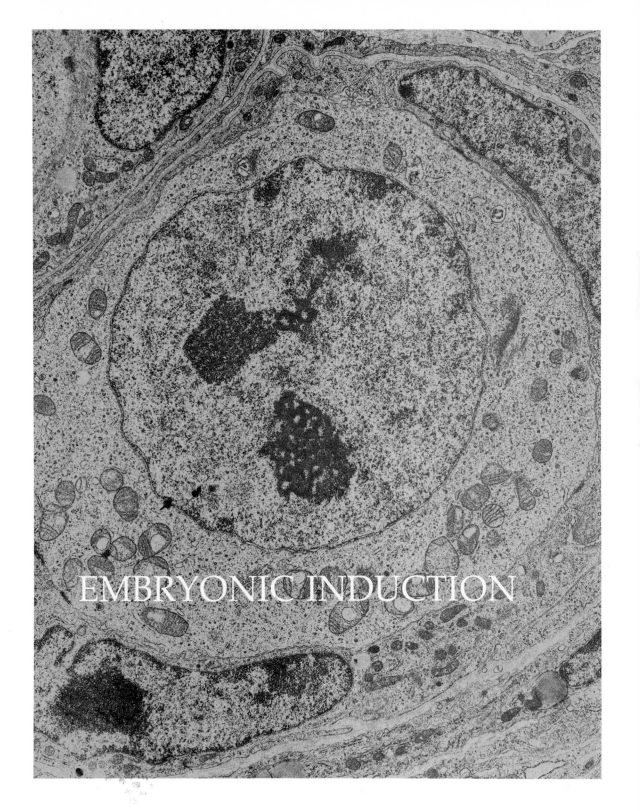

EMBRYONIC INDUCTION

embryonic induction

THE EARLY EMBRYO is a mass of similar cells that eventually become different. How can cells, all derived from the single fertilized egg, all containing identical genes, become different? This is a major area of interest in developmental biology today. We have already touched upon this fascinating problem in previous chapters. In the chapter on cleavage, we saw that certain special types of cytoplasm or surface regions can be segregated to specific cells during the cleavage process. We mentioned that interaction of cells with neighboring cells plays an important role in causing cells to differentiate. This was especially seen in neural tube differentiation, where it was noted that prospective neural ectoderm must interact with the archenteron roof for neural differentiation to occur. This interacting system will be a major topic of this chapter. It is in this system that the concept of embryonic induction originated. Embryonic induction simply means that one tissue (inducing tissue) interacts with another tissue (responding tissue), causing differentiation of the responding tissue. This sort of interaction was also seen in the discussion of organogenesis. Remember that the tip of the optic vesicle induces prospective lens ectoderm to differentiate into lens. The tip of the growing ureter induces metanephrogenic mesoderm to differentiate into metanephric kidney tubules. In short, inductive interactions between cells play a major role in causing cells to differentiate. It is believed that these inductive interactions often involve a chemical inducer that passes from the inducing tissue to the responding tissue. This inducer may act by activating specific genes that code for the specific proteins required for cell differentiation. In this chapter, we will focus upon the nature of the interactions involved in embryonic induction. Here we'll look at some factors responsible for "turning cells on" to differentiate. This will serve as an appropriate introduction to the later study of differentiation. In Chapters 10, 11, and 12, we'll examine what actually goes on during the differentiation process at the nucleic acid, protein, and organelle levels.

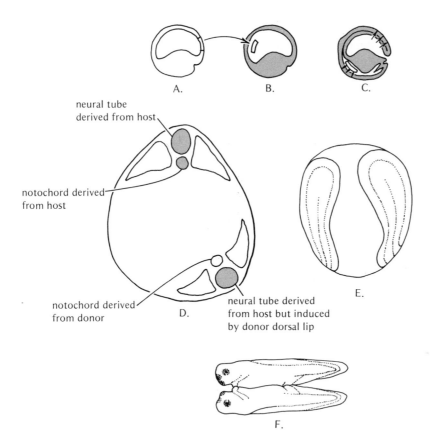

neural tube
derived from host

notochord derived
from host

notochord derived
from donor

neural tube derived
from host but induced
by donor dorsal lip

A.

B.

C.

D.

E.

F.

Figure 8–1.
Transplantation of dorsal lip of blastopore. Dorsal lip from species (a) transplanted to blastocoel of species (b). (c) Shows dorsal lip from host (archenteron roof) and donor interacting with overlying ectoderm. (d) and (e) Two neural systems formed in host. (f) Two embryos formed in some experiments. After Spemann and Mangold, *Arch. Mikrosk. Anat. Entwmech.* 100:599-638.

Origin of the Induction Concept

Let us relate a few experiments that led to the concept of embryonic induction. The first of these experiments involves separating embryos at the 2-cell stage. Hans Driesch separated the sea urchin embryo at the 2-cell stage and found that each blastomere gave rise to a complete, normal (smaller) larva. E. B. Wilson did the same experiment with the *Amphioxus* embryo and got the same results. Each cell was capable of giving rise to an entire embryo.

Hans Spemann separated the amphibian embryo at the 2-cell stage. He found that in about half the cases, at least one of the two blastomeres formed a complete embryo. In many cases, both blastomeres gave rise to a complete embryo. It turned out that if the first cleavage plane passed through the gray crescent region of the embryo, when the embryo was separated each cell blastomere often gave rise to a complete embryo. The gray crescent is a surface region of the embryo that is exposed soon after fertilization by movement of the outer pigmented portion of the embryo towards the point of sperm entry. We will return to discussing the gray crescent later in the chapter.

What did this mean? The results suggested that the gray crescent contains something that is essential for normal embryonic development. The gray crescent gives rise to the dorsal lip of the blastopore. The dorsal lip of the blastopore invaginates into the embryo during gastrulation to form the roof of the archenteron (prospective notochord) that comes to lie under the prospective neural tube. Neural differentiation fails to occur if the archenteron roof does not underlie the prospective neural ectoderm. This is seen if one removes the outer membranes from the amphibian egg and grows the embryo in a solution of lithium chloride or in hypertonic solutions. Under these conditions, instead of the normal invagination process, the prospective archenteron roof grows outward. The archenteron roof does not underly the prospective neural ectoderm and the prospective neural ectoderm fails to differentiate into the neural tube.

That the interaction of the archenteron roof with the prospective neural tube is essential for neural differentiation to occur was beautifully shown by the experiments of Hans Spemann and Hilde Mangold in 1924 that led to the award of a Nobel Prize. These workers transplanted the dorsal lip of the blastopore of one species of newt early gastrula into the blastocoel of another species of newt early gastrula. The species differed in pigmentation so it was possible to determine the origin of specific structures forming in the host embryo. The host embryo developed two neural systems and in some experiments two entire embryos formed from the host embryo (Figure 8–1). The experiment was also done by transplanting the dorsal lip of one embryo near the lateral lip of the blastopore of another embryo. The transplant invaginated and induced a second neural tube as found in the previous experiments. The differences in pigmentation between the host and dorsal lip donor embryos permitted Spemann and Mangold to determine that both neural tubes in the host embryo were derived nearly totally from host tissue. Thus, the transplanted dorsal lip did not give rise to a neural tube. Instead it induced the formation of a neural tube in the host embryo. In cases where whole second embryos formed, although the neural tubes were of host origin, the donor dorsal lip did differentiate into a variety of tissues that normally form from this region. Spemann and Mangold called the **organizer** dorsal lip the organizer because it could cause formation of a complete second embryo.

Gray Crescent

About thirty minutes after fertilization in the frog embryo, a shift of the pigmented surface of the egg occurs towards the point of sperm entry, exposing a region of underlying grayish colored surface cytoplasm. This occurrence is a symptom of dramatic reorganization that is occurring within the zygote. The exposed region of gray cytoplasm is called the gray crescent. The gray crescent, as previously mentioned, is the precursor of the dorsal lip of the blastopore that, in turn, is the precursor of the archenteron roof. We have seen that the archenteron roof (prospective notochord) induces neural differentiation in the overlying ectoderm. An interesting question now comes to mind. Can the gray crescent be transplanted into another embryo to induce neural differentiation? In other words, is the area that gives rise to the dorsal lip capable of serving as the inducing tissue?

gray crescent

The answer to this question is yes. Adam Curtis transplanted the gray crescent from an 8-cell stage frog embryo to the prospective belly ectoderm region of a 1-cell stage embryo. Cleavage occurred, and after gastrulation a second embryo with neural tube and notochord developed at the site of the transplanted gray crescent (Figure 8–2). This gray crescent formed a dorsal lip and archenteron roof and induced formation of a second neural tube (in addition to the neural tube induced by the embryo's own dorsal lip). Since the gray crescent represents only the surface region of the embryo, it becomes clear that important information is contained in this surface area. Such information appears to be essential in controlling differentiation. Recall that this idea is not unique to this system. In Chapter 3, we showed that the surface of sea urchin micromeres is different from that of the mesomeres and macromeres, and these differences may be important in differentiation.

Before we move on to examining the nature of the induction process, let us note that what we have described for the amphibian also occurs in all the other vertebrates. That is, induction of the neural tube by the dorsal lip of the blastopore (archenteron roof) occurs in the prim-

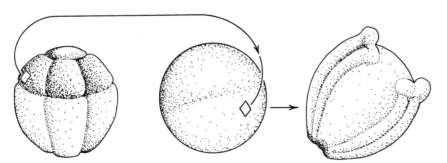

Figure 8–2. Gray crescent transplant. Gray crescent transplanted from eight cell stage amphibian embryo (*Xenopus*) to the ventral margin of the one-cell stage induces a second embryo axis. After A. S. G. Curtis, *Endeavor* 22:134 (1963)

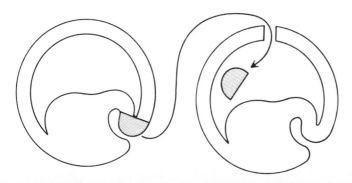

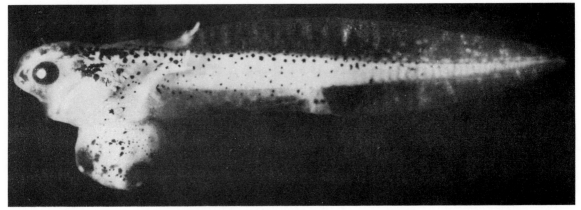

itive chordate, *Amphioxus,* cyclostomes, bony fish, and amphibians. In reptiles, birds, and mammals, the structure that corresponds to the dorsal lip is a portion of the primitive streak, which serves the same inducing function for these groups of vertebrates. Thus, embryonic induction is a phenomenon of general importance in vertebrates and other chordates.

Tissue, Stage, and Regional Specificity in Neural Induction

A question that now comes to mind is: Will any tissue respond to the archenteron roof and differentiate into neural structures? The answer to this question is no. If a piece of dorsal lip is transplanted onto endoderm,

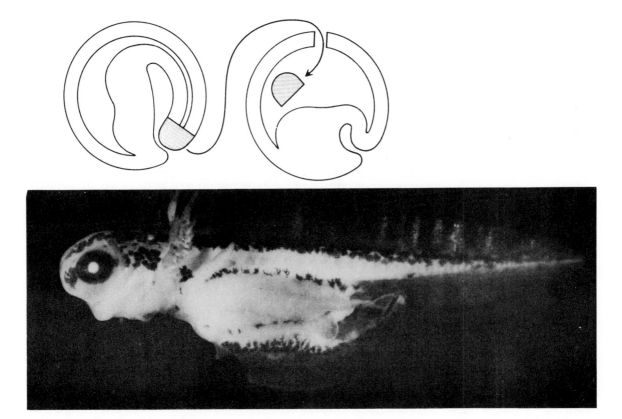

Figure 8–3. Regional inducing specificity of the dorsal lip. (opposite page) Early dorsal lip (early gastrula from amphibian) induces a secondary head in recipient embryo. (above) Late dorsal lip (late gastrula from amphibian) induces a secondary trunk and tail in recipient embryo. Courtesy of Dr. L. Saxen.

no neural differentiation occurs in the endoderm. Only the ectoderm forms neural structures as a result of interaction with the dorsal lip. We can say that only the ectoderm is competent to respond to induction by the dorsal lip. So a tissue-specific competence for neural induction exists.

tissue-specific competence

What about the developmental stage? Can ectoderm from any embryonic stage respond to the inducer by differentiating into neural structures? The answer to this question is no. Only gastrula stage ectoderm is competent to respond to induction. A dorsal lip will not induce neural differentiation if grafted into a neurula stage. If a dorsal lip is transplanted to a blastula stage embryo, the ectoderm differentiates (induced by dorsal lip) into neural structures only after gastrulation is completed, at the time at which the normal host neural system develops.

stage-specific competence

Thus, stage-specific competence for neural induction exists. Only gastrula stage ectoderm is competent to respond to the inducer.

Does a dorsal lip taken from an early gastrula have the same inducing capacity as a dorsal lip taken from a later gastrula? Normally, the early dorsal lip invaginates and comes to lie beneath the anterior part of the prospective neural tube. A later dorsal lip comes to lie beneath the posterior part of the prospective neural tube. An early dorsal lip or anterior portion of the archenteron roof transplanted into another embryo does indeed induce mostly head neural structures in the ectoderm (forebrain, eyes, nose rudiments, hindbrain, and ear vesicles). A late dorsal lip or posterior portion of the archenteron roof induces mostly posterior structures, such as spinal cord and associated trunk and tail organs.

regional inducing specificity

Thus, the dorsal lip or archenteron roof has regional inducing specificity (Figure 8–3).

Chemical Nature of the Neural Inducer

Before we examine the work that attempts to identify the chemical nature of the neural inducer, let us say at the beginning that the molecular nature of the inducer, if it exists, is unknown. It is very worthwhile, however, to look at some of the interesting approaches that have been taken to identify the elusive neural inducer. This will not be an exercise in futility because although the inducer has not been definitively identified, a far better understanding of the nature of embryonic induction has grown out of the investigations in this area.

Niu and Twitty have attempted to isolate the neural inducer by growing dorsal lips in small drops of saline solution (Figure 8–4). These dorsal lips differentiate into mesodermal structures in these drops, so it can be concluded that the conditions in the drops are conducive to normal development of the dorsal lip tissue. Dorsal lips were cultured in these drops for about ten days. They were then removed. It was hypothesized that during the incubation time, the neural inducer was released by the lips into the drops. After removing the dorsal lips, ectoderm was placed in these drops that had been "conditioned" by the dorsal lips. The ectoderm differentiated into neural structures. Induction occurred.

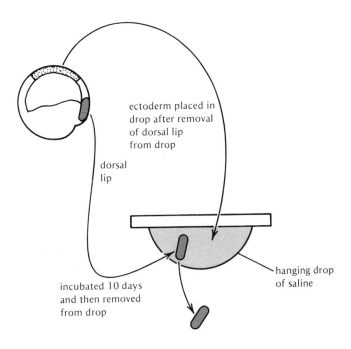

ectoderm placed in
drop after removal
of dorsal lip
from drop

dorsal
lip

incubated 10 days
and then removed
from drop

hanging drop
of saline

Figure 8–4. Attempt at isolating the neural inducer. Based upon experiments of Niu and Twitty, *Proc. Natl. Acad. Sci. U.S.*, 39:985-989 (1953).

Control experiments involved incubating other tissues, such as ectoderm, instead of dorsal lips in the drops prior to placement of the responding ectoderm into the drops. Neural induction did not occur in these cases. Only the dorsal lips released a factor into the drops that led to neural differentiation in the responding ectoderm. What was this factor? Was it the natural inducer? A variety of enzymes were used to identify the chemical nature of the isolated "inducer". Only proteolytic (protein degrading) enzymes such as trypsin completely destroyed the inducing capacity of the drop. In some experiments, other enzymes such as ribonuclease partially inactivated inducing activity. It was concluded that the material in the drops that appeared to be able to induce neural differentiation in responding ectoderm was a protein, possibly a ribonucleoprotein. Unfortunately, although these experiments showed much promise in solving the puzzle of the elusive inducer, little else has come out of this work. We must keep in mind that in order to purify and characterize a molecule, enough material must be available for the biochemical analyses. It would be very difficult to obtain enough inducer using the drop culture method to carry out the critical biochemistry experiments. Modern microchemical methods, however, may offer means of purifying and characterizing the neural inducer even if large quantities of material are unavailable.

Abnormal Inducers

What at first looked like a monkey wrench was thrown into the induction problem by further findings. As more work was done in this area it became clear that many apparently unrelated substances could induce neural differentiation in responding ectoderm. Everything from, for example, killed dorsal lip, ground quartz, dragonfly lymph, killed guinea pig liver or bone marrow, to live or dead human He La cells (cervical cancer cells) was able to induce neural differentiation in responding ectoderm. Many of these experiments used the sandwich technique, in which ectoderm is wrapped around the potential inducer to be tested, and cultivated in fluid. What did all this mean?

Holtfreter found what appeared to be a solution to this enigma. He noted that a piece of isolated prospective ectoderm could differentiate into neural structures when exposed for a short time to saline solutions that would slightly damage the tissue. If the pH of the solutions was lower than 5 or higher than 9.2, or if the ions in the solutions were altered, ectoderm often differentiated into neural structures. This work **sublethal cytolysis** led to the concept of sublethal cytolysis as a cause of induction. This concept suggests that damage to cells in some way induces these cells to differentiate. It may be that many of the so-called abnormal inducers may act in this way. It is very possible that there are many ways to stimulate neural differentiation in responding ectoderm. The normal process of induction by the archenteron roof in the embryo may occur by a different means than sublethal cytolysis, since there is no indication of cell damage in the natural system. As you can see, our understanding of the chemical nature of neural induction is still in its infancy.

Some investigators feel that a simple mechanism may be responsible for neural induction in all normal and abnormal cases. One such mechanism that has been proposed is the sodium uptake model. This model suggests that anything that causes neural induction does so by promoting sodium ion uptake by the ectoderm. Evidence in favor of this model includes the requirement of sodium in the medium for induction to occur. If sodium ion is omitted from the medium or reduced (from 88mM to 44mM), nerve cells fail to differentiate in ectoderm in the presence of dorsal lip. It remains to be seen if ion uptake is a key first step to the induction puzzle. If it is, what do the ions do?

To sum up, many things appear able to cause neural induction in responding ectoderm. The natural inducer may be protein or ribonucleoprotein. Many abnormal inducers may work by damaging the cells. It could be that uptake of specific ions is an important first step in the induction process. The future should provide us with more definitive answers to these problems.

Mechanism of Inducer Action

We have not, as yet, definitively identified the chemical nature of the neural inducer. The chemical nature of the molecule(s) is not the only important aspect of induction. Other key questions include: What is the nature of the response to the inducer in responding ectoderm? What physical conditions are needed for induction to occur? Is contact between the archenteron roof and prospective neural tube needed for induction to occur? Let us examine these problems.

What is the nature of the response to the inducer? The response is controlled by genes in the responding ectoderm. It is unlikely that active genetic information from the dorsal lip plays a role in the response. In other words, the inducer provides permissive rather than instructive information to the responding ectoderm. Permissive refers to turning on the responding tissue's genes. Instructive refers to the provision of genetic information from the inducing tissue to the responding ectoderm. These conclusions are supported by the following experiment.

permissive information

instructive information

That the response to the inducer is a result of genes in the responding ectoderm is shown by experiments with different amphibian embryos. Each species used has markers, such as certain head structures not found in the other species. One of these structures is called a balancer. Only some salamander larvae have balancers.

The balancers are induced in the epidermis by the anterior portion of the archenteron roof. Some salamander species do not possess balancers. In these species, a balancer can be induced if a piece of ectoderm from a species possessing a balancer is transplanted to replace the ectoderm of the host that can not form a balancer. Thus, the archenteron roof of the species without a balancer can induce a balancer in ectoderm from a species competent to form the balancer. The reverse experiment of transplanting ectoderm from species without a balancer to the balancer-forming area of a species that forms a balancer was also performed. In this case, a balancer can not be induced in ectoderm that is not competent to form a balancer. The ectoderm must possess the genetic code to be activated by the inducer. Thus, the inducer does not alter the genetic response of the responding tissue.

Other experiments that show that genes in the responding tissue are activated by the inducer include those using the RNA synthesis inhibitor actinomycin D. If newt gastrula ectoderm is exposed to an inducer, then separated from the inducer and immediately treated with actinomycin D, neural differentiation fails to take place. The actinomycin D–treated ectoderm remained healthy and formed epidermal structures, but not neural structures. This experiment suggests that RNA (presumably messenger RNA) synthesis must occur by the genes of the responding ectoderm if neural differentiation is to take place in response to the inducer.

Cellular Contact in Embryonic Induction

What physical conditions are needed for induction to occur? Is contact between the archenteron roof and prospective neural tube required for induction to occur? The question of the requirement or lack of requirement for cell-cell contact in normal neural induction still remains an enigma.

One approach to investigating the need for contact between cells in the archenteron roof and cells in the prospective neural tube has been to separate these two tissues with different artificial barriers. If induction is caused by a molecule that is passed from the archenteron roof to the overlying ectoderm, then a filter placed between the two tissues that allows molecular passage but not cell passage should not inhibit induction. Material that does not allow even passage of small molecules between the inducing and responding tissues should prevent induction.

Early experiments involved the placement of membrane barriers between the archenteron roof and ectoderm. Cellophane, which does not allow passage of molecules, did indeed prevent induction. If permeable membrane filters were used instead, early experiments indicated that induction still occurred (Figure 8–5). These results suggested that cell contact is not needed for induction, and that the key was rather the passage of molecules between the inducing and responding tissue. Later on, however, when the filters that had been placed between the archenteron roof and ectoderm were examined with the electron microscope, thin cell processes were sometimes observed in the filter pores. Thus, the cell contact issue was not really resolved.

Figure 8–5. Separation of dorsal lip and responding ectoderm by porous filter permits induction to occur.

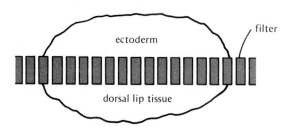

Toivonen and Saxen recently reinvestigated this problem, using filters with the tiny straight pores (0.1μ) produced by shooting neutron beams through the filters. These filters were inserted between the inducing and responding tissues for a very brief time. The tissues were then separated. The ectoderm was cultured separately. Neural differentiation did indeed occur in this ectoderm. Electron microscopic analysis of the filter pores indicated that no cell processes seemed able to enter the filter. This experiment is perhaps the best yet that supports the notion that neural induction does not require contact between the cells of the archenteron roof and prospective neural tube. The induction may occur by diffusion of molecules from the archenteron roof to the responding ectoderm.

Induction Gradients

Although the chemical nature of the neural inducer is not known, indirect evidence suggests that two types of inducers may be active in inducing neural structures. Let us briefly look at some evidence that suggests that there is a "double gradient" of substances in the archenteron roof that is involved in the induction process.

When a variety of abnormal materials were tested for inducing ability, it was found that some materials, such as killed guinea pig liver, would induce mainly forebrain structures in responding ectoderm. Other materials, such as killed guinea pig bone marrow, induced mainly trunk and tail structures in the responding ectoderm (Figure 8–6). These results suggested that more than one type of inducer, with regional inducing specificity, may exist.

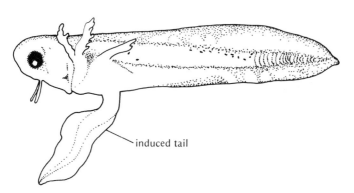

induced tail

Figure 8–6. Induction by abnormal inducer. Alcohol-treated (killed) guinea pig kidney tissue induces tail structures when transplanted into the blastocoel of a gastrula. After experiments by L. Saxen and S. Toivonen.

Figure 8–7.
Implantation of a head
structure inducer (H)
plus a trunk-tail inducer
(T-T) into newt gastrula
yields head, trunk-tail
and middle structures.
After Toivonen and
Saxen, *Ann Acad. Sci.
Fenn. A.* 30; 1-29 (1955).

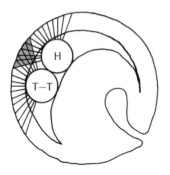

If a mixture of forebrain inducer and tail inducer was placed in contact with responding ectoderm, not only were forebrain and trunk-tail structures induced, but also hindbrain and spinal cord structures were formed (Figure 8–7). Thus, a mixture of a forebrain and trunk-tail inducer could induce intermediate structures. These experiments (by Toivonen and Saxén) suggested that in the natural system, perhaps two types of inducers exist in the archenteron roof. The forebrain inducer may be concentrated anteriorly in the archenteron roof. The trunk-tail inducer may be concentrated posteriorly in the archenteron roof. Where the two substances join in the middle of the roof, middle structures can be induced in the middle area of the responding ectoderm (Figure 8–7).

Figure 8–8.
Hypothetical model of
inducer action. (a)
Inducer acts on
responding cell. (b)
Message reaches
nucleus to begin
transcription of specific
messenger RNA
(required, for example,
for neural protein
synthesis). (c) Specific
messenger RNA enters
cytoplasm. (d) Synthesis
of specific protein
required for
differentiation.

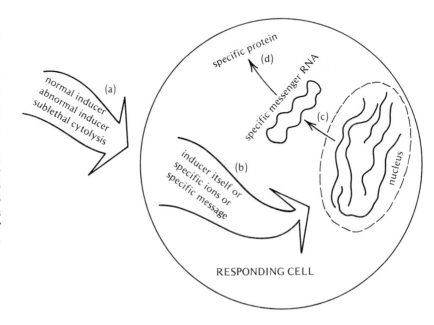

Tiedemann's group isolated a mainly middle structure (hindbrain) inducer. They separated this material by chromatography on DEAE cellulose into several fractions. One of the active fractions only induced forebrain structures in responding ectoderm. Another active fraction only induced trunk-tail structures. No fraction induced much in the way of hindbrain structures. Thus, an inducer that causes hindbrain differentiation can be separated into components that induce forebrain and trunk-tail structures. This is added evidence for the double gradient idea. There is no conclusive proof, however, that two or more inducers are present in the archenteron roof. Time will tell if this is indeed the case.

Summary

In summary, neural induction is an example of how one cell interacts with another cell to cause differentiation. It is likely that the inducer acts by turning on specific genes required for differentiation in the responding tissue (Figure 8–8). We saw that only ectoderm responds to dorsal lip induction (tissue specificity). Only gastrula ectoderm is competent to respond to the inducer (stage specificity) and specific regions of the dorsal lip induce specific structures in the responding ectoderm (regional specificity). The chemical nature of the natural inducer is unknown, although evidence suggests that it may be protein or ribonucleoprotein. Induction may not require cell contact between inducing and responding tissues. Agents that damage cells and other abnormal agents act as inducers. One model that has been proposed to explain the action of the variety of inducing agents is that these agents act by causing the responding tissue to take up specific ions such as sodium. Our understanding of the components of embryonic induction is still in its infancy. The future should shed much light on this intriguing area.

Now that we have see how interactions with neighboring cells can cause differentiation, we can turn to an examination of the process of differentiation in chapters 10, 11, and 12. First we will look at the nature of the forces that shape the embryo. It is these forces that get the cells to the right place at the right time so that processes such as induction and differentiation can proceed.

Readings

Barth, L. G., and L. J. Barth, Ionic Regulation of Embryonic Induction of Cell Differential in *Rana pipiens. Develop. Biol.* 39:1–23 (1974).

Callers, J., Primary Induction in Birds. *Adv. Morphogen.* 8:149–180 (1971).

Curtis, A. S. G., The Cell Cortex. *Endeavour* 22 (No. 87):134–137 (1963).

Driesch, H., Entwicklungsmechanische Studien I–II. *Z. wiss Zool* 53:160–182 (1891).

Holtfreter, J., Mesenchyme and Epithelia in Inductive and Morphogenetic Processes. In *Epithelial-Mesenchymal Interactions,* R. Fleischmajer, ed., William and Wilkins, Baltimore (1968).

Nieuwkoop, P. D. The Organization Center of the Amphibian Embryo: Its Origin, Spatial Organization and Morphogenetic Action. *Adv. Morphogenesis* 10:1–39 (1973).

Niu, M. C., and V. C. Twitty, The Differentiation of Gastrula Ectoderm in Medium Conditioned by Axial Mesoderm. *Proc. Natl. Acad. Sci. U.S.* 39:985–989 (1953).

Saxen, L., and S. Toivonin, *Primary Embryonic Induction.* Logos Press, London, (1962).

Spemann, H., *Embryonic Development and Induction.* Yale University Press, New Haven (1938).

Tiedmann, H., Factors Determining Embryonic Differentiation. *J. Cell Physiol.* 72 (Supplement 1) :129–144 (1968).

Toivonen, S., D. Tarin, and L. Saxén, The Transmission of Morphogenetic Signals from Amphibian Mesoderm to Ectoderm in Primary Induction. *Differentiation* 5:19–25 (1976).

Weiss, P., Perspectives in the Field of Morphogenesis. *Quart. Rev. Biol.* 25:177–198.

Wilson, E. B., Experimental Studies on Germinal Localization. *J. Exp. Zool.* 1:1–72 (1904).

Yamada, T., The Inductive Phenomenon as a Tool for Understanding the Basic Mechanism of Differentiation. *J. Cell. Comp. Physiol.* 60 (supplement): 49–64 (1962).

CHAPTER 9

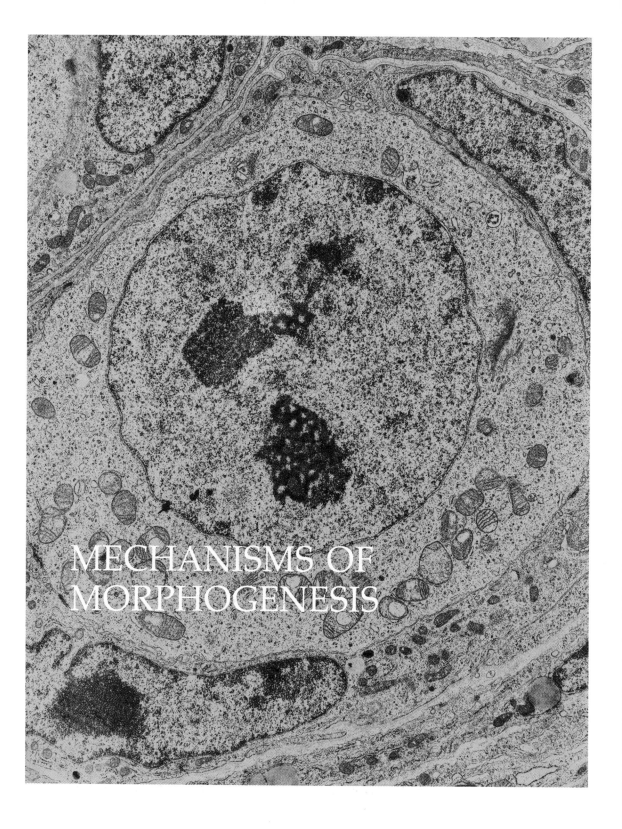

MECHANISMS OF
MORPHOGENESIS

for the surface proteins of given cell types could be produced. These stains were coupled to fluorescent dyes. Two different cell types (a human cell and a mouse cell) were fused together (using Sendai virus which causes cell fusion) and the surface proteins specific for each cell type were observed using these complex stains with the aid of a fluorescence microscope. By watching the movement of each fluorescent stain bound on the surface of the man-mouse cell, Frye and Edidin showed that the human cell surface proteins and mouse cell surface proteins intermixed. The rate of intermixing of these proteins was consistent with the notion that the proteins diffused laterally (moved sideways) in the plane of the cell surface (Figure 9–2).

Numerous subsequent experiments supported the results of Frye and Edidin that led to the conclusion that the cell surface is fluid in nature. We must stress, however, that under certain conditions (for example, low temperatures that solidify the lipid phase of the membrane) movement of proteins in the membrane becomes restricted. In addition, evidence is beginning to accumulate that suggests that certain cell surface proteins are anchored by cytoskeletal elements (microtubules and microfilaments) that appear to restrict their movement.

peripheral membrane proteins

integral membrane proteins

freeze fracture technique

Some of the proteins associated with the cell membrane are loosely associated with the membrane and are easily removed, while others are imbedded deeply in the lipid bilayer, sometimes spanning the entire thickness of the bilayer. The former are called peripheral membrane proteins and the latter are integral membrane proteins. Integral membrane proteins can often be visualized using the freeze fracture technique. This technique involves splitting the membrane down the middle and peeling away one of the lipid bilayers. The surface of the cleaved membrane is coated with heavy metal that forms a replica of the contours of the fractured surface. The replica is examined with the electron microscope. Integral membrane proteins are observed as particles on the replica surface.

To sum up, the cell surface has fluid properties and is a mosaic consisting of lipid with proteins interspersed in the lipid. The proteins can move laterally in the lipid.

Surface Sugars

Figure 9–1 indicates that some of the proteins and lipids of the cell surface have sugar chains attached. These sugar chains, and especially the terminal sugar residues of the sugar chains, reach furthest away from the cell surface and therefore offer a first line of contact with approaching cells and surfaces. Some of the molecular models that have been proposed to explain adhesive recognition of cells in embryonic systems, as will be seen, place a great deal of importance on the role of the surface sugars in adhesion mechanisms.

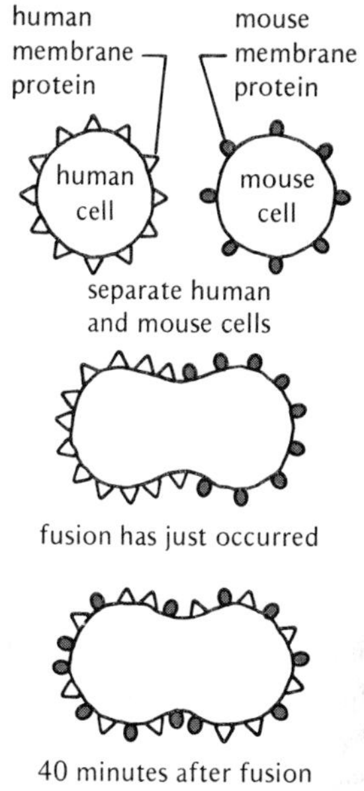

Selective Cell Adhesion and Tissue Assembly

We have seen specific examples of how specific cell adhesion plays a role in developmental events. You may recall, for example, that sperm-egg recognition appeared to depend upon specific adhesion or sticking between the gametes. Also remember that the filopodia of the secondary mesenchyme cells in the sea urchin gastrula preferentially adhere to the animal end of the embryo and play an important role in the formation of the gut tube. How does one begin to investigate the role of cell adhesion in embryonic development? What happens if embryos are dissociated into single cells and the cells are permitted to reassociate? In other words, can we study the mechanisms involved in controlling embryonic cellular rearrangements by taking embryos apart and seeing if they can put themselves back together? The answer to this question is yes. Let us see what these sorts of experiments can tell us about the mechanisms of morphogenesis. We will begin with a model system in sponges and then move on to embryonic systems.

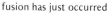

human membrane protein — mouse membrane protein

human cell   mouse cell

separate human and mouse cells

fusion has just occurred

40 minutes after fusion

Figure 9–2. Lateral mobility of proteins in the membrane lipid. Frye and Edidin showed that by 40 minutes after man and mouse cells were fused, their membrane proteins had completely intermixed.

Sponge Cell Selective Adhesion

H.V. Wilson found that sponges could be disaggregated into single cells by pressing through silk cloth. When these single sponge cells were allowed to settle in a dish of sea water, they began to move back together to form a new sponge. If cells from the red sponge (*Microciona prolifera*) were mixed with cells from the purple sponge (*Haliclona occulata*), the red cells sought out other red cells and formed a red sponge, while the purple cells formed a purple sponge. Seldom did red cells stick to purple cells. Thus, there appeared to be species-specific recognition of sponge cells for each other. This observed specificity of sponge cell reaggregation served as a model system to study adhesive recognition in developing systems.

Moscona and Humphreys continued investigations into the nature of specific reaggregation in the sponge system. Using a rotary shaker method developed by Moscona, these workers found that mixtures of red and purple sponge cells, when rotated together, still formed aggregates consisting mainly of only one type of cell. Red stuck to red, purple to purple (Figure 9–3). These results suggested that the major reason for the formation of species-specific sponge cell aggregates was that each type of cell preferentially stuck or adhered to its own kind. The shaker method allowed cell aggregates to form as a result of adhesive stability only and not directed movement (migration) that would occur in a stationary (non-rotated) system.

Moscona and Humphreys sought to determine the molecular nature of this species specific adhesion of sponge cells. They found that sponges could also be disaggregated into single cells by immersing them in calcium-magnesium-free sea water. These cells, however, unlike cells obtained by the silk cloth method, did not reaggregate if returned to sea water and rotated at low temperature (5° C). They did, however, reag-

Figure 9–3. Species-specific sponge cell re-association (reaggregation) as demonstrated by Moscona and Humphreys.

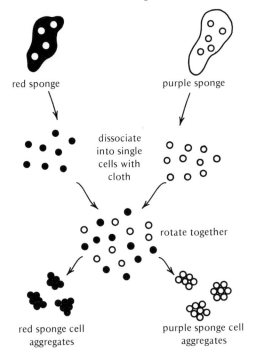

gregate if the calcium-magnesium-free sea water disaggregation super-natant (the solution left over after disaggregated cells are removed), plus calcium, was added back to the cells. The supernatant (or supe) from red sponges promoted reaggregation of only red sponge cells. The superna-tant from the purple sponges only promoted purple sponge cell reaggre-gation (Figure 9–4). It was concluded that a molecule responsible for species specific sponge cell adhesion was released upon sponge disag-gregation in calcium-magnesium-free sea water. These molecules could be added back to the cells from which they were removed, allowing them to stick back together. It was thought that cells could also resynthesize this molecule, because reaggregation of calcium-magnesium-free sea water disaggregated sponge cells could occur, to some extent, at higher temperatures (24° C) but not at low temperatures (5° C). At higher temperatures (24° C) cellular synthetic reactions could occur.

The adhesion-promoting molecules from sponge cells were purified by Moscona's group and Humphrey's group. The molecules appear to

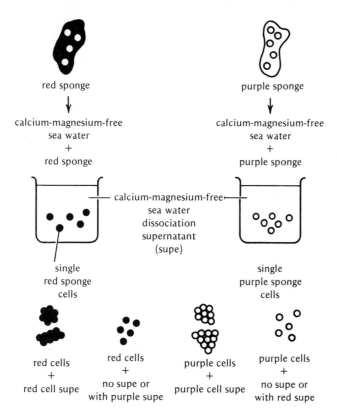

Figure 9–4. Species-specific sponge aggregation factors. From experiments of Moscona and Humphreys.

Figure 9–5. Sorting out of reaggregated amphibian embryo cells. A piece of the medullary plate and a piece of prospective epidermis are excised and disaggregated by means of alkali. The free cells are intermingled (epidermal cells are indicated in black). Under readjusted conditions the cells reaggregate and subsequently segregate so that the surface of the explant becomes entirely epidermal. From Townes and Holtfreter, *J. Exp. Zool.* 128:53–120 (1955).

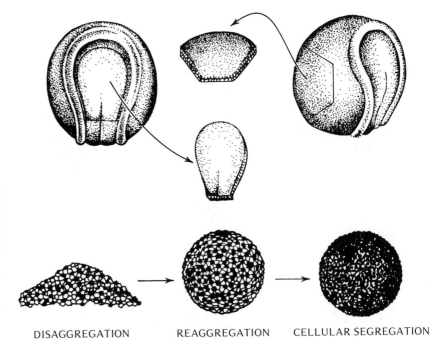

DISAGGREGATION   REAGGREGATION   CELLULAR SEGREGATION

be high molecular weight glycoproteins or proteoglycans (proteins associated with carbohydrate). Turner and Burger found that the sugar D-glucuronic acid contained on the adhesion-promoting molecule of *Microciona prolifera* may be responsible for binding to the receptor site on the sponge cell surface. The nature of the chemical bonds between readhering sponge cells is still, however, not well understood. However, molecules have been isolated, purified, and characterized that appear to be responsible for allowing sponge cells to adhere to their own kind.

Selective Adhesion in Embryonic Systems

The methods used in studying sponge cell adhesion have also been used to study embryonic cell adhesion. Townes and Holtfreter, Moscona, and Steinberg have performed some of the pioneering adhesion experiments with embryonic cells. Embryos of all types can be disaggregated into single cells using alkaline solutions, calcium-magnesium-free solutions, or solutions containing proteolytic (protein degrading) enzymes such as trypsin. Embryonic cells disaggregated by such means can be rotated using the Moscona shaker method. The cells from a whole embryo or

MEDULLARY PLATE +
EPIDERMIS

MEDULLARY PLATE +
EPID. + FOLD

MEDULLARY PLATE
ON ENDODERM

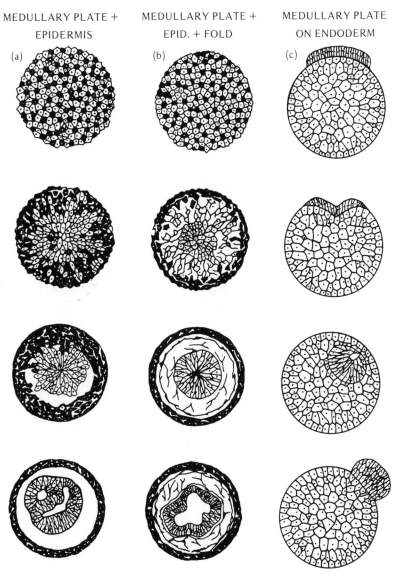

Figure 9–6. Sorting out of reaggregate amphibian embryonic cell combination. Diagrammatic sections through successive stages of composite reaggregates. (a) Randomly arranged cells of epidermis (black) and medullary plate (white) move in opposite directions and re-establish homogeneous tissues. (b) Cells from the neural fold were added. The latter move to occupy the space between neural tissue and epidermal covering. (c) A piece of medullary plate or of larval forebrain first moves into, then out of a matrix of endoderm. From Townes and Holtfreter, *J. Exp. Zool.* 128:53–120 (1955).

from a variety of embryonic tissues will recombine into aggregates of many cell types. After a day or two, however, cells tend to "sort out" into groups of only one specific cell type. Cells usually seek their own kind, and the mixed cell aggregate becomes transformed into an aggregate containing several areas, each composed of only one cell type (Figures 9–5 and 9–6). For example, ectoderm cells seek out ectoderm cells, retina cells stick to retina cells, pre-heart cells stick to pre-heart cells. The nature of this "sorting out" phenomenon is unclear but it appears to involve relative strengths of adhesiveness between the different cell types. The most tightly adhesive cells squeeze together towards the center of the aggregate, pushing the less adhesive cells to the periphery. Recently McGuire and co-workers found that embryonic cells, if disaggregated from the embryos very gently, also behaved similarly to the sponge cells. That is, embryonic liver cells, when rotated with other embryonic cell types, mainly adhered immediately to other liver cells and not to cells from other embryonic tissue cell types. Thus, there appears to be a tissue specific adhesive recognition between like cells from embryos. This adhesive recognition may play a key role in governing embryonic cell rearrangements.

Let us now turn to some experimental evidence that suggests that specific adhesive recognition between cells in embryos plays an important role in morphogenesis. Roth's group (Barbera, Marchase, and Roth) utilized the chick embryo retina-tectum system to accomplish this end. Nerves grow from the retina of the eye to specific regions of the optic tectum, the brain center that interprets visual impulses. Nerves from the dorsal portion of the retina always grow to the ventral portion of the

Figure 9–7. Retino-tectal adhesive recognition. Barbera, Marchase, and Roth found that dorsal retina cells adhered best to ventral tectum and ventral retina cells to dorsal tectum. The time needed for retina cell adherance to tectum varied in these experiments.

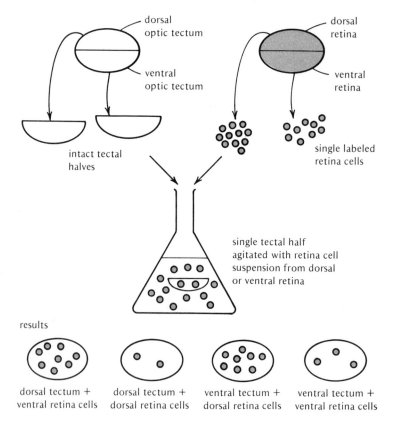

optic tectum (not the dorsal portion). Roth and co-workers sought to determine whether adhesive recognition plays a major role in these specific nerve connections. These workers cut retinas into dorsal and ventral halves and tecta into dorsal and ventral halves. Retina cells from each half of the retina were labeled with radioactive precursors (molecules that become incorporated into cells) or by their own pigment. Single labeled dorsal or ventral retina cell suspensions were agitated with either dorsal tectal halves or ventral tectal halves. This technique of shaking chunks of tissue with more than one cell type was developed by Roth and Weston to measure adhesion of like to like and unlike to unlike cell types. It was found, in general, that dorsal retina cells were picked up by ventral tectal halves to a greater extent than were dorsal retina cells picked up by dorsal tectal halves. Ventral retina cells, on the other hand,

Figure 9–8. Adhesion of ventral retina cells to dorsal and ventral tectal halves. Right: dorsal tectum. Left: Ventral tectum. Retina cells are from pigmented retina and appear as black dots on the tecta. From Barbera, Marchase, and Roth, *Proc. Natl. Acad. Sci. U.S.*, 70:2482–2486 (1973), courtesy of Stephen Roth.

were picked up to a greater extent by dorsal tectal halves than by ventral halves (Figures 9–7, 9–8). Roth and co-workers concluded that there is a specific adhesive recognition between dorsal retina and ventral tectum and ventral retina and dorsal tectum. This adhesive recognition may play a role in determining where nerves from the retina become located on the tectum. After all, in the embryo, dorsal retina nerves innervate ventral tectum and ventral retina nerves innervate dorsal tectum. The nerve endings may feel the surrounding area and get to their final resting place as a result of specific adhesion to that area.

Molecular Models of Adhesive Recognition in Embryonic Systems

In the previous sections we saw that molecules could be isolated that promote the adhesiveness of sponge cells in a species specific way. Other research has indicated that molecules can also be isolated from embryonic cells that appear to be involved in their adhesion to each other. Aron Moscona and Jack Lilien have isolated a molecule from a culture medium in which chick embryo neural retina cells were grown. This molecule appears to promote adhesiveness of embryonic neural retina cells, but not cells from other embryonic organs. If chick embryo retina cells are rotated with the adhesion-promoting factor, the cell aggregates formed over time are much larger than if the adhesion factor is absent. Similar adhesion promoting factors have been isolated, for example, from mouse teratoma cells by Oppenheimer and Humphreys, from cellular slime molds by Rosen and Barondes, and from sea urchin sperm by Vacquier and colleagues (as described in Chapter 2). Thus the isolation, purification, characterization, and determination of the functional groups of these adhesion factors may help elucidate the nature of the specific cell adhesion at the molecular level.

adhesion promoting factors

It is unlikely that the mechanisms controlling selective cell association are the same in all systems. A couple of models, however, have recently been proposed that attempt to explain cell adhesion in dynamic systems such as developing embryos and tumors. One model, the cell surface glycosyl transferase–carbohydrate acceptor model, was proposed by Roseman and developed further by Roth and others. This model suggests that embryonic cell adhesion and "de-adhesion" during cellular rearrangements in the embryo result from cell surface enzymes called glycosyl transferases, which attach to or let loose of the carbohydrate chains of glycoproteins and glycolipids at the cell surface (Figures

glycosyl transferase– carbohydrate acceptor model

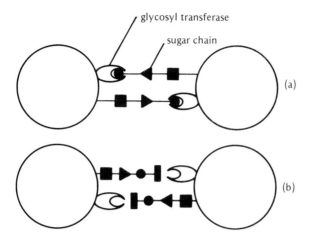

Figure 9–9.
Hypothetical model of
embryonic cell
adhesion. (a) Adhesion
takes place when the
appropriate glycosyl
transferases and sugar
chains are present. (b)
Adhesion fails to occur
because the glycosyl
transferases and sugar
chains are not
appropriate for
interaction to occur.
After Roseman, *Chem.
Phys. Lipids* 5:270 (1970).

9–9, 9–10). Glycosyl transferases are enzymes that catalyze the transfer
of single sugar residues from nucleotide-sugars to the ends of the carbohydrate chains of glycoproteins and glycolipids. Thus these enzymes are
responsible for the synthesis of sugar chains. Are these enzymes present
on the cell surface? Most experiments suggest that they are. Are the
sugar chains present on the cell surface? The answer to this question is
definitely yes. So it makes sense that a glycosyl transferase on one cell
may grab a sugar chain on an adjacent cell. Many such enzyme-substrate
complexes would lead to adhesion between the cells. Release of the
sugar chains would lead to de-adhesion.

The beauty of this hypothesis is that it can explain the complexity of
specific cell associations and disassociations in embryos. Each transferase is specific for a given receptor chain and for the sugar it transfers.
Thus, certain cells could stick together if the right transferases and the
right sugar chains are present. It should be emphasized that this model
only seeks to explain initial cell contacts. It does not deal with more
stable associations that may result after secretion of cementing substances in differentiated tissues. Experimental evidence from several
systems is accumulating that supports the glycosyl transferase hypothesis. Some evidence, however, does not support the hypothesis. It is still
too early to say if this model is indeed a correct explanation of cell adhesion in embryonic systems.

The evidence that supports a role for cell surface carbohydrates in
cell associations is somewhat more substantial. Experiments from laboratories such as those of Roseman, Roth, Oppenheimer, Steinberg,
Moscona, Burger, Rosen, and Barondes suggest that cell surface carbohydrates are involved in the adhesion process. If embryonic cells are
treated with purified glycosidases (enzymes that catalyze removal of
sugar residues from carbohydrates) their adhesion and aggregation rates

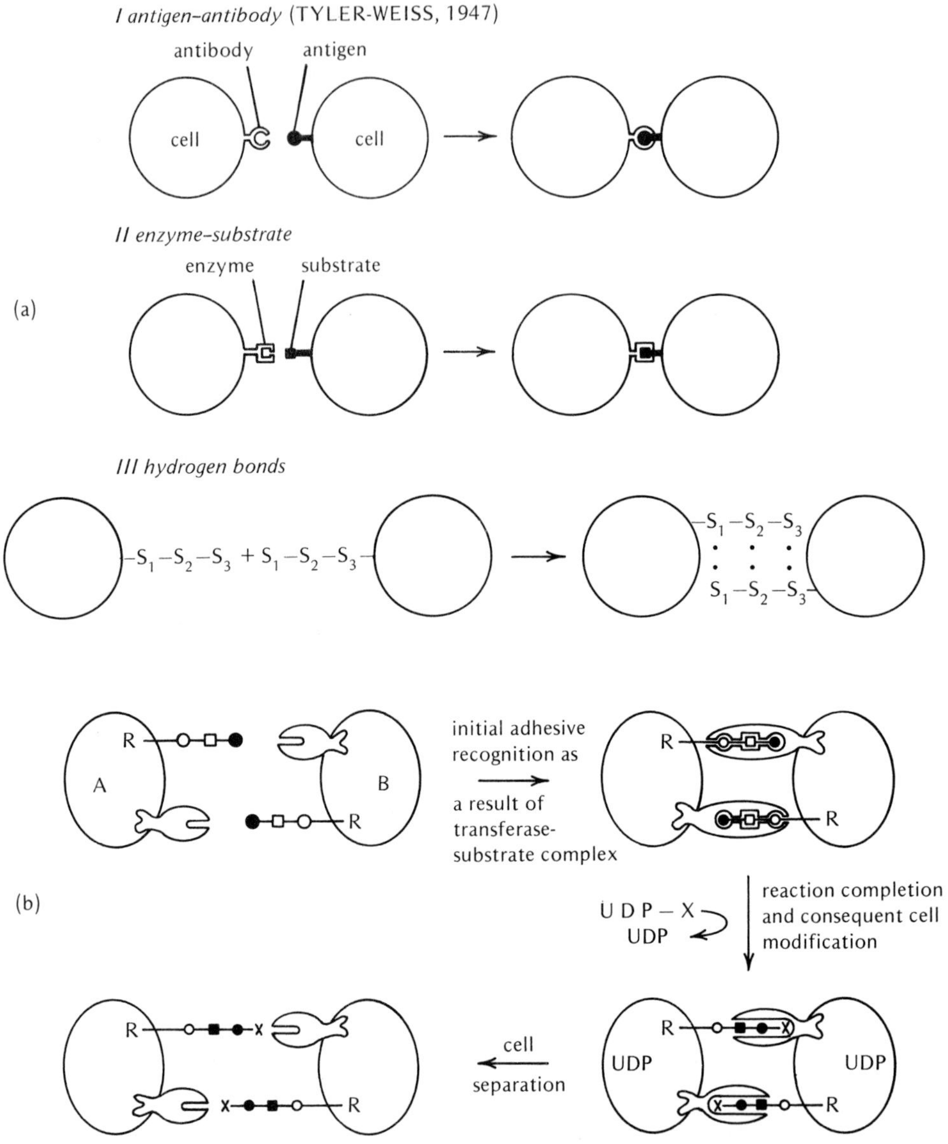

Figure 9–10. Hypotheses of cell-cell adhesion: antibody-antigen, enzyme substrate, hydrogen bonds, modification of cell-cell adhesion by glycosyl transferase reaction. In the last hypothesis, the cells adhere as a result of the binding of a complex carbohydrate chain to the corresponding glycosyl transferase. Once bound, an internally generated sugar-nucleotide (UDP-X) provides X to the enzyme on the internal face of the membrane. The transferase utilizes X to complete the reaction, and in the third step (cell separation) the product of the enzymatic reaction dissociates from the enzyme. From: Roseman, The Biosynthesis of Complex Carbohydrates and their Potential Role in Intercellular Adhesion, in A. A. Moscona, (ed.) *The Cell Surface in Development*, John Wiley and Sons, (1974) pp. 255–272.

change. Aggregation-promoting molecules are often inactivated when treated with glycosidases. In addition, specific carbohydrate binding proteins have been isolated from slime mold and embryo cells. These proteins promote association of the cells if the cells are rotated with these molecules. The proteins appear to bind specific sugar residues on the cell surface. Thus, protein-carbohydrate interaction may be of key importance in controlling cell-cell associations.

carbohydrate binding proteins

In summary, the molecular nature of cell associations in embryos is not well understood. It does appear, however, that cell adhesion plays an important role in the cellular rearrangements that occur in developing embryos. Carbohydrate-protein interaction may be the molecular mechanism that controls embryonic cell associations in some systems. Since the carbohydrate chains of cell surface glycoproteins and glycolipids reach farthest away from the cell, these chains must at least play a role in initial contact with approaching cells and substrates.

Other Mechanisms

Specific adhesive recognition among cells in embryos is clearly not the only so-called mechanism that controls the development of form in embryos. We have treated cell adhesion in length because this area has been the subject of intensive investigation in recent years, culminating in specific testable models of the molecular nature of adhesive recognition among cells. Many other specific mechanisms appear to be important in controlling morphogenesis in embryos. Let us briefly examine some that we have not previously discussed in great detail.

Cytoskeletal Elements in Morphogenesis

Throughout the text, we have mentioned that cytoskeletal elements such as rod-shaped proteins, microtubules, and microfilaments—the "microskeleton" of the cell—appear to be involved in a variety of embryonic cell activities. Cytoplasmic microtubules are associated with elongating cells.

During embryonic cellular rearrangements and in the development of form in embryonic organs, cell elongation is often observed to occur. When cell elongation takes place, microtubules appear lined up in the long axis of the cell. For example, microtubules are observed in elongating cells around the amphibian blastopore, in the neural plate, in cells moving into the primitive groove in bird embryos, and in outgrowing nerve cells. Cell elongation is prevented if a drug called colchicine, which specifically disrupts microtubules, is present. Colchicine prevents nerve cell outgrowth and inhibits cell elongation in the chick primitive streak and in the neural plate. Microtubules thus appear to play a role in embryonic cell elongation because colchicine is not known to affect anything other than microtubules. These rigid structures may directly force the cells to elongate, or they may allow cytoplasm to flow in the direction of elongation by serving as tracks for cytoplasmic flow.

Cytoplasmic microfilaments may cause cell narrowing. Microfilaments appear to be a simple type of cellular contractile system. Morphogenesis of tubular glands, for example, may be aided by bundles of microfilaments that could contract and exert force on the sides of cells. This could result in cell narrowing at one end. If a sheet of cells narrows at one side, the sheet will bend. Other hypotheses, however, have been advanced to explain how certain cell layers (epithelia) bend or fold (Figure 9–11). For example, if the outer surface cells of the layer are very tightly attached to each other so that they cannot move laterally at all and cell division takes place in the layer, the cell sheet will buckle inward. This would result because the outer surface of the sheet is unable to expand laterally. Microfilaments may also act here by causing the outer cells to become less elastic. Bundles of microfilaments that may act in such a way have been observed in surface regions of embryonic pancreas, lens, thyroid, and lung.

Extracellular Materials and Morphogenesis

glycosaminoglycans

collagen

Two groups of extracellular molecules appear to play an important role in morphogenesis. These are glycosaminoglycans and collagen. Glycosaminoglycans are sugar polymers consisting of uronic acids and amino sugars and are often linked to proteins, forming proteoglycans. Recall that one of the sponge cell adhesion factors is a proteoglycan. Collagen is a long structural protein that is fibrous and tough. Collagen is often associated with proteoglycan.

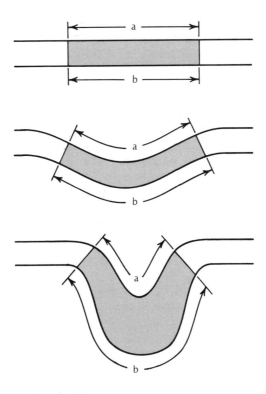

Figure 9–11. Epithelial folding: a hypothetical mechanism. Surface (a) (outer surface) is tightly adhering. Cell division takes place within the layer and the sheet buckles. Modified from Zwann and Hendrix, *Amer. Zool.* 13:1039 (1973).

Collagen may exert a specific morphogenetic stimulation on certain cells. For example, Hauschka and Konigsberg have shown that collagen stimulates prospective muscle cells to fuse together to form multinucleated cells that will form the adult muscle. Collagen is present in embryonic regions where skeletal muscle forms, and may play a role in the development and location of muscle. Also, an oriented collagen pattern appears to be needed in feather development. If an oriented pattern of collagen is absent, normal, organized feather pattern development does not occur.

Glycosaminoglycans also appear to be of importance in morphogenesis. These molecules seem to serve in maintaining the shape of branching epithelial structures. For example, if embryonic salivary gland, lung, or kidney epithelia are treated with enzymes that remove surface glycosaminoglycans, normal morphogenesis is inhibited and the structures tend to round up, to form a ball.

In summary, in this section we examined some of the factors involved in morphogenesis. Cytoskeletal elements such as microtubules and microfilaments, localized cell division, and extracellular substances such as collagen and glycosaminoglycans appear to play important roles in developing epithelial structures.

Cell Death in Morphogenesis

Before concluding our discussion on some of the mechanisms that appear to influence the development of form in embryos, let us briefly mention another important factor that helps to shape certain structures. Some structures are molded as a result of cell death that occurs in specific regions of these structures. For example, at an early stage in embryonic development our hands and feet once resembled paddles or flippers. Our fingers and toes formed as a result of cell death in the regions between the prospective fingers (or toes on the foot) as shown in Figure 9–12. Certain abnormalities, in which this cell death fails to occur, lead to webbed hands or feet. Cell death occurs in many developing embryonic structures. We saw for example, that during development of the urogenital system, certain structures in each sex degenerate after the indifferent stage of sexual development. The tadpole tail regresses during metamorphosis as a result of cell death.

Increased levels of lysosomal enzymes (for example, acid phosphatase, ribonculease, cathepsins, etc.) are observed in many embryonic regions that are "programmed to die." In the urogenital system, sex hormones may control death in certain structures. For example, androgens may trigger cell death (by influencing lysosomal enzymes) of Müllerian ducts in males. In amphibians, on the other hand, regression of the tail and gills appears to be controlled by thyroid hormone. If the thyroid gland is removed from tadpoles, metamorphosis does not occur—the tadpole does not change to become a frog. Tails and gills remain. If thyroid hormone is given to the "thyroidless" tadpoles, metamorphosis proceeds. Lysosomal enzymes increase in the areas that die. It should be stressed that specific hormones such as thyroid will only initiate cell death in some amphibian cells (tail, gills, and horny teeth) while in others (for example, limb bud) it causes cell proliferation. The hormone may cause certain genes to be "turned on" in some cells and other genes in other cells.

Pattern Morphogenesis

Pattern refers to the orderly arrangement of parts of the organism, and results from many types of morphogenetic and differentiative events. In other portions of this text we discuss pattern formation in the context of

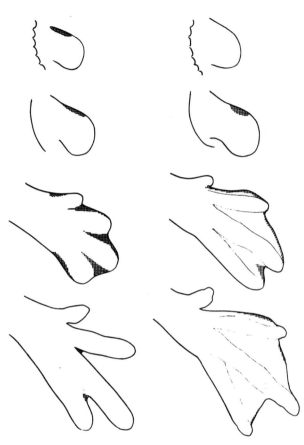

Figure 9–12. Cell death in development of digits in chick and duck embryo hindlimb. Shaded areas die in forming limb. From Saunders and Fallon in M. Locke (ed.) *Major Problems in Developmental Biology,* 289–314 Academic Press (1967). Courtesy of J. W. Saunders.

topics such as nerve growth to end organs, limb morphogenesis and regeneration, finger and toe development, plant embryo polarity, mosaic cleavage patterns, and cytoplasmic segregations. In this section no attempt will be made to discuss this topic in the context of all or even most of the systems that have been examined by hundreds of investigators concerned with the development of pattern. Instead, we will deal with a few simple systems in which some of the fundamental findings have been made. How does an ordered arrangement of parts come about?

Generally when we think about pattern formation we think of several sorts of mechanisms that could be involved in controlling this process. In order for an orderly arrangement of parts to come about there must be gene action. That is, genes that control specific differentiations such as hair bristle and pigment formations must be turned on so that the differentiated aspects of the parts that we are concerned with come about.

The second important factor is the environment in which the cells that form the patterned elements reside. We have seen in many parts of this text that cells often interact with other cells or with non-cellular environments. This interaction initiates selective gene activation or aids functioning in responding cells.

A third set of events that is sometimes involved in pattern formation is cell migration and selective adhesion, described earlier in this chapter. These events, in many systems, get the cells that eventually form the pattern to the future sites of pattern development.

In the fruit fly *Drosophila*, some of the mechanisms mentioned above come into play in the formation of bristle patterns that appear in the epidermis. It was found that the presence or absence of a bristle at a specific site in the epidermis depended upon the genotype of that portion of the epidermis. If, at a given locus, the tissue is mutant preventing bristle formation, no bristle forms even if this section of tissue is surrounded by nonmutant, bristle-forming epidermis. If the tissue is wild type at that locus, bristle formation occurs even if the tissue is surrounded by mutant tissues. Such results indicate that specific genes do indeed control bristle formation in tissue. In other cases, however, surrounding tissue appears to play a role in the formation of patterned differentiations.

Melanocytes are pigment-forming cells, most of which are derived from the neural crest. Only the melanocytes in the pigmented retina of the eye are derived from the neural tube itself. The neural crest melanocyte precursors migrate from the neural crest to a variety of sites in the body such as skin and hair follicles. Pigment pattern in mammals is a form of pattern that can be studied in depth because the genes that control such pigment pattern are known in mammals such as mice. One can transplant a piece of skin from a black mouse onto the back of a yellow mouse or visa versa. When yellow skin is placed on the back of a black mouse, the melanocytes from this skin move out into adjacent host tissue hair follicles, and begin to produce black pigment despite their own genotype. In the reciprocal experiment, the melanocytes from the black transplant migrate into the hair follicles of the yellow host and give rise to yellow pigment. Thus, formation of yellow or black pigment appears to be controlled by the hair follicle cells and not only by the melanocytes. Sulfhydryl compounds, such as glutathione, may be the hair follicle factors that stimulate yellow pigment formation in the melanocytes. That is, it may be that glutathione present in yellow hair follicles inhibits black pigment synthesis while stimulating yellow pigment synthesis. This suggestion is supported by the finding that *in vitro* cultured, isolated yellow melanocytes will form black pigment. If sulfhydryl compounds are added to the culture medium, however, these melanocytes will revert to synthesizing yellow pigment.

The most dramatic examples of pigment patterns on animals are the spotted coats of many mammals and birds. What about an animal with white spots on a dark background? How does such a pattern form? Cytological examination of hair follicles in the white spots of such an animal indicates the absence of differentiated melanocytes in these areas. Are the hair follicles in the white spots capable of permitting melanocytes to form pigment? The answer to this question is yes. Melanocytes from a black skin transplant can migrate into a white spot region, enter the hair follicles and make pigment. Thus, the white spotted areas are capable of sustaining differentiation of melanocytes that are already mature. It is likely that the white spotting pattern, therefore, may be caused by some factors present in pre-white spots that prevent melanocyte differentiation. Even if melanocyte precursors enter the white spot areas, they do not differentiate.

The case of pigment pattern formation appears, therefore, to involve all three of the mechanisms that we mentioned earlier. These are

1. migration of the pre-melanocytes from the neural crest to the skin and hair follicles

2. interaction of the melanocytes with the environment (that is, products produced by the hair follicles cells)

3. specific gene activation in the melanocytes that results in pigment synthesis

Pattern formation, therefore, is not a unique phenomenon but results from mechanisms that operate in many embryonic processes. A complete understanding of the formation of organized arrangements in organisms, that is, patterns, will come about only when we understand the molecular nature of cell-cell interactions and differentiation.

Morphogenesis in a Simple System: Cellular Slime Molds

Morphogenesis in vertebrates is complex and difficult to study. In this chapter we see that we do not know a great deal about the mechanisms that control morphogenesis in vertebrate embryos. Slow progress in this area is being made, but the major problem in such studies is that they usually must be performed with isolated embryo cells *in vitro*. Thus, for example, studies of cell-cell interactions involved in morphogenesis are often done in culture in which cells are in an environment that is not the same as that in the embryo. There are, however, a few simpler systems that have been studied that have helped to increase our understanding of the types of forces that shape the organism. One of these systems is the cellular slime mold. Let us briefly examine this system and how it has helped us to understand some of the mechanisms of morphogenesis.

What are cellular slime molds? These organisms at one stage in their life cycle are amoebae. At another stage in their life cycle, the amoebae aggregate and eventually form a multicellular fruiting body, composed of a stalk and a mass of spores (that give rise to the amoebae) at the tip of the stalk. We have here a system that is fairly simple and goes through morphogenetic processes that resemble some of those occurring in higher organisms. There is an aggregation phase in which the free amoebae congregate together to form an integrated migratory cellular mass (slug). This process resembles some of the cellular rearrangements that occur in vertebrate embryos. Differentiation occurs in the cells in the aggregated mass (or slug) and morphogenesis occurs in which the slug becomes transformed into the fruiting body (sporocarp). So the same sorts of events, namely morphogenesis and differentiation, that occur in higher embryos occur in this rather simple system. What is known about the mechanisms that are responsible for slime mold morphogenesis?

The aggregation phase in several slime mold systems such as *Dictyostelium discoideum* has been examined by several groups. The free amoebae continue to divide and remain single as long as there is an adequate supply of the bacteria they feed on. When the bacteria are exhausted, the amoebae in dense cultures begin to aggregate, forming the multicellular slug. What causes this aggregation to occur? Certain cells secrete the nucleotide cyclic AMP (cyclic adenosine monophosphate). The amoebae move in the direction of increasing concentration of cyclic AMP. Such movement of cells towards an increasing concentration of a chemical is **chemotaxis**   called chemotaxis. So, such specific chemotaxis is an important mechanism that controls one phase in slime mold development, the aggregation phase.

Cyclic AMP in some way causes amoebae to move together. Other work by Rosen's and Barondes' groups suggests that the mechanism by which the amoebae stick together is by protein-carbohydrate interaction. **lectins**   These workers have isolated carbohydrate-binding proteins (lectins) from slime mold cells that are competent to aggregate. The carbohydrate-binding proteins are thought to occur on the cell surfaces of amoebae. These molecules bind specifc sugar residues on carbohydrate chains also apparently present on the amoebae cell surfaces. It is proposed that cyclic AMP gets the amoebae together while the carbohydrate binding protein-sugar chain interactions are responsible for amoebae sticking together.

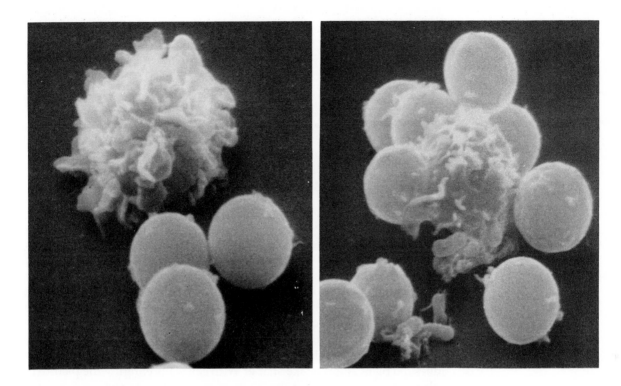

Figure 9–13. Scanning electron micrographs of slime mold cell (*Polysphondylium pallidum*) and fixed sheep erythrocytes. (left) In the presence of D-galactose, cell-cell interaction is inhibited. (right) D-glucose does not inhibit cell-cell interaction. Courtesy of Steven Rosen.

What causes the mass of amoebae, the slug, to differentiate into the fruiting body? Cells at the leading end of the migratory slug become prestalk cells while the remainder become prespore cells. The position of the cells in the slug therefore appears to play a role in determining the fate of the cells. The fate of cells is reversible, because if the slug is divided into the prestalk cells and prespore cells, each part can produce a complete fruiting body. Thus some cells shift their fates to produce the missing cell types. It is position with respect to other cells and the external environment that likely plays a major role in the shift of fates of cells in a cut slug just as it does in an intact slug. Different cells are now at the tip of the cut slug. These cells were internal in the intact slug. New positions may determine new fates.

In summary, specific factors that control development in the cellular slime molds, such as cyclic AMP, have been investigated. Other factors such as carbohydrate-binding proteins appear to play a role but are somewhat less understood at this time. Cell position also appears to govern differentiation in slime molds, but it is not at all well understood how position "works" at the molecular level. Still, simple systems such as this one are helping us understand mechanisms that may operate in the morphogenesis of higher organisms.

Morphogenesis in Plants

Throughout the text we have stressed that cell-cell interactions play a key role in development. We saw that animal embryos take form as a result of cells and cell layers moving over and under one another. Such morphogenetic events take place as the result of intimate contacts made between the surfaces of the embryonic cells, which sometimes result in stable associations. At other times, contacts are only transitory in nature and the cells move on to establish other more permanent contacts. Plants, to many of us, are rather static structures in which the cells are encased in thick cellulose cell walls. How then do plants develop? How does the plant embryo take shape? With cell walls entering the picture in plants, it is apparent that the types of movements and contacts associated with animal embryos might not be possible in plant embryos. As we will see, the plant embryo generates its form as a result of cell formation and growth. Cell migration does not play a role in plant embryo morphogenesis.

Before we describe plant embryonic development, it might be useful to summarize some of the principal differences between animal and plant development in general terms. In plants, cells develop in place, where they are formed. There is no gastrulation-like process that shapes the embryo as a result of cellular rearrangements. There is no real cleavage stage in plant embryos following fertilization. Instead, immediate differentiations occur.

Development in *Fucus*

Some of the key aspects of plant development can best be observed in plants that possess embryos that develop outside the parent. Recall that sea urchins and amphibians are prime candidates for studying early

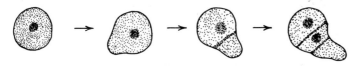

Figure 9–14. Early development of *Fucus* embryo. From Jaffe, *Adv. Morph.* 7:295–328 (1968).

development in animals because they are easily observable in the laboratory, there is a large number of gametes available, and they develop in a simple media. *Fucus* is brown seaweed that offers many of the same advantages for study. Large numbers of gametes can easily be observed in the laboratory. About 15 hours after fertilization, a striking event occurs in the zygote. At one side of the zygote an outgrowth appears. This tubular outgrowth, the rhizoid, occurs only on one side of the zygote. Thus an early polarity is established in the form of a specific differentiative event. About 9 hours later, a nuclear division occurs and a wall is formed perpendicular to rhizoid outgrowth (Figures 9–14, and 9–15).

rhizoid

In *Fucus,* therefore, we now have two cells that are clearly not equal in appearance. Each cell has a different developmental fate. Each cell divides, and derivatives of the rhizoid cell produce a holdfast, a root-like anchor, that attaches the plant to the substrate. The derivatives of the other cell (terminal cell) form a leaf-like body. What causes this polarity in the *Fucus* zygote? What causes a rhizoid to develop at one side of the zygote?

A rhizoid forms in zygotes in which nuclear division is blocked by colchicine, which blocks assembly of the spindle fiber microtubules (the spindle components that are involved in separation of the chromosomes

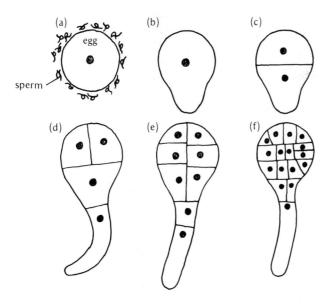

Figure 9–15. Early development in *Fucus.* From experiments of Nienburg, 1928, 1933; Smith, 1955).

Figure 9–16. Polarity in *Fucus*. From data of Kniep, 1907; Whitaker, 1931, 1939; Olson, 1937; Jaffe, 1956.

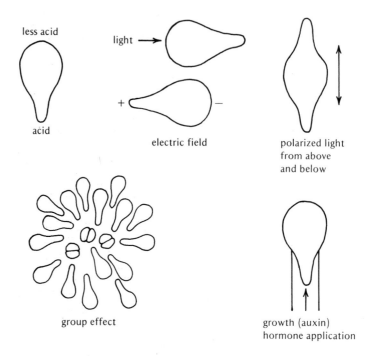

during cell division). The formation of the rhizoid therefore appears to depend on differences in the cytoplasm rather than differences in early nuclei. What factors influence rhizoid development? Environmental influences appear to be key in setting up polarity in the zygote. Rhizoids will form on the warm side of temperature gradient if one is set up in cultures. Rhizoids will form on the more acid side of a pH gradient set up in cultures. Rhizoids will form on the shaded side of a white light gradient set up in cultures. These environmental factors seem to trigger some cytoplasmic rearrangements that account for the development of polarity (Figure 9–16).

What is known about the mechanism by which environmental factors influence polarity in the zygote? Studies examining the ultrastructure of zygotes exposed to illumination from only one side show that the nuclear surface becomes polarized before any visible rhizoid appears. Finger-like projections on the nucleus radiate towards the site of rhizoid formation. These projections apparently are nuclear membrane extensions. Ribosomes, mitochondria, and fibrillar vesicles also concentrate in the region where the rhizoid will form. These changes appear at about 12 hours after fertilization, before the rhizoid appears.

Within fifteen minutes after illumination with unilateral light, *Fucus* zygotes already show some signs of polarity. For example, if one lyses (dissolves) the cells with hypotonic (dilute) solutions, lysis will begin on the shaded side, on the side where the rhizoid develops fifteen hours later! It seems that very early change occurs in response to unilateral illumination.

Other changes that may contribute to the initiation of polarity in the *Fucus* zygote include the accumulation of the sulfated polysaccharide fucoidin in the area of the cell wall at the rhizoidal end of the cell. Also, electric currents are generated by *Fucus* embryos in a capillary tube that is illuminated at one end. The end at which the rhizoids form becomes electronegative compared to the other end. It seems that current moves through the zygotes as a result of sodium, calcium, and chloride ion movements through the embryos.

Jaffe suggests that the current that moves through the zygote plays a major role in the development of polarity. He proposes that the current causes the movement of charged cytoplasmic macromolecules to specific points in the cell. For example, the current may cause fucoidin build-up at the rhizoidal end of the cell. These suggestions are supported by the finding that polarity in *Fucus* zygotes can be determined by imposing a current through the cells by artificial means. Studies with the simple plant embryo of *Fucus* have helped us to understand a few of the many factors that appear to influence early differentiation in plants. Let us now turn to a brief general description of plant embryonic development. We should keep in mind the differences between early plant and animal development noted earlier.

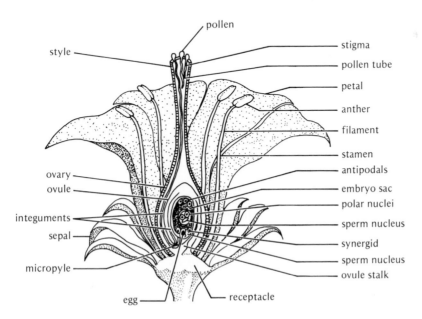

Figure 9–17. Parts of a flower. Fertilization has occurred in the ovule.

Figure 9–18. *Capsella* ovule just prior to fertilization. From Ebert & Sussex, *Interacting Systems in Development,* Holt, Rhinehart & Winston, 1970, p. 114.

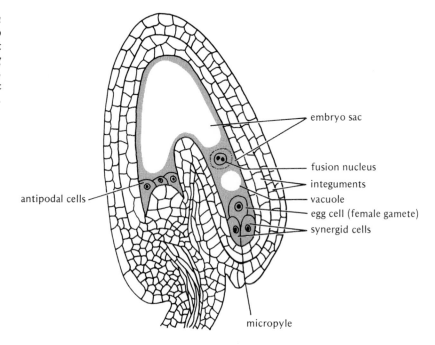

antipodal cells

embryo sac

fusion nucleus

integuments

vacuole

egg cell (female gamete)

synergid cells

micropyle

Early Plant Embryo

The embryo of the plant *Capsella bursa-pastoris,* a flowering plant called shepherd's purse, has been the subject of a great deal of study. We will summarize the development of this plant embryo in this section and then move on to the interesting story of plant embryo culture. The embryo, like the embryo of other flowering plants, develops in the flower, in the *ovule* (Figures 9–17 and 9–18). So, it is inaccessible at early stages to direct observation. Early development has been reconstructed by light and electron microscopic observation of sectioned embryos (Figure 9–19).

The egg of *Capsella* appears to be metabolically inactive. It does not contain a great deal of aggregated ribosomes, endoplasmic reticulum, Golgi apparatus, or reserve nutrients. It is a small cell contained in the embryo sac in the ovule. The haploid nucleus is situated at the broad end of the pear-shaped egg. The male gamete consists of a haploid nucleus in cytoplasmic sheath.

Fertilization occurs after transport of the male gamete from the stigma of the flower via the pollen tube to the vicinity of the egg. Activation of the egg occurs after fertilization. New ribosomes are synthesized. Ribosomes come together to form polysomes, aggregates of ribo-

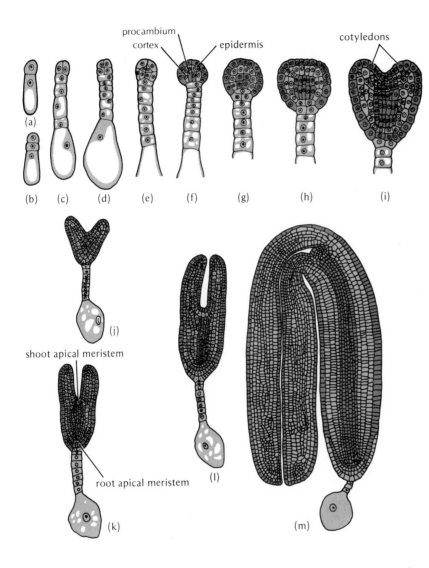

Figure 9–19.
Development of *Capsella*
embryo. From A. W.
Haupt, *Plant
Morphology*, McGraw
Hill (j)–(m) From
Schaffner, M., *Ohio
Naturalist* 7:6.

somes attached to messenger RNA. Golgi and endoplasmic reticulum
increase in abundance. Thickening of the thin cell wall that surrounds
the egg also occurs.

The first nuclear division separates the embryo into two unequal
cells. One cell is small and at the rounded end of the embryo. The other
is vacuolated, large and elongated. Thus, as in *Fucus*, an early differentiation has occurred. Each of these two cells has a different fate. The
elongated cell divides and gives rise to a linear suspensor that anchors
the embryo, that consists of five to seven large vacuolated cells. The
small terminal cell gives rise to the globular part of the embryo, forming

all of the embryonic organs. By about the 50-cell stage in the globular mass, three primary tissue systems differentiate:

1. a surface epidermis

2. a central core of elongated procambium which gives rise to the vascular system

3. a cylinder of cells that will form the cortex, situated between the two other tissues

shoot apical meristem

root apical meristem

The globular mass becomes heart-shaped (heart stage) as two hemispherical mounds form the cotyledons (embryo leaves). Right between the cotyledons, at one pole, a group of cells remains relatively undifferentiated. These cells form the shoot apical meristem. At the other pole, between the cotyledons and near the suspensor, another group of cells forms the root apical meristem. The meristems remain embryonic (Figure 9–19). The meristem itself continues to divide, giving rise to the other tissues; after seed germination, they will give rise to cells that differentiate into stem and root tissues. The embryo continues to enlarge in the ovule cavity. The rate of embryo growth then declines and finally stops. The embryo is now part of a dormant seed that will not grow again until germination.

We see, therefore, that embryonic development in plants is different from that in animals. No fertilization membrane forms. What prevents polyspermy in plants such as *Capsella* is unknown. The shaping of the plant embryo occurs by formation of cells and growth of these cells, not by cell migrations as in animals. No gastrulation occurs in plant embryos. The cells of plant embryos develop in place, where they form. The shoot and root complexes are only represented by the meristems and no vegetative leaves, shoots or roots exist in the plant embryo. Such development only occurs after germination.

The type of embryonic development that we have seen in *Capsella* is generally characteristic of most plant embryos. Some differences exist. For example, in gymnosperms (cone-bearing plants, whose seeds are not enclosed in an ovary) the early nuclear divisions are not followed by cell wall formation. The young embryo is thus a mass of cytoplasm with many nuclei (coenocytic). A globular cell mass and a region of larger suspensor cells also develop in these embryos. In ferns and mosses, the embryo does not develop in an ovule but develops in a free living gametophyte stage. In mosses and ferns, early embryonic differentiation occurs. A region of large vacuolated cells called the foot develops, resembling the suspensor of *Capsella*. Development continues to the adult stages in the fern and moss without going through an intervening dormant seed period.

Plant embryos are different from animal embryos in their more static development, lacking cellular migration. The end is the same, however; an adult organism eventually forms.

Plant Embryo Culture and Totipotency

Exciting experiments were performed by Steward and others that shed a great deal of light upon plant totipotency, the ability of plant cells to form tisues other than those that they have already formed. In many parts of the text, we examine animal systems with respect to the factors that govern differentiation. We note that in animal systems, cells in embryos often are not fixed in their fates. If such cells are transplanted to another part of the embryo they often differentiate into the type of tissue that newly surrounds them. Position or environment, therefore, appears to govern differentiation. There is little evidence, however, that adult animal tissues generally can differentiate into other types of tissues, although specific conditions may eventually be found that will allow for such reversals to occur. Since all cells in the body in general possess the same genes, and differentiation is believed, in most cases, to represent selective gene expression in specific types, it is not unreasonable to assume that differentiation may be reversible if certain genes can be turned off and others turned on.

Plants provide an important answer to the problem of the reversibility of the differentiated state. Steward and others have cultured fragments of phloem tissue from adult carrot roots under artificial conditions. Phloem is the tissue that transports nutrients from the leaves to the stem and roots. The carrot fragments grow as undifferentiated masses of tissues in a salts solution with glucose and supplemented with coconut milk that contains growth-promoting factors. If the cell masses are dissociated into single cells and cultured as free cells in a suspension containing coconut milk, a remarkable process occurs. The cells, that seem to be the progeny of differentiated phloem tissue, divide and begin to form embryo-like structures called embryoids. These embryoids **embryoids** develop shoots and roots like normal carrot embryos, and can develop into complete carrot plants that bear flowers and seeds.

Pollen grains and mature plant tissues other than root phloem have yielded embryoids in culture. These tissues include flower buds, leaves, and stems. It is possible that the cells that give rise to the embryoids are not really derived from the differentiated tissues, but instead may de-

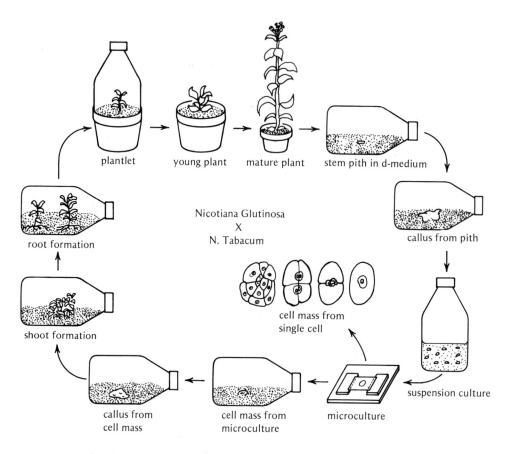

Figure 9–20. Culture of complete plants from single cells. From Vasil and Hildebrandt, *Planta* 75:139 (1967) Springer-Verlag.

velop from the actively proliferating meristem tissue that is often associated with differentiated tissues. There is evidence, however, that differentiated plant cells that are large and highly vacuolated actually alter their appearance and dedifferentiate. These cells can be observed to lose their vacuoles, enlarge their nuclei and nucleoli, and begin increasing DNA and ribosomal RNA synthesis. In tobacco plants, single differentiated cells will dedifferentiate into callus tissue. Such callus, under appropriate culture conditions, can give rise to entire tobacco plants (Figure 9–20).

The nature of the growth factors that lead to dedifferentiation and differentiation of plant cells in culture has been investigated. Specific concentrations of the plant hormones auxin (indolacetic acid) and kinetin appear to control root and shoot development of callus tissue in culture. It is clear that propagation of new plants from cells of other plants has important implications for agriculture. Agriculture has been making use of such propagation methods for many years already and it is likely that this field will grow substantially in the years ahead.

One can isolate single plant cells in culture and remove their cell walls mechanically or with enzymes. These wall-free cells are called protoplasts. New cell wall synthesis will occur in the presence of the hormones kinin or auxin. Whole plants can arise from such protoplasts derived from tobacco, carrot, petunia, belladonna, and asparagus.

That wall-free plant cells can give rise to entire plants has very important implications for plant breeders. With wall-free plant cells, one can fuse one cell with another in culture. Already, hybrid species of tobacco and petunia have been produced by fusing single protoplasts of different species with each other in culture. In addition, chloroplasts (cell organelles involved in photosynthesis) from one species have been introduced into protoplasts of other species. Nuclear transplants have also been made possible by protoplast technology. Work is currently in progress to introduce nitrogen-fixing bacteria into protoplasts of plants that cannot normally fix nitrogen. If it were possible to transfer nitrogen-fixing ability to plants which usually require large amounts of nitrogen fertilizers, the need for such fertilizers could be lessened or eliminated!

In summary, plant development and animal development differ in certain respects. Plant embryo cells generally develop in place. Animal embryo cells rearrange to form organ rudiments. Like animals, plant systems offer some unique qualities for the study of development. The question of the totipotency of cells, for example, has been studied in plant cell culture. New developments in protoplast plant culture are leading to important advances in the fields of plant breeding and agriculture.

Lectins: Tools for Studying the Role of the Cell Surface in Morphogenesis and Malignancy

In many sections of this text, we have mentioned that lectins, which are carbohydrate-binding proteins or glycoproteins, have helped us to understand certain cell surface phenomena. These molecules have, in the last decade or so, been major tools used by cell biologists in the study of the cell surface. Let us spend a little time here to discuss these widely used molecules in more detail, in the context of morphogenesis. We will sometimes stray a bit from the theme of morphogenesis in order to present the lectin story in an uninterrupted manner.

Lectins have been isolated from the seeds of plants and also from extracts of sponges, snails, crabs, fish, slime molds, and vertebrate cells. A first question we might ask is: What is the function of lectins *in vivo*? We have already mentioned that lectins may be important in sperm-egg recognition. Sea urchin sperm bindin isolated by Vacquier appears to be a lectin that recognizes carbohydrate groups on the cell surface of eggs, and thus appears to function in sperm-egg recognition. Lectins isolated from slime molds by Rosen and Barondes' groups may play a role in recognition and adhesiveness of slime mold amoebae as they aggregate to form a multicellular slug, as described previously. Lectins, therefore, may be important mediators of the cell recognition aspects of morphogenesis.

What about the function of lectins in plants, which are a major source of the lectins that have been studied to date? Lectins may function as plant antibodies to protect the plant against microbial attack. Lectins can agglutinate harmful microbes and can inhibit fungal cell wall hydrolysing enzymes. Lectins may also help plants store and transport sugars, and may bind plant enzymes into multi-enzyme systems that function together in metabolic reactions. Lectins may also control cell differentiation and mitosis, especially during seed germination. Animal cells such as lymphocytes can be induced to differentiate by lectins. The lectins bind and cross-link specific carbohydrate-containing cell surface receptors. Such cross-linking appears to set into motion a series of reactions that lead to differentiation and cell division.

An important function of lectins in plants appears to involve the binding of nitrogen-fixing bacteria to the roots of legumes such as beans. Legumes possess nodules of the nitrogen-fixing bacterium *Rhizobium* on their roots. These bacteria fix nitrogen into a form usable to the plants and thus are of major importance to the plants. Lectins appear to be secreted on the root surfaces of such plants. Specific lectins have been shown to specifically bind only the strains of *Rhizobium* found on the roots of the plant from which the lectin is extracted. For example, soybean agglutinin, a lectin isolated from soybean plants, has been shown to bind only strains of *Rhizobium* found on the roots of soybeans. This agglutinin does not bind other *Rhizobium* strains that form nodules on the roots of plants other than soybeans.

Before moving on to the question of how lectins are used to study the role of the cell surface in morphogenesis and malignancy, a brief description of the different types of lectins is in order. Lectins bind terminal sugars of the cell surface oligosaccharide chains or groups of sugars in the chains. The following is a list of lectins and the sugar(s) they preferentially bind.

Concanavalin A (Con A), isolated from the jackbean, binds alpha-D-mannose, D-glucose and beta-D-fructose groups.

Wheat germ agglutinin binds N-acetyl-D-glucosamine-like residues.

Ricinus communis agglutinin, isolated from the castor bean, binds D-galactose and N-acetyl-D-galactosamine-like groups.

Soybean agglutinin binds N-acetyl-D-galactosamine and D-galactose-like residues.

Garden pea lectin binds D-mannose groups.

Lentil lectin binds D-mannose and N-acetyl-D-glucosamine-like groups.

Lotus lectin binds L-fucose groups.

These are just a few of the many lectins that have been isolated. We should stress that these lectins tend to preferentially bind the given sugars. The sugar specificity is seldom absolute, however, and most lectins will bind many sugars but perhaps with less tenacity than the sugars they show preference for. The fact that there is some sugar specificity of lectin binding, has however, been a key reason for the successful use of lectins for the study of the cell surface.

How have lectins been used to study cell surfaces in morphogenesis and malignancy? We have already presented one such experiment in the chapter on cleavage. Recall that Oppenheimer's group found that sea urchin embryo micromeres and mesomere-macromeres differ with respect to their reaction to the lectin Concanavalin A. We noted that although the micromeres, mesomeres, and macromeres bound concanavalin A, only the micromeres showed a lectin-induced capping of concanavalin A cell surface receptor sites. In other words, the cell surface of the micromeres allowed the lectin to move the lectin receptor sites into a cap at one end of the cell. This suggested that the sugar-containing concanavalin A receptor sites are more mobile in the cell surface of the micromeres than in the surface of the macromeres or mesomeres. What does this have to do with morphogenesis? We know that the micromeres form the migratory primary mesenchyme cells that play a role in gastrulation of the sea urchin embryo. The above-mentioned studies provide evidence that the cell surfaces of these pre-migratory cells are indeed different from the surfaces of the other cell types. It may be that the difference in lectin receptor site mobility in the cell surface directly aids cell migration during morphogenesis. In any case, the use of lectin in this experiment showed that cell surfaces of known populations of embryonic cells are very different in terms of the mobility of specific receptor sites.

There are many, many other studies using lectins. Let us just deal with some of them here that relate to morphogenesis and malignancy. In 1888 Stillmark found that extracts from castor bean seeds agglutinated (clumped) human erythrocytes (red blood cells). Since then lectins have been extracted from the seeds of over 800 species of flowering plants, especially the legumes. Lectins have been found not only in seeds but in leaves, roots, and bark of plants and also in extract from many animal cell types. Numerous studies have shown that, in general, malignant tumor cells or transformed cells in culture and early embryo cells are agglutinated with rather low concentrations of many lectins, while normal adult cells are seldom agglutinated at these concentrations. Why are embryonic cells and tumor cells agglutinated, while normal cells are not?

Studies using lectins labeled with fluorescent dye, ferritin, or a radioactive isotope show that, in general, most cells bind lectin. That is, normal adult cells or culture cells, tumor cells and embryonic cells have lectin receptor sites in their cell surfaces. Using the labeled lectins it has also been shown, however, that those cells that are agglutinated with lectins (tumor and embryo cells, for example,) tend to have mobile lectin receptor sites in their cell surfaces so that capping and clustering of these sites occurs in the presence of lectin.

Lectins are multivalent molecules that can cross-link several surface receptor sites, leading to a clustering or capping of the sites. This clustering or capping of the lectin receptor sites may explain how lectins agglutinate cells. A buildup of lectin-bound receptor sites in specific regions of the cell surface may facilitate lectin-bound cell agglutination. It has been proposed that early embryo cells and malignant cells have altered cytoskeletal elements (microtubules and microfilaments) that fail to restrict movement of cell surface lectin receptor sites. In normal adult cells, it is suggested that these cytoskeletal elements are attached to the inner membrane surface, serving to restrict movement of surface receptor sites. These suggestions are supported by the finding that drugs that disrupt cytoskeletal elements, such as colchicine, local anesthetics, and cytochalasin B dramatically alter the mobility and distribution of cell surface lectin receptor sites in a variety of cell types. Lectins, therefore, have been useful in identifying specific differences in the surfaces of tumor cells, normal cells, and embryonic cells.

Extensive studies with lectins have led to several additional conclusions. Moscona's group has shown that young chick embryo cells were agglutinated by concanavalin A while older cells were not. Such studies were extended to other systems by several investigators and the findings in general suggest that young, motile embryo cells are agglutinable with lectins, while older embryo cells and those which don't move much are less agglutinable. This is generally true in many systems. In those systems in which follow-up studies with labeled lectin were carried out, the results show that agglutinable embryo cells display mobile surface lectin receptor sites as found in the sea urchin micromeres.

In summary, we can say that lectins have helped us identify cell surface sugar-containing molecules and have led to the conclusion that active cells in embryos and tumors possess mobile lectin receptor sites on their surfaces. It remains to be seen whether or not this characteristic plays a key role in facilitating the types of movements and interactions that occur during morphogenesis and malignant spread (Chapter 13).

In this chapter we have examined some of the mechanisms that appear to play a role in morphogenesis. We have also looked at a variety of different systems and different techniques that have been used to investigate the problem of morphogenesis. With this information behind us, let us now turn to a study of how specific parts of the embryo become different.

Readings

Armstrong, P. B., Light and Electron Microscope Studies of Cell Sorting in Combinations of Chick Embryo Neural Retina and Retinal Pigment Epithelium. *Wilhelm Roux's Archiv.* 168:125–141 (1971).

Barbera, A. J., R. B. Marchase, and S. Roth, Adhesive Recognition and Retinotectal Specificity. *Proc. Nat. Acad. Sci. U.S.* 70:2482–2486 (1973).

Burnside, B., Microtubule and Microfilaments in Newt Neurulation. *Develop. Biol.* 26:416–441 (1971).

Etkin, W., Hormonal Control of Amphibian Metamorphosis. In *Metamorphosis,* W. Etkins and L. Gilbert, eds., Appleton-Century-Crofts, New York (1968).

Frye, L. D. and M. Edidin, The Rapid Intermixing of Cell Surface Antigens After Formation of Mouse-Human Heterokaryons. *J. Cell Sci. 7:319–335 (1970).*

Hauschka, S. K., and I. R. Konigsberg, The Influence of Collagen on the Development of Muscle Clones, *Proc. Natl. Acad. Sci. U.S.* 55:119–126 (1966).

Humphreys, T., Chemical Dissolution and *in vitro* Reconstruction of Sponge Cell Adhesions. *Develop. Biol.* 8:27–47 (1963).

Karfunkel, P., The Role of Microtubules and Microfilaments in Neurulation in *Xenopus. Develop. Biol.* 25:30–56 (1971).

Moscona, A. A., Studies on Cell Aggregation and Demonstration of Materials Selective Cell-Binding Activity. *Proc. Natl. Acad. Sci. U.S.* 49:742–747 (1962).

Moscona, A. A., ed., *The Cell Surface in Development.* John Wiley & Sons, New York (1974).

Oppenheimer, S. B., Cell Surface Carbohydrates in Adhesion and Migration. *Amer. Zool.* 18:13–23 (1978).

Roseman, S., The Biosynthesis of Complex Carbohydrates and Their Potential Role in Intercellular Adhesion. In *The Cell Surface in Development*, A. A. Moscona, ed., John Wiley & Sons, pp. 255–272 (1974).

Rosen, S. D., S. H. Barondes, and P. L. Haywood, Inhibition of Intercellular Adhesion in a Cellular Slime Mold by Univalent Antibody against a Cell Surface Lectin. *Nature* 263:425 (1976).

Roth, S., and J. Weston, The Measurement of Intercelluar Adhesion. *Proc. Natl. Acad. Sci. U.S.* 58:974–980 (1976).

Saunders, J. W. and J. F. Fallon, Cell Death in Morphogenesis. In *Major Problems in Developmental Biology*, M. Locke, ed., Academic Press, New York, 289–314 (1967).

Schroeder, T. E., Cell Constriction: Contractile Role of Microfilaments in Division and Development. *Amer. Zool.* 13:949–960 (1973).

Singer, S. J., and G. L. Nicolson, The Fluid Mosaic Model of the Structure of the Cell Membrane. *Science* 175:720–731 (1972).

Steinberg, M. S., Does Differential Adhesion Govern Self-Assembly Process in Histogenesis? Equilibrium Configurations and the Emergence of a Heirarchy Among Populations of Embryonic Cells. *J. Exp. Zool.* 173:395–434 (1970).

Spooner, B. S., and N. K. Wessells, An Analysis of Salivary Gland Morphogenesis: Role of Cytoplasmic Microfilaments and Microtubules. *Develop. Biol.* 27:38–54 (1972).

Townes, P. L., and J. Holtfreter, Directed Movement and Selective Adhesion of Embryonic Amphibian Cells. *J. Exp. Zool.* 128:53–120 (1955).

Vacquier, V. D., and G. W. Moy, Isolation of Bindin: The Protein Responsible for Adhesion of Sperm to Sea Urchin Eggs. *Proc. Natl. Acad. Sci. U.S.* 74:2456–2560 (1977).

Wessells, N. K., and J. Evans, The Ultrastructure of Oriented Cells and Extracellular Materials Between Developing Feathers. *Develop. Biol.* 18:42–61 (1968).

Wessells, N. K., *Tissue Interactions and Development.* W. A. Benjamin, Menlo Park, California (1977).

Wilson, H. V., On Some Phenomena of Coalescence and Regeneration in Sponges. *J. Exp. Zool.* 5:245–258 (1907).

Pattern Morphogenesis

Crick, F., Diffusion in Embryogenesis. *Nature* 225:420–442 (1970).

Lock, M., The Development of Patterns in the Integument of Insects. *Adv. Morphogen* 6:33–88 (1967).

Markert, C. L., and H. Ursprung, *Developmental Genetics.* Prentice Hall, Englewood Cliffs, New Jersey (1971).

Poole, T. W., Dermal Epidermal Interactions and the Site of Action of the Yellow (Ay) and Non-Agouti (a) Coat Color Genes in the Mouse. *Develop. Biol.* 36:208–211 (1974).

Stern, C., and Tokunaga, C., Non-Autonomy in Differentiation of Pattern-Determining Genes in *Drosophilia.* I. The Sex-Comb of Eyeless-Dominant. *Proc. Nat. Accad. Sci. U.S.* 57:658–664 (1967).

Ursprung, H. The Formation of Patterns in Development. *Major Problems in Developmental Biology,* M. Locke ed., Academic Press, New York, pp. 177–216 (1966).

Tokunaga, C., Autonomy or Non-Autonomy of Gene Effects in Mosaics. *Proc. Nat. Acad. Sci. U.S.* 69:3283–3286 (1972).

Wolpert, L., J. Hicklin, and A. Hornbruch, Positional Information and Pattern Regulation in Regeneration of *Hydra. Symp. Soc. Exp. Biol.* 25:391–415 (1971).

Morphogenesis In Plants

Ebert, J. D., and I. Sussex, *Interacting Systems in Development*, 2nd. ed. Holt, Rinehart & Winston, New York (1970).

Jaffe, L. F., Localization in the Developing *Fucus* Egg and the General Role of Localizing Currents. *Adv. Morphogen.* 7:468 (1968).

Pollock, E. G., and W. A. Jensen, Cell Development During Early Embryogenesis in *Capsella* and *Gossypuim*. *Amer. J. Botany* 51:915 (1964).

Quatrano, R. S., An Ultrastructural Study of the Determined Site of Rhizoid Formation in *Fucus* zygotes. *Exp. Cell Res.* 70:1 (1972).

Quantrano, R. S., Rhizoid Formation in *Fucus* Zygotes: Dependence on Protein and Ribonucleic Acid Synthesis. *Science* 162:468 (1968).

Skoog, F., and C. O. Miller, Chemical Regulation of Growth and Organ Formation in Plant Tissues Cultured *in vitro*. *Symp. Soc. for Exptl. Biol.* 11:118 (1957).

Steward, F. C., M. O. Mapes, A. E. Kent, and R. D. Holsten, Growth and Development of Cultured Plant Cells. *Science* 143:1 (1964).

Vasil, I. K., The Progress, Problems, and Prospects of Plant Protoplast Research. *Adv. Agron.* 28:119 (1976).

Vasil, V., and A. C. Hildebrandt, Differentiation of Tobacco Plants from Single Isolated Cells in Microcultures. *Science* 150:889 (1965).

Vasil, I. K. and V. Vasil, Totipotencey and Embryogenesis in Plant Cell and Tissue Cultures. *In Vitro* 8:117 (1972).

Wordlau, C. W., *Embryogenesis in Plants*. John Wiley and Sons, New York (1955).

CHAPTER 10

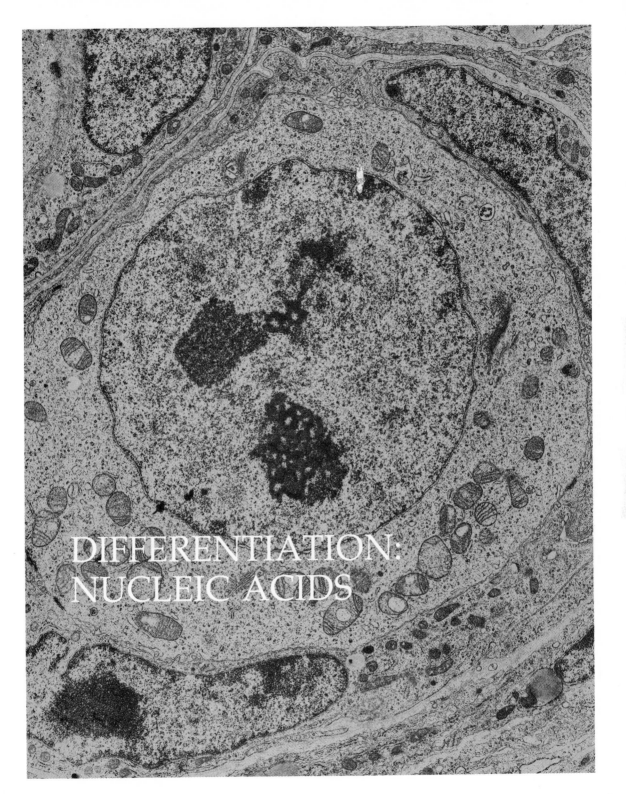

DIFFERENTIATION:
NUCLEIC ACIDS

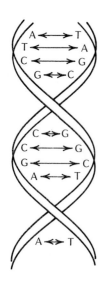

Figure 10–1. Section of DNA double helix showing nucleotide pairing. A pairs with T, and C pairs with G, by specific hydrogen bonds.

ALL CELLS IN THE EMBRYO are derived from a single cell—the fertilized egg. How then do muscle cells, blood cells, nerve cells, pigment cells, and gland cells develop if they are all derived from the same original cell and presumably contain identical genes? We know that what makes these cells different is mainly specific proteins that are required for specialized function. This will be the topic of the next chapter. We also know that the structure of these proteins is coded by the sequence of nucleotides in the DNA of the chromosomes of the cells. It makes sense that if cells with identical genes become different, only certain genes in certain cells are activated. This would lead to the synthesis of specific proteins in the specific cell types.

In previous chapters we saw that it is the environment in which the genes lie that appears to be responsible for activating or repressing certain genes. We saw that messages that tell cells to activate their genes for nerve-specific proteins, for example, appear to be passed from an inducing to a responding tissue. We will expand upon this theme later in this chapter. In this chapter we will examine differentiation at the nucleic acid level, since this is the level that plays a major role in controlling protein synthesis. We will examine evidence that supports the contention that during differentiation, specific genes are indeed activated at specific times in specific cells. In addition we will look at possible means by which these genes may be repressed or activated. Let's briefly first turn to a look at some of the techniques developed to examine differences in cellular nucleic acids, which reflect activation of specific genes.

Research Techniques

Techniques have been developed to help determine if the nucleotide sequences of two batches of DNA or a batch of DNA and a batch of RNA are similar. This is the sort of information needed to compare differences and similarities in nucleic acids in the differentiation process. These techniques are based upon the known specific pairing that occurs between the bases of nucleotides in nucleic acids. We know that adenine (A) on one chain of DNA pairs with thymine (T) on the other chain, while guanine (G) pairs with cytosine (C). We also know that RNA nucleotides contain uracil (U) instead of thymine. Uracil in an RNA chain pairs with adenine on a neighboring DNA chain. These specific pairings occur by hydrogen bonds that form between the bases of the nucleotides (Figure 10–1.)

The strands of the DNA double helix can be separated by mild heating that breaks the hydrogen bonds between the nucleotides. The strands will reassociate (anneal) if the mixture is slowly cooled. The strands will pair up correctly so that As bind Ts and Cs bind Gs. This is the basis for the techniques we will consider. Since one can easily separate double DNA helices and then allow the strands to reassociate again, DNA from different organisms or tissues could be compared for similarities and differences in base sequence. Since association of two strands of DNA will only occur if the sequence of bases are complementary, different DNAs can be compared by allowing reassociation between them to occur. If reassociation is rapid and complete, the DNAs have similar or identical base sequences. If reassociation does not occur under the carefully controlled conditions of the experiment, then the base sequences in the two types of DNA are not similar. These experiments are called hybridization experiments and are used to determine if two DNAs have similar or different base sequences.

anneal

hybridization experiments

There are many variations of the basic hybridization experiment that have been developed. Some involve binding of RNA to DNA in the presence of other types of RNA. These experiments help determine if nucleotide sequences in two types of RNA are similar. If two types of cells have similar or different RNA in terms of base sequence, we will know if the same or different genes have been activated in these cell types. These experiments will be described shortly.

The experimental protocols used in nucleic acid hybridization experiments are complex, and the conditions under which reassociation of nucleic acid strands occurs must be carefully controlled. No attempt will be made in this introductory text to describe the details of these experiments. For these, the student is referred to the references at the end of the chapter. Let us, however, get a general idea of how these experiments are performed and what they can tell us about differentiation at the nucleic acid level.

The basic DNA-DNA hybridization experiment often involves production of radioactive DNA by injecting a radioactive label, such as "hot" phosphate, into the organism. Nuclei with radioactive DNA are isolated by centrifugation of the homogenized tissue, and nucleic acids

Figure 10–2. (a) DNA-DNA hybridization experiment. (b) RNA-DNA competition hybridization experiment.

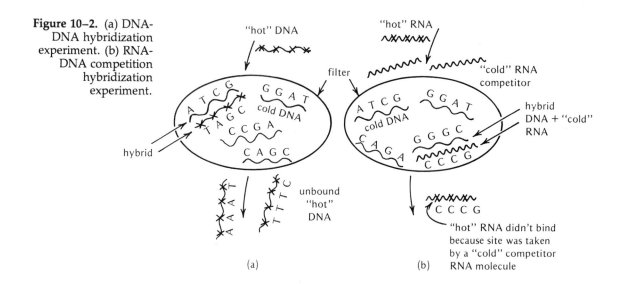

are chemically extracted from the nuclei. RNA is enzymatically removed from the DNA preparation. In this way we have extracted radioactively labeled DNA. Nonradioactive DNA is also used in these experiments and is extracted from other specimens in the same way as given above. The nonradioactive DNA is heated so that the hydrogen bonds between the nucleotides are broken. The DNA helices thus dissociate into single strands as previously described. These nonradioactive single pieces of DNA are attached to a filter or imbedded in a column of gel. Next, the radioactive single stranded DNA segments are passed through the filter or column containing the nonradioactive single stranded DNA segments. If the base sequences in the DNA on the filter and those of the radioactive DNA are similar, base pairing will occur and the radioactive DNA will anneal with the DNA on the filter. The amount of radioactive DNA bound to the DNA on the filter can easily be measured with a scintillation counter that measures radioactivity (Figure 10–2). If the two DNAs are not similar in terms of base sequence, the radioactive DNA will pass through the filter without binding to the DNA on the filter. In this way, therefore, one can determine if two DNAs have similar or different base sequences. Such DNA-DNA hybridization experiments sug-

species-specific differentiation of DNA

gest that there is species-specific differentiation of DNA. That is, DNA base sequences differ between species. Evolutionary relationships between different species have been confirmed using the DNA-DNA hybridization technique. In general, closely related species appear to have more DNA base sequences in common that do distantly related species. In addition, DNA-DNA hybridization experiments suggest that there is no tissue-specific differentiation of DNA base sequence in cells of the same organism. This makes sense because all cells in an organism are derived from the fertilized egg and should possess identical genes.

Major advances in nucleic acid technology have been made rather recently. These include the technique of gene cloning. In this technique, purified messenger RNA that has been isolated is used to make many copies of the gene (DNA) that codes for the messenger RNA. This is accomplished with the enzyme reverse transcriptase (RNA-dependent DNA polymerase). This enzyme copies RNA templates to make complementary DNA strands (cDNA). Many copies of a given gene can thus be cloned in this manner. Such "gene clones" are extremely useful in the determination of whether a given species of messenger RNA is present in a specific cell at a specific time. This is accomplished by measuring how much of the RNA under study is complementary to the specific cDNA using DNA-RNA hybridization techniques. In this way cDNA is used to directly determine the presence or absence of a particular species of RNA. The study of differentiation at the messenger RNA level, therefore, has become far more direct through use of cDNA probes.

gene cloning

reverse transcriptase

Tissue-Specific Differentiation of RNA

Are different RNA molecules present in different tissues? In other words, is there tissue-specific differentiation of RNA? Intuitively, we would guess that, yes, since different tissues possess given tissue-specific proteins, these proteins must be coded for by specific RNA molecules. We would guess that although all cells in an organism possess identical genes, only some genes are active in given tissues. In this way cells in tissues can differentiate by synthesizing specific proteins required for their specialized functions. Let us briefly describe an experiment that provides evidence for the contention that different tissues do indeed contain different RNAs. Note again that it is the base sequence in messenger RNA (and DNA) that determines the amino acid composition of specific proteins. Differences and similarities in base sequence of nucleic acids can be estimated using a modification of the hybridization techniques described previously. If pieces of DNA or a segment of DNA and a segment of RNA hybridize with each other, the base sequences are complementary. If they don't, the base sequences are not similar.

In order to determine if tissue-specific differentiation of RNA exists, a modification of the hybridization technique can be used. Segments of

tissue-specific differentiation of RNA

DNA are heated and cooled or chemically treated so that they become single-stranded. These strands are adsorbed onto a nitrocellulose membrane filter. Then, radioactively labeled RNA from a specific tissue is incubated with DNA from the same organism on the filter. This radioactive RNA is produced by injecting radioactive RNA precursors into the organism. Radioactive RNA is then extracted from the specific tissues such as spleen, liver, and kidney. In addition to radioactive RNA from a specific tissue, the incubations are carried out in the presence of increasing concentrations of unlabeled (nonradioactive) RNA from different tissues. If unlabeled RNA from a given tissue significantly inhibits the binding of the labeled RNA to the DNA, this indicates that the base sequence in this unlabeled RNA is similar to that of the radioactive RNA. This is the case because the unlabeled RNA competes for hybridization sites on the DNA template. Fewer labeled RNA molecules are then able to bind to the DNA because the sites have been taken by the similar unlabeled RNA molecules (Figure 10–2). On the other hand, if unlabeled RNA from a given tissue does not inhibit binding of the radioactive tissue RNA to the DNA on the filter, then the base sequence of the unlabeled RNA is not similar to the base sequence of the radioactive RNA. This is the case because the unlabeled RNA, in this situation, binds to DNA strands different from those to which the labeled RNA binds. The student is referred to the references cited at the end of this chapter for a more in-depth analysis of the many parameters and problems associated with these experiments.

competition-hybridization experiment

Figure 10–3 gives an example of the results of this type of so called competition-hybridization experiment in which labeled RNA from mouse kidney is incubated on the mouse DNA filter together with unlabeled RNA from mouse kidney, spleen, or liver. As can be seen, kidney

Figure 10–3. Tissue-specific differentiation of RNA. Labeled mouse kidney RNA is inhibited from binding to mouse DNA by incubation of the DNA filters in the presence of unlabeled RNA from different mouse tissues. The best inhibitor (competitor) is kidney RNA followed by spleen RNA. Liver RNA is the least effective competitor. From McCarthy, B. J. and Hoyer, B. H., *Proc. Natl. Acad. Sci., U.S.* 52:915–922 (1964).

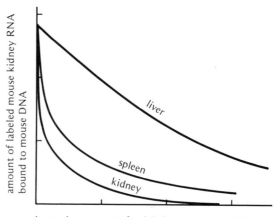

increasing amount of unlabeled competitor RNA

RNA is the most effective competitor for the hybridization reaction between kidney RNA and DNA. Spleen RNA is less effective and liver is the least effective competitor. These results suggest that the three tissues do not contain identical sets of RNA, but that different genes are active in different tissues. Thus, these sorts of experiments provide evidence for the idea that differential gene activation is a key aspect in the differentiation process. These results, taken together with the known protein differences of specialized tissues (next chapter), strongly support the contention that different genes are active in different tissues.

Stage-Specific Differentiation of RNA

Are different RNA molecules present at different developmental stages? In other words, is there stage-specific differentiation of RNA? To attempt to answer this question, we will again use the competition-hybridization method already described. Recall that the sequence of bases in RNA (and DNA) is the genetic code. It is this nucleotide sequence that codes for each specific protein. So, if we talk about similarities or differences in RNA during different developmental stages, we are very concerned with base sequence. Similarities and differences in base sequence between nucleic acids can be determined using hybridization techniques. If segments of DNA or a segment of DNA and a segment of RNA hybridize with each other, the base sequences are complimentary. If they don't, the base sequences are not similar.

stage-specific differentiation of RNA

Now, to determine if there is stage-specific differentiation of RNA during early embryonic development, the competition-hybridization technique is used as follows. Segments of DNA are heated and cooled or chemically treated so that they become single-stranded. These strands are adsorbed on a filter. Then, radioactively labeled RNA from a specific developmental stage of the embryo is incubated with the DNA (from the same organism) on the filter. These incubations, however, are carried out with the added presence of increasing concentrations of unlabeled RNA from the same or different developmental stages of the organism. If a specific batch of unlabeled RNA from a given developmental stage significantly inhibits the binding of the labeled RNA to the DNA, this indicates that the base sequence in this unlabeled RNA is similar to that of the labeled RNA. This is so because the unlabeled RNA binds to the

Figure 10–4. Stage-specific differentiation of RNA. The graph shows that labeled sea urchin prism RNA is inhibited from binding to sea urchin DNA by incubation of the DNA filters in the presence of a variety of unlabeled RNAs from different sea urchin developmental stages. The best inhibitors (competitors) are urchin prism and urchin gastrula RNA. Starfish RNA does not show concentration-dependent inhibition of binding. From Whitely, A. H., McCarthy, B. J., and Whitely, H. R., *Proc. Natl. Acad. Sci., U.S.,* 55:519–525 (1966).

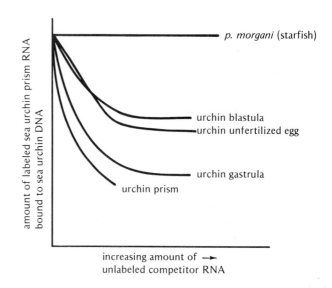

complementary DNA strands on the filter leaving no places for the labeled RNA to bind (Figure 10–2). If, on the other hand, the unlabeled RNA from a given developmental stage does not inhibit binding of the labeled RNA to the DNA on the filter, then the base sequence of the unlabeled RNA is not similar to the base sequence of the radioactive RNA. This is so because the unlabeled RNA, in this case, binds to DNA strands different from those to which the labeled RNA binds.

Figure 10–4 gives an example of this type of competition-hybridization experiment in which labeled RNA from the prism (post-gastrula) stage of the sea urchin embryo is incubated on the sea urchin DNA filter together with unlabeled RNA from different sea urchin embryonic stages. As can be seen, unlabeled prism RNA is the best competitor of labeled prism RNA binding to the DNA on the filter. The next best competitor is unlabeled gastrula RNA, with unlabeled blastula and unfertilized egg RNA being significantly poorer competitors. Unlabeled *P. morganii* (starfish) RNA does not compete at all. From this experiment, we can conclude that it appears that the sea urchin gastrula has many RNA sequences in common with the sea urchin prism stage. The sea urchin blastula and unfertilized egg have fewer RNA sequences in common with the prism stage. Starfish RNA is not similar to sea urchin prism RNA in terms of base sequence. This work suggests that there can indeed be stage specific differentiation of RNA. This makes a lot of sense, in that it seems reasonable that at different developmental stages different genes may be activated to form the RNA needed to code for the specific and sometimes different proteins that may be required for development to proceed from stage to stage.

Before leaving the technique of nucleic acid hybridization, let's mention some important information that has been discovered using this technique concerning the nature of the genome of the cells of higher organisms (eucaryotic cells). Eucaryotic cells, all cells above the bacteria and blue-green algae, have a well defined nucleus. Procaryotic cells (bacteria and blue-green algae) generally contain a single naked DNA double helix. Most eucaryotic cells contain thousands of times the amount of DNA found in procaryotic cells. This large amount of DNA may be in part required to code for and regulate the tremendous amount of cellular diversity that occurs during the differentiation process in higher organisms.

eucaryotic cells

procaryotic cells

Using the DNA-DNA hybridization technique, it has been shown that some single strands of DNA from eucaryotic cells reanneal with each other rather quickly at low DNA concentrations. Other pieces of the DNA, however, don't reanneal with each other until the DNA concentration and lengths of time increase (Figure 10–5). It has been suggested from this sort of experiment that the strands that reanneal quickly (at low concentrations) have many base sequences in common. The more slowly annealing segments that hybridize at higher DNA concentrations represent DNA strands that appear to be present in only a few or single copies. The more quickly DNA strands hybridize at low DNA concentrations, the greater the number of identical DNA copies present. This is the case because it is not difficult for one strand to find and bind with a

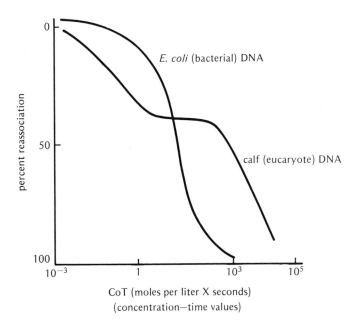

Figure 10–5. Reassociation of procaryote (bacterial) DNA and of eucaryote (calf) DNA. The calf DNA curve shows early annealing sequences (and low concentrations of DNA needed) and late annealing sequences (and high concentrations of DNA needed). The E. coli curve is S shaped, suggesting single copies of gene. From Britten, R. J., and D. E. Kohne, Repeated Sequences in DNA, *Science* 161:529–540, Fig. 3, (August 9, 1968). Copyright 1968 by the American Academy for the Advancement of Science.

complementary strand if there are many copies of a given sequence of DNA. The DNA strands that recombine only under high DNA concentrations and at long incubation times are present in fewer copies and thus have difficulty in colliding with a complementary sequence.

Thus, the eucaryotic cell has some of its DNA present in single copies of a base sequence, while many copies of other sequences exist. Procaryotic cells appear to possess mostly single copies of given genes (Figure 10–5).

In summary, the eucaryotic genome contains unique and repeated genes. Eucaryotic cells probably require regulatory genes and other types of genes in multiple copies in order to be able to undergo the complex sequence of developmental events that is characteristic of the differentiation process.

Other Approaches to Studying Nucleic Acid Differentiation

In the next chapter, we will examine differentiation at the protein level. Here we are looking at nucleic acids—the molecules that contain the genetic code and thus represent a step that precedes protein synthesis. So, what happens at the nucleic acid level in a cell is something that we should consider before we examine differentiation in terms of proteins. Recent advances have been made that allow us to examine nucleic acids in developing systems. We have already considered one of these methods—nucleic acid hybridization. Using this method we have seen that there appears to be tissue-specific and stage-specific differentiation of RNA during development. This technique has also been used to home in very specifically on the messenger RNA molecule. We have only examined total cellular RNA in the experiments described previously. It is also possible to identify specific messenger RNA molecules using the new technology mentioned earlier. This is accomplished by taking purified molecules of messenger RNA and making complementary DNA to this **reverse transcriptase**   message by using the enzyme reverse transcriptase. This enzyme uses RNA as a template to make DNA and is discussed more fully in Chapter 13. Now we have available DNA complementary to specific messenger **complementary DNA**   RNA molecules. This complementary DNA (cDNA) is a very potent tool **(cDNA)**   that can now be used to determine if the specific messenger RNA (that was used to produce it) is present in cells. Single strands of this DNA can be adsorbed to a filter, and then batches of RNA can be passed through the filter. Any specific messenger RNA that is complementary to the cDNA will hybridize with the cDNA. These hybrids can be detected by a variety of methods previously described. Thus, we can determine if a specific message for a specific protein is present in a population of cells. We can determine the amount of this message and therefore can say a lot about specific gene activation at the nucleic acid level.

A second method that has been developed to examine nucleic acid changes during differentiation consists of using the RNA synthesis inhibitor actinomycin D. Differentiating cells are incubated with this drug at various time. If actinomycin D treatment prevents synthesis of a specific protein, this suggests that transcription of the messenger RNA for that protein was blocked by the drug. If, on the other hand, actinomycin D treatment does not block synthesis of a specific protein, it is likely that the message for that protein was already present in the cell before actinomycin D treatment. In this way, we can ascertain whether a specific gene becomes activated shortly before protein synthesis occurs or whether stable messenger RNA molecules were produced at some time prior to incubation with actinomycin D. Whether protein synthesis immediately follows transcription of messenger RNA by genes or whether proteins are synthesized utilizing stable message that have been in the cell for some time are important questions in understanding differentiation at the nucleic acid level.

actinomycin D

A third approach to investigating the role of changes in nucleic acids during differentiation is to identify specific messenger RNA molecules by their ability to direct the synthesis of specific proteins. This can be accomplished by incubating the RNA to be tested with the necessary components of the protein synthesis system that can be prepared from bacterial cells. These components include ribosomes, amino acids, transfer RNAs, and a variety of factors and enzymes. We can determine if messenger RNA molecules specific for a given protein are present by observing the synthesis of that protein with such a cell-free protein synthesizing system.

These techniques have been used in many developing systems. For example, in developing red blood cells, muscle cells, and pancreas cells, there comes a point when actinomycin D no longer prevents the onset of specific protein synthesis. It therefore appears that stable messenger RNA molecules accumulate in these cells in preparation for the extensive synthesis of cell specific proteins characteristic of the differentiated state.

In the mouse, synthesis of hemoglobin in prospective red blood cells is blocked by actinomycin D before day 10 of gestation. After day 10, actinomycin D does not block hemoglobin synthesis. Thus, any hemoglobin message made before day 10 is stable and remains able to synthe-size the protein after day 10.

We might now ask the question: How does messenger RNA become stabilized? It may just be that the rate of message synthesis is increased before day 10 so that there are too many message copies around to be totally degraded. Also, messages may be synthesized and additional nucleotides may be attached (such as polyadenylic acid). These messages with attached polyadenylic acid may combine with some protein to protect them from degradation. It has been proposed that perhaps most or all genes in the genome are transcribed in eucaryotic cells, but only some are protected from degradation. It may be that this sort of regula-tion plays a major role in determining which genes actually form mes-sages that eventually are active in synthesizing protein—many mes-sages may be transcribed, but only some may stay around long enough to reach the cytoplasm to synthesize protein. This concept proposed by Wilt and others may be one important piece of the differentiation puzzle.

The chromosomes of higher cells are complex. Before we move on to differentiation at the protein level, let's look at another interesting proposal that may help account for the observation that during differen-tiation certain genes seem to be turned on while others appear to be turned off.

Control of Gene Transcription during Differentiation

Chromosomes in the cells of higher organisms contain a great deal of protein associated with the DNA. This protein may be a key factor in differentiation because it may play an important role in repressing certain genes, while allowing others to become active. There are two major types of chromosomal proteins, histone proteins and non-histone

histone proteins

proteins. Histones are low molecular weight (10,000–20,000) proteins that are rich in basic amino acids such as lysine and arginine. Histones from different organisms and different cell types appear to be structur-ally similar. Non-histone proteins associated with DNA are low moleclar

non-histone proteins

weight (7,000–15,000) and are structurally much more heterogeneous than the histones. Most of them are phosphorylated and are acidic instead of basic.

Some experiments have been performed that suggest that histone proteins non-specifically bind to the DNA of the chromosomes causing compaction of the DNA, preventing binding of the enzyme RNA polymerase that allows RNA synthesis (transcription) to occur. It has been suggested that the non-histone proteins may protect regions of DNA from histone binding or may remove histones from specific regions of DNA so that RNA synthesis can proceed at these regions. Several experiments support these notions. Rabbit thymus chromosomes were deproteinized. That is, all of the histone and non-histone proteins were removed. Histones were then added back to the rabbit thymus DNA. Thymus non-histone proteins were added to one portion of this partially reconstituted thymus chromatin. Bone marrow non-histone proteins were added to another portion of the thymus chromatin. The chromatin with the thymus non-histone protein added transcribed RNAs similar to those normally present in thymus tissue. The chromatin with the bone marrow non-histone protein added transcribed RNAs similar to those found in bone marrow. Thus, the non-histone proteins appear to be very important in controlling which genes are activated in specific tissues. It may be that the acidic, negatively charged non-histone proteins bind to the positively charged histones, removing them from specific regions of the DNA. In this way, certain genes may be activated.

It is known that hormones can often activate specific genes in specific tissues during differentiation. How do these hormones work, and how is hormone action related to the proposed role of non-histone proteins mentioned above? Estrogen treatment of chick oviduct cells stimulates RNA synthesis within two minutes after hormone treatment. In a series of elegant experiments, O'Malley and Means showed that the estrogen complex (estrogen attached to specific cell receptors) appears to bind to specific non-histone chromosomal proteins. It was shown that binding of the estrogen complex was specific for oviduct non-histone protein containing chromatin. Oviduct reconstituted chromatin containing red blood cell non-histone proteins bound far less hormone complex than did such chromatin containing oviduct non-histone proteins. This suggests that hormone complexes may "turn on" genes by binding to non-histone protein on the chromosomes.

Additional experiments utilizing DNA-RNA hybridization techniques indicated that estrogen-stimulated oviduct chromatin produced eight times as much oviduct specific ovalbumin messenger RNA as chromatin from oviducts in which estrogen treatment was halted. The non-histone protein fraction from estrogen-treated oviduct chromatin may be added back to DNA plus histones from oviducts in which estrogen treatment was halted. This chromatin now is able to transcribe amounts of specific ovalbumin messenger RNA characteristic of estrogen-treated oviduct tissue. Non-histone from estrogen-halted chromatin lowers the amount of ovalbumin messenger RNA produced by DNA (plus histones) from estrogen treated oviducts.

Thus, in summary, inducer complexes, hormone complexes, and other molecules that appear to control differentiation may in some way activate genes by binding to non-histone proteins on chromosomes. This absorbing area of research is moving rapidly. It is probable that the exact means by which gene action is regulated during differentiation of eucaryotic cells may be completely understood in the near future.

The Eucaryotic Chromosome

Mention should be made of the structure of the eucaryotic chromosome and recent models regarding its functioning. We know that there is a large quantity of histone protein associated with chromosomes. We also know that non-histone protein and a small amount of RNA are associated with chromosomal DNA. What is the arrangement of these molecules in the chromosome and how does the chromosomal unit function?

nucleosomes   Recent work in the area of chromosome structure has led to the finding that eucaryotic chromosomes are composed of particles called nucleosomes. Although the structure of these units is still being investigated, evidence suggests that the nucleosomes are 70–100 Å in diameter and are usually composed of 4 histone proteins and about 140 base pairs of DNA. The nucleosomes are like beads on a string. Between the beads, the piece of "string" consists of about 30–60 base pairs of DNA. It appears likely that the DNA is wrapped over the histones rather than embedded within the nucleosomes, because nucleosome DNA is accessible to chemical reagents and can be transcribed. One type of histone (very lysine-rich H-1) does not appear to be present within the nucleosome particles. Instead, H-1 appears to be associated with the strand of DNA between the nucleosomes. The other four types of histones, which are less lysine-rich than H-1, are present in the nucleosomes themselves.

The length of DNA in a chromosome, if it were to exist as a single stretched-out strand, would not fit into a cell. In a typical human chromosome, for example, the DNA would stretch out to about 3 centimeters. The length of such a chromosome in reality, however, is only about 5 micrometers (0.0005 centimeter). The packing of the DNA into nucleosomes reduces the length by a factor of six or seven. This is still too long, so some other sort of packing must additionally occur in chromosomes. It is likely that the nucleosome beaded string is itself coiled to form a fiber of about 300 Å in diameter. This would account for the thicker structures that are seen with the electron microscope, and for the modest chromosome length that is also observed.

We note in this chapter that although the ratio of histone protein to DNA is relatively constant from cell type to cell type, the amount of non-histone chromosomal protein is very variable. Cells with very active genes tend to have more non-histone protein than cells with inactive genes. In addition, there are many types of non-histone proteins, unlike the histones. Thus, the non-histone proteins are candidates for playing a specific role in gene activation. For example, if brain chromatin is dissociated into its major components (DNA, histone, and non-histone proteins) and reassembled, but using a red blood cell non-histone protein fraction instead of that from the brain, this chromatin was able to produce globin messenger RNA. Globin is the protein of hemoglobin and is not usually produced by brain cells, but is usually only produced by red blood cell precursors. Here, however, addition of non-histone protein from red blood cells allows brain chromatin to produce globin message. It appears that non-histone proteins activate specific genes, possible by altering the histones that may mask a given gene.

Exactly how the non-histone proteins interact with the basic nucleosome structure of the chromosomes is not known. It is likely, however, that molecules such as hormones bind to specific non-histone proteins and the protein-hormone complex may then interact with the genes. This is one way in which genes may be specifically activated, resulting in tissue-specific differentiation patterns. More about this is presented in another section of this chapter.

Before leaving the topic of the eucaryotic chromosome, it should be mentioned that each structural gene in the cell, such as that for globin, probably has associated with it a series of control genes that can turn on the structural gene. The products of control genes may be the specific non-histone proteins. A given activator, such as a hormone, could bind to the specific non-histone protein. This complex may then bind directly to the genes. The hormone, in other cases, may bind to the cell surface and cause reactions that result in the accumulation of small secondary molecules such as cyclic AMP, that may bind to the non-histone proteins which in turn activate genes.

That hormones act in the ways described above has been shown in several systems. For example, steroid sex hormones in mammals penetrate the cell membrane and bind to specific protein receptors in the cytoplasm. The receptors are absent in cell types which are not stimulated by the given hormone. The activated receptor penetrates the nucleus and interacts with the chromatin to activate specific genes. Specific receptors for estrogens, progesterones, androgens, and cortisol have been isolated. The estrogen receptor, for example, is a protein of about 80,000 daltons that binds estrogen, but not similar molecules without estrogen activity.

Direct observation of the effect of hormones on chromosomes is seen in the fruit fly, *Drosophila*. Within five or ten minutes after addition of the insect hormone ecdysone, six major new "puffs" are observed to appear in the chromosomes. These puffs are localized regions of chromosomal uncoiling that appear to represent areas of active gene transcription. An additional one hundred puffs or so form later on that apparently result from activation caused by regulator molecules produced by the earlier puffs. The nature of eucaryotic chromosome structure is being elucidated, and the complex picture of the components involved in differential gene activation is becoming a bit clearer. The complete answer to the question of how eucaryotic chromosome structure relates to function will probably be understood in the not too distant future.

Ribosomal RNA Genes in the Amphibian Embryo

We have mainly considered differences and similarities in messenger RNA during development because this type of RNA directly codes for the tissue-specific proteins that are important in differentiation. Before moving on to an examination of these proteins let us briefly consider a type of RNA that is rquired for the assembly of ribosomes, the seats of protein synthesis.

anucleolate mutant of *Xenopus*

The synthesis of ribosomal RNA is controlled by specific genes that appear to become activated at specific developmental times. This is nicely shown by the anucleolate mutant of *Xenopus* (the African clawed toad—a frog). Homozygous mutant individuals, obtained by mating two heterozygous individuals, die after hatching. These homozygous individuals do not make ribosomes and have no nucleoli. Such mutants can be produced by mating two 1-nu individuals, (each of whom has 1 nucleolus in their cells). They produce offspring of the following types: homozygous wild type (2-nu—two nucleoli in their cells), heterozygous (like the parents—1-nu) and the homozygous mutant (0-nu—no nucle-

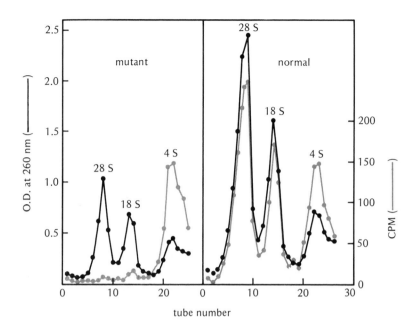

Figure 10–6. RNA synthesis in normal and mutant *Xenopus* embryos at neurula-tailbud stage. Light lines indicate radioactivity incorporated into newly synthesized RNA. Dark lines show where the specific types of RNA appear in tubes of fractions from a density gradient centrifugation experiment. The anucleolate mutant does not synthesize ribosomal RNA (28S + 18S RNA), while the normal embryo does. From D. D. Brown, *J. Exp. Zool.* 157:101–114, 1966.

oli). The anucleolate mutants (0-nu) do, however, possess ribosomes that were made by the heterozygous mother frog. These maternal ribosomes enable these mutants to synthesize proteins for some time during early development. New ribosomes, however, seem to be required by the time the embryo hatches. Since the mutants can't make new ribosomes, they die by the early swimming tadpole stage.

Let's look at some specific evidence that indicates that ribosomal RNA synthesis is indeed controlled by specific genes that are active at specific developmental stages. Figure 10–6 shows that ribosomal RNA is synthesized in normal *Xenopus* from the neurula to the tail-bud stages. The figure also shows that ribosomal RNA is not synthesized in the mutant, although other RNA is still made. As can be seen from the figure, these results were obtained by incubating neurula stage embryos with a radioactive label that is incorporated into RNA. Some of the different types of RNA synthesized are separated from each other on the basis of their sedimentation (S) (based on size of the molecule) in a gradient of a substance such as sucrose. Ribosomal RNA includes 28S, 18S and 5S size classes. 4S RNA, also shown in the figure, is soluble RNA. 5S ribosomal RNA is not shown in the figure.

Does the anucleolate mutant possess the genes for ribosomal RNA, but for some reason these genes are not producing ribosomal RNA? Or are the ribosomal RNA genes in the anucleolate mutant absent? The DNA-RNA hybridization technique was used to answer this question. Nonmutant *Xenopus* DNA was dissociated into single strands and incu-

Figure 10–7. Hybridization of ribosomal RNA with DNA. 2-nu DNA from normal *Xenopus* + ribosomal RNA. 1-nu DNA from heterozygous *Xenopus* + ribosomal RNA. 0-nu DNA from a nucleolate mutant *Xenopus* + ribosomal RNA. From: Markert, C. L., and H. Ursprung. *Developmental Genetics*, Prentice Hall, (1971) p. 12. Reprinted by permission of Prentice-Hall, Inc., Englewood Cliffs, New Jersey.

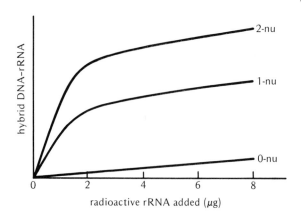

bated with purified radioactive nonmutant ribosomal RNA. The ribosomal RNA hybridized with the ribosomal RNA genes of the DNA. The amount of such hybridization was determined by measuring the radioactivity bound to the DNA on the filter, as previously described. When DNA from the anucleolate mutant was used, however, no radioactive ribosomal RNA bound to the DNA (Figure 10–7). Thus, anucleolate mutant DNA does not contain ribosomal RNA genes. The anucleolate mutant, therefore, does indeed lack the genes that code for ribosomal RNA. This can be visually confirmed by noting that the anucleolate mutant lacks certain secondary constrictions that are usually present on two chromosomes of nonmutant cells. These constrictions were found to be the sites at which nucleoli originate and where ribosomal RNA is produced (Figure 10–8). So we see that ribosomal RNA genes really exist and that they must be activated before the early swimming stage if the embryo is to survive.

Throughout this text we have discussed examples of how cytoplasmic or environmental factors and inducers appear to be able to activate specific genes at specific times. Let us conclude this chapter by looking at the nature of this sort of activation in the frog embryo system before we turn to an examination of differentiation at the protein level.

Figure 10–8. Drawing of mitotic chromosomes of normal *Xenopus* larva. N = Nucleolar organizer region.

Nuclei from a variety of embryonic cells can be transplanted into enucleated amphibian eggs so that scientists can research the questions: Can a nucleus from a later developmental stage still support development of the early embryo? What factors influence the turning on and off of genes at various developmental stages?

Gurdon and others have transplanted nuclei from differentiated intestinal epithelium cells, tail fin epithelial cells, or cultured skin cells of *Xenopus* tadpoles into enucleated frog eggs. Some of these nuclei were able to support completely normal frog development (Figure 10–9). A gut cell nucleus that, for example, synthesized ribosomal RNA stopped

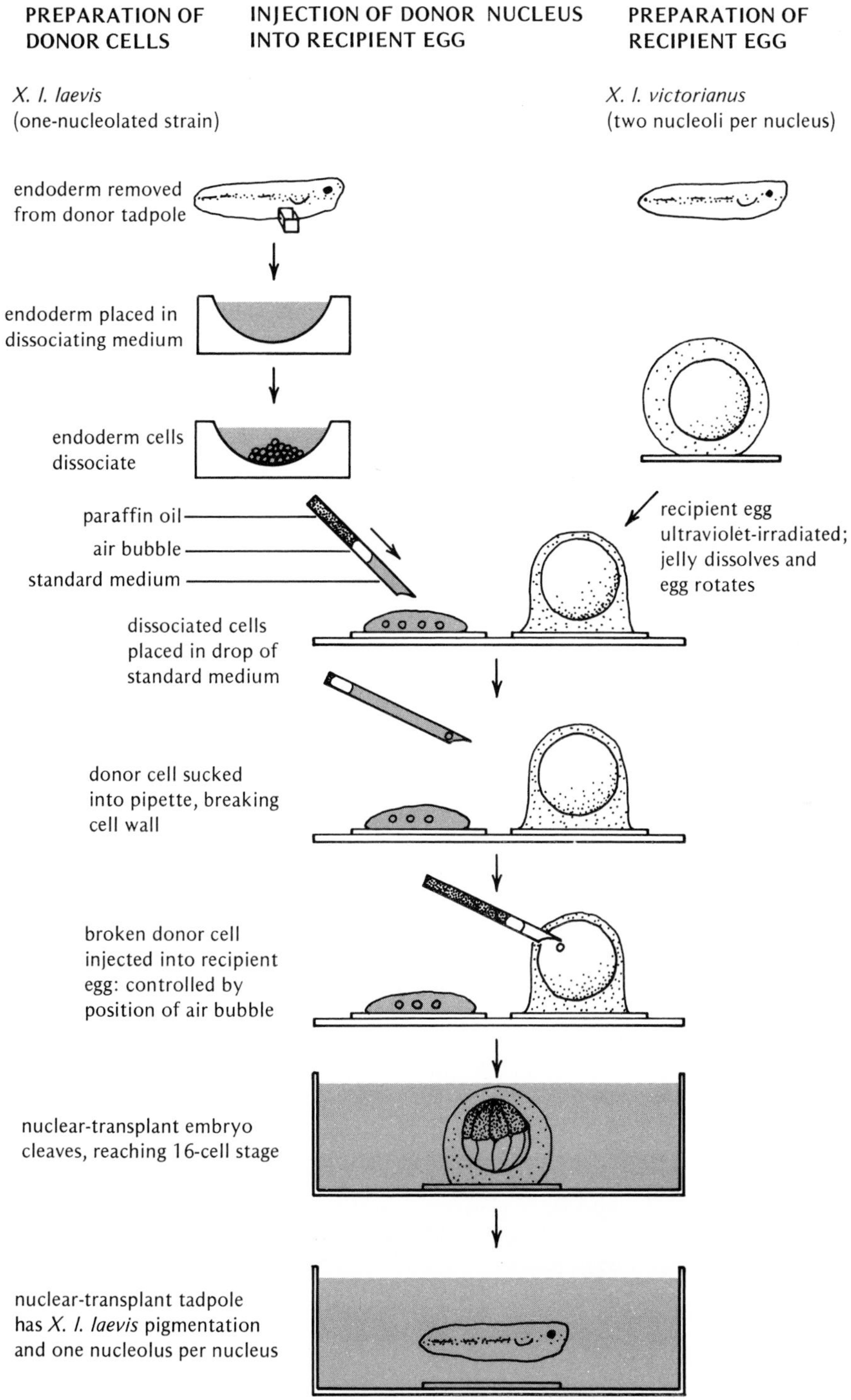

PREPARATION OF DONOR CELLS	INJECTION OF DONOR NUCLEUS INTO RECIPIENT EGG	PREPARATION OF RECIPIENT EGG

X. l. laevis
(one-nucleolated strain)

X. l. victorianus
(two nucleoli per nucleus)

endoderm removed from donor tadpole

endoderm placed in dissociating medium

endoderm cells dissociate

paraffin oil ——
air bubble ——
standard medium ——

recipient egg ultraviolet-irradiated; jelly dissolves and egg rotates

dissociated cells placed in drop of standard medium

donor cell sucked into pipette, breaking cell wall

broken donor cell injected into recipient egg: controlled by position of air bubble

nuclear-transplant embryo cleaves, reaching 16-cell stage

nuclear-transplant tadpole has *X. l. laevis* pigmentation and one nucleolus per nucleus

Figure 10–9. Nuclear transplantation in *Xenopus*. Donor nuclei from one strain are transplanted to recipient eggs of another strain. The tadpoles from such transplants show characteristics of the nuclear parent, showing that the nuclei of the tadpoles are derived from the transplanted nucleus. From Gurdon, *Endeavour* 25:96 (1966).

synthesizing this RNA when transplanted into the egg. At gastrulation, when normal ribosomal RNA synthesis increases rapidly, nuclei derived from the transplanted intestinal nucleus begin normal ribosomal RNA synthesis. Thus, it appears that RNA synthesis is regulated by the type of cytoplasm in which the nucleus lies. Nuclei from many types of differentiated cells still can return to the embryonic state by turning on "old" genes that may have been active only in the early embryo, if the cytoplasmic conditions that control such a change should reappear. More will be said about this intriguing notion when we discuss cancer and embryology in the last chapter of this text. Now let's turn to proteins—the molecules that make differentiated cells truly different. These are primary products of the nucleic acid events that we examined in this chapter.

Readings

Britten, R. J. and D. E. Kohne, Repeated Sequences in DNA. *Science* 161:529–540 (1968).

Braun, D. D., RNA Synthesis During Amphibian Development. *J. Exp. Zool.* 159:101–113 (1964).

Callan, H. G., Chromosomes and Nucleoli of the Axolotl, *Ambystoma mexicanum. J. Cell Sci.* 1:85–108 (1966).

Davidson, E. H., and R. J. Britten, Organization, Transcription and Regulation in the Animal Genome. *Quart. Rev. Biol.* 48:565–613 (1973).

Gurdon, J. B., *The Control of Gene Expression in Animal Development.* Claredon Press, Oxford (1974).

Markert, C. L., and H. Ursprung, *Developmental Genetics.* Prentice Hall, Englewood Cliffs, New Jersey (1971).

McCarthy, B. J., and B. H. Hoyer, Identity of DNA and Diversity of mRNA Molecules in Normal Mouse Tissues. *Proc. Natl. Acad. Sci. U.S.* 52:915–922 (1964).

Miller, O. R., Jr., and R. R. Beatty, Visualization of Nucleolar Genes. *Science* 164:955–957 (1969).

O'Malley, B. W., and A. R. Means, The Mechanism of Steroid Hormone Regulation of Transcription of Specific Eucaryotic Genes. *Prog. Nucleic Acid Res. Mol. Miol.* 19:403–419 (1976).

Stein, G. S., T. C. Spelsberg, and L. J. Kleinsmith, Nonhistone Chromosomal Proteins and Gene Regulation. *Science* 183:817–824 (1974).

Tsai, S. Y., S. E. Harris, M. J. Tsai, and B. W. O'Malley, Effects of Estrogen on Gene Expression in Chick Oviduct. *J. Biol. Chem.* 251:4721 (1976).

Wessells, N. K., and W. J. Rutter, Phases in Cell Differentiation. *Sci. Amer.* March (1969).

Whiteley, A. H., B. J. McCarthy, and H. R. Whiteley, Changing Populations of Messenger RNA During Sea Urchin Development. *Proc. Natl. Acad. Sci. U.S.* 55:519–525 (1966).

Wilt, F. W., The Beginnings of Erythropoiesis in the Yolk Sac of the Chick Embryo. *Ann. N.Y. Acad. Sci.* 241:99–112 (1974).

CHAPTER 11

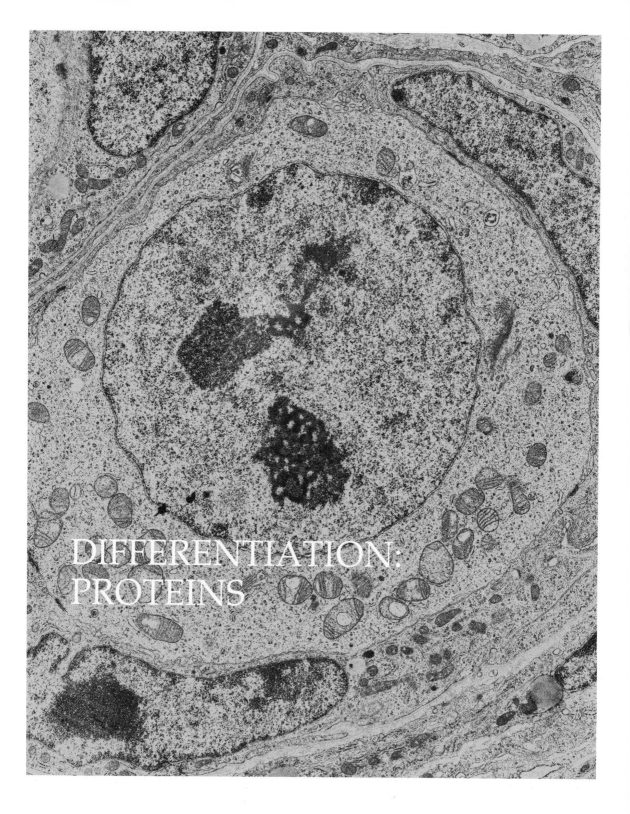

DIFFERENTIATION:
PROTEINS

IN THE LAST CHAPTER, we saw that the RNA in cells changes during development. We also noted that it is extremely difficult to interpret some of these changes in nucleic acids because of the uncertainties in determining which species of RNA are functional. Since proteins are the important products of nucleic acid messages, and since the presence or absence and synthesis or lack of synthesis of specific proteins can be definitively determined, many investigators have turned to studying proteins as markers of development and differentiation. In this chapter we will get a glimpse of proteins in development. We will look at protein synthesis in early development, methods of analysis in the protein field, and finally we'll examine some of the specific proteins that make cells differentiated.

Protein Synthesis in Early Development

The sea urchin embryo, as mentioned earlier in the text, is ideal for studying development because of the vast number of embryos that can easily be obtained, and because normal development occurs in sea water. We can add radioactive amino acid (such as radioactive methionine) to the sea water in which sea urchin embryos are developing and measure the amount of amino acid that becomes incorporated into newly synthesized protein. We can incubate the embryos with the "hot" amino acid for a short time, such as an hour, and then place the embryos in sea water without the radioactive amino acid. If we do this at various development stages, we can determine the amount of newly synthesized protein at each specific stage by measuring the amount of radioactive protein formed at each stage after administration of each short pulse of "hot" amino acid. Such experiments allow us to measure the amount of new protein synthesized with reasonable accuracy if we make sure that the radioactive amino acid is being taken up by the cells at all the stages. We also must be sure that the intracellular supply (pool) of the amino acid available for protein synthesis remains constant during the stages being examined. These and other factors must be considered if we are to reliably determine the relative amount of protein synthesis at different developmental stages.

Figure 11–1 is an example of the type of results that have been obtained by investigators examining protein synthesis in the sea urchin embryo. As can be seen, protein synthesis rapidly increases for the first 3 or 4 hours after fertilization and then begins to decline, only to increase again at about the gastrula stage. Radioactive methionine was used in this experiment and was indeed shown to be taken up by unfertilized eggs and other stages. So, amino acid uptake was not a problem in these experiments. Also, the supply (pool) of intracellular methionine in these experiments was shown not to change appreciably in the different stages examined. Thus the pool of methionine was also not a problem in this work. These experiments suggest that differential amounts of total protein are synthesized at various stages of embryonic development.

Figure 11–1 also shows that if embryos are treated with the RNA synthesis inhibitor actinomycin D, protein synthesis still increases after fertilization but does not undergo the second increase at gastrulation. These results suggest that the proteins synthesized during early development use messenger RNA that has been formed in the egg prior to fertilization. At gastrulation, however, new species of messenger RNA may be transcribed and this would be blocked by actinomycin D. Recall that we discussed this sort of finding in the previous chapter. More recent work has, however, shown that some specific proteins such as histones are synthesized in the early hours of sea urchin embryonic development utilizing newly formed messenger RNA as well as that stored in the egg prior to fertilzation.

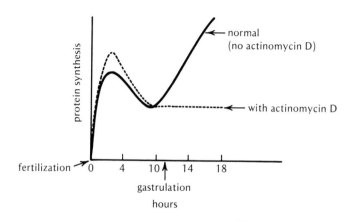

Figure 11–1. Protein synthesis at various stages of sea urchin embryo development. From H. Ursprung and K. D. Smith, *Brookhaven Symp. Biol.* 18:1–13 (1965).

We have described synthesis of total protein at various developmental stages. What about specific proteins? Do all proteins in the cell follow the same synthesis pattern as that for total protein shown in Figure 11–1? Figure 11–2 shows that if individual proteins are separated chromatographically at different developmental stages after the embryos are treated with a pulse of radioactive amino acid, differential amounts of a single protein are synthesized at different developmental stages. Note that a lot of protein A is synthesized at the first hour after fertilization in the sea urchin, while only a little bit of protein D is synthesized at this time. At the eighth hour after fertilization, however, less protein A is synthesized while a lot of protein D is synthesized. These results show that different proteins are synthesized in different amounts at different developmental stages. The experiment shown in Figure 11–2 does not identify the natures of proteins A and D. In fact, these proteins may be similar and may represent different numbers of identical subunits or association of the protein with co-factors. Later in the chapter we will examine very specific proteins and reaffirm the notion that given proteins are often synthesized in differential amounts at different developmental stages. In this section we have just looked at a few experiments in one model system to get an idea of the ways in which some of the early experiments in the protein synthesis area have been performed. We have looked at protein synthesis in whole embryos. This serves as an introduction to the examination of specific proteins in specific cells during differentiation.

Measurement of Specific Protein Synthesis

We have just seen that protein synthesis is generally measured by incubating embryo cells in solution containing radioactive amino acid such as ^{14}C methionine or ^{14}C leucine. The cells are then homogenized and large molecules are separated from the amino acids and other small molecules by acid precipitation and chromatography on columns of gel that separate molecules according to size and/or charge. The amount of radioactive amino acid that has been incorporated into each of these separated large molecules is easily determined by using a scintillation counter. Thus the amount of newly synthesized protein can be determined by measuring amino acid incorporation into this material.

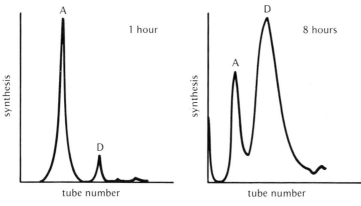

Figure 11–2. Specific proteins synthesized at different stages of sea urchin embryo development. Protein synthesis is measured as counts per minute of radioactive amino acid incorporated into protein in each column fraction. After C. H. Ellis, *J. Expt. Zool.* 163:1–22 (1966).

We mentioned that specific proteins can be separated by techniques such as gel chromatography. Using gel chromatography in conjunction with amino acid incorporation, we can determine the rate of synthesis of the specific proteins separated on the gels. We can often identify specific proteins on the gels by comparing their profiles on the gels with those of characterized protein standards. Electrophoresis (separation of proteins in electric fields) is also used to compare one protein sample with a known protein standard.

Immunological methods are also widely used to identify specific proteins. A specific protein can be injected into rabbits or goats, which then produce antibodies against the protein that can then be isolated from the blood of the animal. The antibodies are used to identify a specific protein in a mixture of many proteins. The antibodies against the specific protein will react with the specific protein, forming a precipitate. Thus the antibodies complex with a protein and can be used to identify that specific protein.

Newly synthesized specific protein can be identified using the immunological technique coupled with the incorporation of the radioactive amino acid method. Cells can be incubated with radioactive amino acid for a short time, homogenized, and the homogenate (or an extract thereof) can then be incubated with antibodies against a single specific protein. The antibodies react with the specific protein, precipitating it out of solution. The amount of radioactivity in the precipitate is then measured. This will tell us how much radioactive amino acid has been incorporated into the specific protein precipitated by the antibody. In this way we are able to determine how much of the specific protein was newly synthesized during the time of incubation of the cells with radioactive amino acid.

This technique has been very useful in identifying the patterns of synthesis of specific proteins during differentiation. This method clearly determines synthesis of new protein. We should remember that by just measuring the activity of a given enzyme or the presence or absence of a specific protein at a given developmental stage, we are not looking at synthesis. Synthesis must be measured by a method such as that described above in which newly formed protein can be determined. The simple presence or absence of a specific protein or the presence or absence of specific enzyme activity is not necessarily a measure of synthesis but may result from activation or inactivation of an enzyme by specific cofactors or from differential degradation of the protein by degradative enzymes present in the cells.

Before moving on to a look at a variety of specific proteins during differentiation, let us see how the immunological technique described above has been used to determine the nature of specific protein synthesis in one model system. The system we will examine is synthesis of the enzyme tryptophan pyrrolase by the rat liver. This enzyme opens the indole ring of tryptophan leading to the synthesis of an important vitamin, nicotinamide. The activity of this enzyme rises rapidly at birth. The activity of the enzyme is also increased by the hormone hydrocortisone and the amino acid tryptophan (Figure 11–3). The immunological technique described previously was used to determine how these substances cause the observed increases in tryptophan pyrrolase activity. Let us briefly examine these experiments because they will not only show how the technique was used, but also will help us to understand a bit about the nature of the way in which protein synthesis and degradation are regulated.

Purified tryptophan pyrrolase was injected into rabbits to produce specific antibody against this enzyme. Rats were inoculated with hydro-

tryptophan pyrrolase

Figure 11–3. Stimulation of tryptophan pyrrolase activity by hydrocortisone, tryptophan or both. Redrawn and modified from: R. T. Schimke, *Natl. Cancer Inst. Monograph* 27:301–314, 1967.

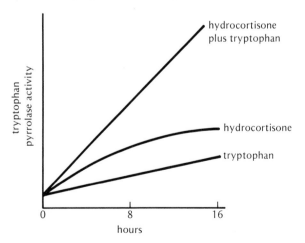

Injected material	Counts per minute of radioactive amino acid in protein precipitated by antibody against tryptophan pyrrolase.
NaCl	1406
Hydrocortisone	9466
Tryptophan	1954

Figure 11–4. Stimulation of tryptophan pyrrolase synthesis by hydrocortisone but not by tryptophan. Modified from Schimke, et al., *J. Biol. Chemistry* 240: 322-331 (1965)

cortisone, tryptophan, or saline and after a few hours radioactive amino acid was administered. The livers of the rats were removed and homogenized. Tryptophan pyrrolase antibody was then added to the homogenate to precipitate out tryptophan pyrrolase and newly synthesized enzyme was determined by measuring the amount of radioactivity in the antibody precipitated material.

Figure 11–4 shows that hydrocortisone appears to increase synthesis of tryptophan pyrrolase. This conclusion is based upon the observed increase in radioactive tryptophan pyrrolase in the hydrocortisone treated livers over that of the controls. There does not appear to be increased tryptophan pyrrolase enzyme synthesis in the tryptophan treated livers, however (Figure 11–4). How then does tryptophan result in the observed increase in tryptophan pyrrolase activity (Figure 11–3)? This question was resolved in another experiment. Liver proteins were labeled as before with radioactive amino acid. Later on, either saline or tryptophan was injected. At a still later time, the livers were homogenized and the amount of radioactive tryptophan pyrrolase that was still present was determined after precipitation by the antibody as before. Figure 11–5 shows that the amount of radioactively labeled tryptophan pyrrolase declines steadily after saline injection. Tryptophan, however,

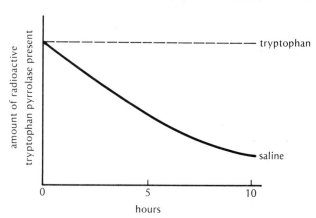

Figure 11–5. Prevention of tryptophan pyrrolase degradation by tryptophan. Tryptophan pyrrolase was made radioactive with radioactive amino acid inoculation. Then, at 0 time and at 4 and 8 hours the animals received injections of saline or tryptophan. The amount of radioactive tryptophan pyrrolase was determined at 3, 6 and 9 hours. From: Schimke, R. T., *Natl. Cancer Inst. Monograph* 27:301–314, 1967.

prevents this loss of labeled enzyme. These results suggest that trypto-phan keeps tryptophan pyrrolase enzyme activity high by protecting tryptophan pyrrolase from degradative enzymes.

In summary, we can identify whether or not a specific protein is being synthesized by measuring incorporation of radioactive amino acid into the protein precipitated by specific antibody. We saw that activity of an enzyme can be regulated by stimulating synthesis of the enzyme or by preventing its degradation. Other work has shown that enzyme activity can also be regulated by inhibitor or activator molecules. Let us turn to an examination of proteins in differentiating cells now that we have an idea of how one goes about measuring protein synthesis and some of the factors that regulate the activity of specific enzymes.

Specific Proteins and Differentiation

Hemoglobin

Figure 11–6. Diagram of hemoglobin showing 4 subunits. A heme group is associated with each subunit.

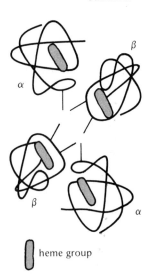

heme group

Hemoglobin is a tetramer—a molecule composed of four folded poly-peptide chains. It also contains a non-protein, iron-containing heme group attached to each polypeptide chain (Figure 11–6). The heme groups of the hemoglobin bind oxygen. The most important function of hemoglobin is to carry oxygen to different parts of the body. Hemoglo-bin is the major protein contained in red blood cells and serves as a distinct marker of differentiation in these highly specialized cell types.

There are different types of hemoglobin. Embryonic hemoglobins, fetal hemoglobins, and normal and abnormal adult hemoglobins exist. Stage-specific differences in hemoglobin must have evolved as a result of the different respiratory needs of embryo, fetus, and adult organisms. For example, in mammals, the fetus obtains oxygen as a result of diffu-sion from the maternal blood across the placenta to the fetus. The hemo-globin of the fetal red blood cells must have a higher affinity for oxygen than that of the mother in order for such an exchange to occur.

What is the nature of these differences in the hemoglobin molecule at different developmental stages? Adult hemoglobin in mammals consists of four polypeptide chains, as mentioned previously. Two of these chains are coded for by one gene while the other two chains are coded for by another gene. Adult hemoglobin can thus be designated as αα ββ indicating the two polypeptide chains coded by gene A and the two polypeptide chains coded by gene B. In the early mouse embryo, other genes also code for other hemoglobin chains. The A gene is active in the embryo along with X, Y, and Z genes. The embryonic hemoglobins do not contain β chains because the B gene is not active until late in fetal development. Thus important changes in hemoglobin result from activation of the B gene so that β polypeptide chains become a part of new hemoglobin molecules.

Abnormal types of adult hemoglobin include hemoglobin S and hemoglobin C. Hemoglobin S is found in individuals having the disease sickle-cell anemia. In this disease the red blood cells are sickle shaped. Hemoglobin S differs from normal hemoglobin in only one of the types of polypeptide chains—the β subunits. The α subunits in sickle cell hemoglobin are identical to those of normal adult hemoglobin. The normal β subunit consists of 146 amino acids. The only difference in the β subunits of normal and sickle-cell hemoglobin is a single amino acid. A valine residue in sickle-cell hemoglobin replaces the number 6 amino acid, glutamic acid, of the normal β chain. This single amino acid change could be caused by a single base change in the messenger RNA coding for the β subunit. Glutamic acid is coded for by the base sequence GAA, while valine is coded for by GUA. So a single base difference in the RNA (and DNA) can result in a disease that can be fatal. Such a base change probably occurred as a result of a single point mutation in the DNA of the chromosome. A single base change resulting in a single amino acid change can alter a protein enough so that a major change in shape of red blood cells occurs, leading to a major disease.

Hemoglobin C, another abnormal hemoglobin, is also the result of a single amino acid change. The same β chain amino acid that is replaced in hemoglobin S is also replaced in hemoglobin C. Instead of valine, however, a lysine substitutes for the glutamic acid residue in the hemoglobin C. The bases AAA code for lysine, while GAA code for glutamic acid. Thus a single base change, an A for a G, can again lead to an amino acid change in the β subunit. Hemoglobin C, however, does not cause any disease in people. Thus, if one specific amino acid out of 146 amino acids in the β subunit is changed to a valine, a major disease results, while a change to a lysine does not cause any disease. Now that we know a little about the nature of the protein hemoglobin, let's turn to a look at the nature of red blood cell differentiation.

Red blood cells in embryos of men and mice first originate in the yolk sac, and then later in development are produced by the liver. After birth, red blood cell production is taken over by the bone marrow.

Important questions we may now ask are: When do cells that give rise to red blood cells begin to differentiate? What sort of events mark the beginning of differentiation? Recall also that embryonic hemoglobin differs from adult hemoglobin. Do the same cells, or cells derived from the same parent cell, give rise to the different types of hemoglobin? Or do different cell lines give rise to the different hemoglobins?

In the mouse, hemoglobin begins to appear in red blood cell precursors in the blood islands of the yolk sac by the eighth day of gestation. On the ninth day these precursor cells enter the circulation of the embryo. Embryonic hemoglobin is synthesized by these cells. These cells divide four times before their nuclei condense and become inactive. On the twelfth day of gestation, red blood cell production begins in the liver and this new population of cells synthesizes adult hemoglobin. Yolk sac red blood cell production still occurs through the fourteenth day so that both types of red blood cells (yolk sac and liver derived), with their own types of hemoglobin, are found in fetal circulation. Spleen and bone marrow take over red blood cell production between day 15 and birth, while liver production declines. Thus it appears that there are at least two populations of red blood cells, one that synthesizes embryonic hemoglobin and another that synthesizes adult hemoglobin. Whether or not a common ancestor cell originally seeded the yolk sac and the liver, the spleen and bone marrow is not known.

In the yolk sac, hemoglobin (globin) messenger RNA is synthesized prior to day 10 of gestation. After day 10, synthesis of hemoglobin is not inhibited by actinomycin D. This suggests that the messenger RNA made prior to day 10 is stable and used later on for the synthesis of the protein chains of hemoglobin. Thus, perhaps the main event that results in the formation of hemoglobin-containing red blood cells is activation of the genes that code for the polypeptide chains of hemoglobin (globin).

Messenger RNA that codes for the polypeptide chains of hemoglobin has been isolated. Such messenger RNA can be identified by its ability to direct the synthesis of globin in a cell-free system and by hybridization experiments as described in the last chapter. Early fetal mouse liver cells do not contain globin messenger RNA. The glycoprotein hormone, erythropoietin, causes these cells to begin to synthesize globin messenger RNA. Hemoglobin synthesis begins shortly thereafter. This hormone is produced by the kidney in response to decreases in tissue oxygen content. It appears to be a key factor in inducing prospective erythrocytes to differentiate into mature red blood cells. It probably acts by turning on globin genes. Exactly how this hormone works is unknown, but it may act in a similar way to that proposed for estrogen (estradiol) in the previous chapter. Perhaps its final action is on specific chromosomal proteins that function to regulate gene action. Exactly how certain cells become determined as red blood cell–forming is unknown. However, the events that immediately lead to differentiation—globin message transcription followed by hemoglobin synthesis—are becoming better understood.

embryonic hemoglobin

adult hemoglobin

erythropoietin

Myosin

Myosin protein is probably present in many cell types. Muscle cells, however, contain a large amount of this protein. Myosin thus serves as a major marker for muscle cell differentiation. Myosin is a part of contractile networks that may be present in all eucaryote cells. Since the main function of muscle cells is contraction, it makes sense that myosin should be enriched in these cell types. Myosin makes up about ten to twelve percent of the fresh weight and nearly fifty percent of the dry weight of adult muscle.

Myosin is a fibrous protein composed of two identical subunits that are wound around each other (Figure 11–7). In muscle, myosin molecules aggregate into filaments. This assembly phenomenon will be examined in the next chapter. Myosin combines with other proteins such as actin. Units composed of myosin and actin associated with other proteins are contractile, and serve as a cellular contractile apparatus. Proteins associated with actin filaments include α-actinin, β-actinin, tropomyosin, and troponin. These proteins serve a variety of functions in permitting the contractile apparatus to remain intact and work smoothly at appropriate times.

Let us look first at developing muscle cells and then zero in on the nature of differentiation in these cell types. In this chapter, we will stress myosin synthesis in this system. In the next chapter, we'll get some insight into how muscle proteins assemble into functional contractile units.

Using fluorescent labeled antibody against myosin, it has been shown that muscle cell contractile proteins are present in many prospective muscle cells before one can observe any organized contractile apparatus in these cells. Prospective muscle cells (myoblasts) are derived from the myotome region of the somite and hypomere mesoderm (Chapter 6). The prospective muscle cells divide to form many additional myoblasts. When these cells stop dividing, they fuse together, forming long muscle tubes (myotubes) containing many nuclei in a common cytoplasm.

In two-day-old chick embryos, somite myotome cells are actively dividing. Some of these cells form spindle shapes containing detectable myosin filaments. By the fourth day most of the myoblasts contain muscle contractile fibrils (myofibrils). All this occurs before fusion of these cells into the multinucleated myotubes. In developing limb muscle, however, myofibrils only appear in the multinucleated myotubes and not in the pre-fusion myoblasts.

That elongated myotubes arise as a result of fusion of single cells rather than by continued nuclear divisions without cell division was

Figure 11–7. Myosin consists of two subunits wound around each other.

myoblasts

myotubes

myofibrils

Figure 11–8. Isocitrate dehydrogenase in myotubes. Model (B) (fusion model) is correct because hybrid enzyme was indeed found during electrophoresis of muscle from allophenic mice (mice resulting from aggregated cells from different mouse strains). From Mintz and Baker, *Proc. Natl. Acad. Sci., U.S.* 58:593 (Fig. 1) (1967).

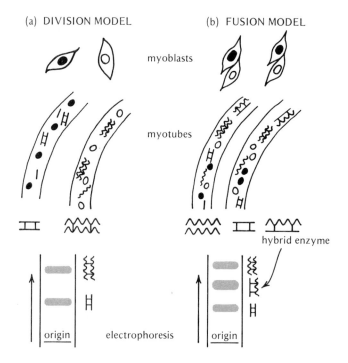

elegantly shown by aggregating cells from two different mouse embryos. Each embryo strain differed in the nature of one enzyme, isocitrate dehydrogenase, made up of subunits. The mosaic embryos were implanted into the uterus of a foster mother mouse, and developed into normal mice consisting of cells from the two different strains of mice. The skeletal muscle of these mice contained some enzyme like that of each parent, but also some hybrid enzyme, consisting of some subunits from one type of parental enzyme and some subunits of the other parent's enzyme. Hybrid enzyme of this type could only have been produced if the myotubes contained nuclei derived from cells of each parent in a common cytoplasm (Figure 11–8). In this cytoplasm, the nuclei from each produce a message that codes for its own enzyme subunits. These subunits, when synthesized, then can polymerize into the hybrid enzyme observed. Such enzymes are easily detected using electrophoresis, which separates proteins on the basis of size and charge. Thus, skeletal muscle forms as a result of fusion of single cells into elongated myotubes.

Myoblast fusion has also been studied *in vitro* in attempts to learn about the specific factors that influence this phenomenon. Myoblasts in culture do not fuse until about six to eight hours after cell division has ended. Presumably these cells are acquiring specific cell surface properties that facilitate the fusion process. Limb myoblasts in culture will synthesize myosin and assemble myofibrils if fusion is prevented by chelating agents that bind divalent cations. Thus, even in limb muscle, in culture, fusion is not absolutely necessary for muscle differentiation to occur.

In culture, it has been shown by Yaffe and others that much of the messenger RNA that codes for myosin is probably synthesized hours before myoblast fusion takes place. Actinomycin D inhibits synthesis of muscle proteins if applied before six hours prior to fusion. If applied within six hours of fusion, actinomycin D has little effect on muscle protein synthesis. Thus, it appears that stable messenger RNA molecules that code for muscle proteins are synthesized long before fusion occurs.

Cell culture experiments have also shown that specific factors may be required for muscle cell fusion and differentiation to occur. Hauschka and Konigsberg found that collagen can stimulate myoblast fusion to form myotubes. Collagen is present in the somite and hypomere where skeletal muscle forms. It may be that this substance plays a role in controlling myoblast fusion and differentiation *in vivo*. Exactly what factors cause certain cells to become muscle cells is, as yet, unknown.

Pancreatic Proteins and Differentiation

Before we examine the differentiating pancreas, let us say a few words about one of the pancreatic proteins—the hormone insulin. This hormone is extremely important in the metabolism of carbohydrates, and without it the transformation of glucose to glycogen does not occur. In the absence of insulin, blood sugar levels build up, eventually killing the organism. This disease, diabetes, results from lack of insulin, and can be controlled with insulin injections. Recent work has shown that the human insulin gene can be incorporated into the bacterial (*E. coli*) genome. Bacteria, in this way, begin to synthesize human insulin. Nobel prizes were recently awarded to Arber, Smith, and Nathans for their work in discovering enzymes that cut DNA, allowing incorporation of new genes such as the insulin gene. Human insulin should be very useful in treating many difficult cases of diabetes that do not respond well to insulin from animals. Virtually unlimited supplies of this hormone may now become available. Such work is leading to major

insulin

Figure 11–9. Insulin. The hormone insulin is composed of two polypeptide chains. Disulfide bridges form between sulfhydryl groups of specific amino acids (cysteine residues).

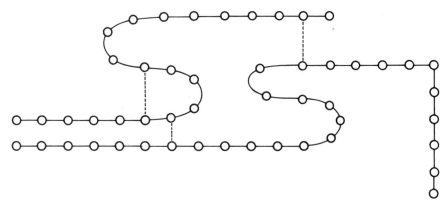

breakthroughs in medicine and this sort of genetic engineering is clearly a new frontier in biology today.

Insulin is a small protein hormone with a molecular weight of 6,000 daltons and consists of two popypeptide chains (Figure 11–9). It is one of many important proteins produced by the pancreas. In the mouse embryo, insulin is first detected in the beginning pancreas rudiment at nine to ten days after fertilization. The cells that produce insulin pinch off from other pancreas tissue, forming separate masses called **Islets of Langerhans** (Figure 11–10). The cells in the islets differentiate into distinct cell populations: **A-cells** which synthesize the hormone **glucagon** and **B-cells** that produce insulin. B-cells are easily distinguished from A-cells by dense cytoplasm and beta granules. Both hormones are involved in carbohydrate metabolism. The cells that produce digestive enzymes are located in the ends of a branched duct system in pouches called **acini** (Figure 11–10). The digestive enzymes produced in the acini include lipase, amylase, trypsin, and chymotrypsin. Acinar cells have extensive endoplasmic reticulum, Golgi apparatus, and storage granules (**zymogen**).

The mouse pancreas rudiment begins to appear as an evagination from the gut tube nine to ten days after fertilization. Pancreatic enzymes and hormones are detectable, but in very small amounts. Then follows a period of rapid cell division in which the enzyme and hormone levels remain low. By the fifteenth day or so, cell division stops and the synthesis of specific enzymes and hormones rapidly increases. During this time, the cells of the pancreas also differentiate cytologically. Endoplasmic reticulum, Golgi apparatus, and zymogen granules become increasingly abundant in the nondividing acinar cells.

Differentiation of the endodermal epithelium of the pancreas appears to be induced by mesenchyme that overlies the gut. The mesenchyme merges with the endodermal pancreatic rudiment as the rudi-

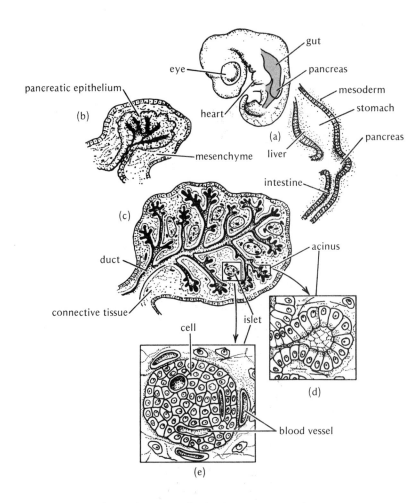

ment evaginates from the gut tube at day 9 or 10 in the mouse embryo. If
an 11-day mouse embryo pancreas rudiment is separated from its
mesenchyme and cultured on a porous filter, proliferation and differentiation of the endodermal epithelium fail to occur. If the mesenchyme is
placed on the other side of the porous filter, however, cell division
followed by differentiation of the epithelium occur. Specific proteins are
synthesized and characteristic tubules develop. Mesenchyme from other
tissues and even from other species also induces differentiation. The
mesenchyme appears to produce a factor that can pass through the filter
pores to stimulate cell division in the epithelium.

   A cell-free extract of whole chick embryos or of mesenchymal tissues
can be added to the medium in which pancreas epithelium is cultured.
These extracts can cause division and differentiation in the epithelium in
the absence of mesenchyme. Cell division is needed for differentiation

FUdR

BUdR

mesenchymal factor (MF)

because if DNA synthesis of pre-dividing cells is inhibited by FUdR (5-fluorodeoxyuridine), differentiation does not occur. BUdR (5-bromodeoxyuridine) is a thymidine analog that replaces thymidine in DNA of dividing cells but does not prevent DNA synthesis or cell division. DNA produced, however, is abnormal. This substance also prevents differentiation of pancreas cells treated during the division period. In the presence of BUdR mitosis continues, some common cell proteins are synthesized, but the increased levels of pancreas-specific proteins and differentiation do not occur. It may be that BUdR prevents production of messenger RNAs that code for cell specific proteins. The RNA synthesis inhibitor actinomycin D also inhibits the protein synthesis increase, if the inhibitor is applied during the cell division period or prior to the transition to the increase in specific protein synthesis. It has no effect once the protein synthesis rate accelerates. Thus stable messages may be produced prior to the increase in specific protein synthesis. To sum up, it seems that a period of normal DNA synthesis must occur before the transition to the high rate of specific protein synthesis can occur.

The mesenchymal factor (MF) that stimulates cell division and differentiation of pancreas epithelium has been isolated by Rutter and colleagues. This factor was covalently bound to inert beads. The beads were cultured with pancreas epithelium cells. The cells attached to the beads and continued DNA synthesis and cell division. Zymogen granules formed in the cells, indicating that differentiation had occurred (Figure 11–11). Since it is unlikely that MF attached to the beads entered the cells, the results suggest that MF stimulates mitosis and differentiation by some sort of action at the surface of the pancreas epithelial cells.

In summary, the pancreas develops as an evagination of the gut tube. The pancreatic rudiment joins mesenchyme surrounding the gut tube. In the presence of mesenchyme, morphogenesis begins and low amounts of specific proteins are synthesized. A cell division period and continued morphogenesis occur. Finally specific protein synthesis increases rapidly and cytological differentiation is completed. Stable messenger RNA coding for the specific proteins appears to be synthesized prior to the period of increased specific protein synthesis. Mesenchyme seems to induce cell division in the pancreas epithelium. Wessels and others have proposed that the increased cell mass in the developing pancreas (and in other systems) may be a key factor in the transition to the state of increased specific protein synthesis and cytodifferentiation. Large masses of cells could alter their immediate environment, creating conditions that turn on high rates of specific protein synthesis. This idea may lead to the design of additional specific experiments that could uncover some of the key control mechanisms operating in differentiating systems such as the pancreas.

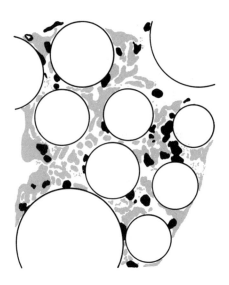

Figure 11–11. Pancreas epithelial cells attached to sepharose beads coated with mesenchymal factor (MF). Cell division is occurring in these cells. Such division seldom occurs if cells are added to beads without MF. After Pictet, R. and Rutter, W. J., *Nature New Biology* 246:49 (1973).

Tissue and Stage-Specific Isozyme Patterns

We have seen that proteins often consist of subunits that are coded for by different genes. Hemoglobin subunits are coded for by several genes. The relative activity of these genes varies at different developmental stages so that fetal hemoglobins and adult hemoglobins are clearly differ-ent with respect to their subunit compositions. These different hemoglo-bins presumably are best adapted to function in the different environ-ments of the embryo and the adult.

Isozymes are enzyme molecules that are functionally similar but which behave differently during electrophoresis, due to differences in the charge of the enzyme molecules. Some forms of the enzymes are more negatively charged than others. These charge differences result from the fact that the enzymes are composed of subunits and, just like hemoglobin, different subunits can combine to form the isozymes. Each different subunit is coded for by a different gene. So, during develop-ment, given genes may be more active than others and the proteins formed by polymerization of the subunits produced by these genes may differ, as in the case of hemoglobin. Isozymes thus serve as indicators of differential gene function in different tissues and at different develop-mental stages.

isozyme

Lactate dehydrogenase (LDH) is an enzyme that exists in multiple molecular forms (isozymes). Different isozymes of LDH predominate in

Lactate dehydrogenase (LDH)

different species, in different tissues, and at different developmental stages. LDH catalyzes the interconversion of pyruvate and lactate. Pyruvate is a key intermediate product in the metabolic oxidation of glucose to carbon dioxide and water, resulting in the release of energy that is essential for various metabolic activities. LDH is a key enzyme that makes it possible for cells to function effectively when oxygen is temporarily unavailable. This is the case because as pyruvate is converted to lactate, NAD is generated from NADH. NAD is essential for the completion of an early step in a metabolic set of reactions called glycolysis. It is assumed that, like hemoglobin, different LDH isozymes function best under different metabolic conditions. Thus, one isozyme may predominate at a specific stage or in a specific tissue because it is best adapted to function under the specific cytoplasmic conditions present. A brief consideration of LDH isozymes during development is appropriate now that we know a bit about the nature of isozymes. Before we do this, let us examine how it was determined that LDH exists as isozymes in the first place.

When crude homogenates of tissues such as mouse skeletal muscle are subjected to electrophoresis, the various proteins present in the homogenate separate on the electrophoresis gels on the basis of their net electrical charge. After a time, the gel block is immersed in a staining solution that contains lactic acid (an LDH substrate), the required cofactor NAD, and a tetrazolium salt. If LDH is present in the gel, lactic acid is oxidized and the tetrazolium salt is reduced to colored formazan that precipitates at the site of LDH activity. Five of these colored zones appear after electrophoresis of a mouse muscle homogenate (Figure 11–12). Each colored zone identifies an LDH isozyme. Five isozymes of LDH have been found in most mammal and bird tissues examined. What causes the formation of the five LDH isozymes?

Figure 11–12. Changing LDH isozyme pattern in heart muscle at different stages of the mouse. LDH-1 migrates towards the anode (+) and LDH-5 migrates to the cathode (−) during electrophoresis. Diagrammatic. (Drawing based upon photo from Markert, C. L. and Ursprung, H., *Developmental Genetics,* © 1971, Fig. 4–7, p. 44. Reprinted by permission of Prentice-Hall, Inc., Englewood Cliffs, New Jersey.)

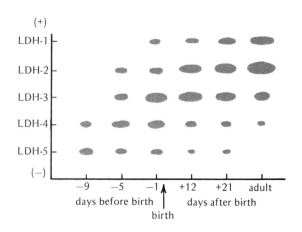

Markert and Ursprung purified LDH by several methods. The pure LDH also electrophoresed as five isozyme bands. By analytical ultracentrifugation in sucrose density gradients, it was determined that the molecular weight of each LDH isozyme is about 140,000 daltons. When the pure LDH isozymes were subjected to agents or conditions that separate proteins into subunits, only two, not five, electrophoretic bands were present that stained for protein. Each band represented a subunit of LDH. Each subunit had a molecular weight of about 35,000 daltons. The two bands were designated LDH subunit A and LDH subunit B.

Separated LDH subunits will recombine to form the five LDH isozymes as follows:

LDH-1   BBBB

LDH-2   ABBB

LDH-3   AABB

LDH-4   AAAB

LDH-5   AAAA

It is believed that such combinations of subunits occur in cells and in this way lead to formation of the five LDH isozymes.

Different genes code for LDH subunit A and LDH subunit B. Thus the amount of LDH-1, LDH-2, LDH-3, LDH-4 and LDH-5 in cells of different tissues or different developmental stages reflects, in part, the amount of A subunit versus the amount of B subunit that is produced at a given time in a given tissue (Figure 11–12 and 11–13). For example,

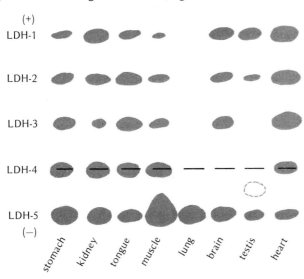

Figure 11–13. LDH isozyme patterns in 8 adult rat tissues. Electrophoretic patterns show tissue specific differences in LDH isozymes. Dotted circle in testis pattern represents an isozyme that may be testis specific. Diagrammatic. (Drawing based upon photo from Markert and Ursprung, *Developmental Genetics*, © 1971, Fig. 4, 6. p. 43. Reprinted by permission of Prentice-Hall, Inc., Englewood Cliffs, New Jersey.)

mouse eggs possess mostly LDH-1 (the B tetramer). Thus, the B gene appears to be most active during oogenesis. As development proceeds, the A gene becomes activated and the B gene is suppressed. LDH-5 (the A tetramer) begins to predominate. By birth, the B gene is activated in many tissues of the body so the isozyme pattern shifts again toward LDH-1. Thus, shifts in LDH patterns reflect shifting relative activities of the A and B genes.

Adult rat tissues also differ with respect to LDH pattern (Figure 11–13). For example, skeletal muscle shows a lot of LDH-5 while heart muscle has more LDH-1, -2, and -3. Here, too, shifting relative activities of the A and B genes appear to occur. It is likely that specific cytoplasmic factors such as pH and ionic species present control the rates of subunit synthesis. Such conditions also may affect subunit accumulation and assembly. We must also remember (recall the work on tryptophan pyrrolase) that the amount of a given protein that is present at a given stage or in a given tissue is not only a function of synthesis of the protein but also is a function of degradation of that protein. Thus, the presence of specific LDH isozymes represents a balance between synthesis and degradation of those isozymes.

To sum up, both stage-specific and tissue-specific patterns of proteins appear during the differentiation process. Such protein patterns result from differential synthesis, degradation, and activation of specific proteins. Now that we have been introduced to differentiation at the nucleic acid and protein levels, we can move on to a look at how larger components of cells come about during the differentiation process.

Readings

Ellis, C. H. Jr., The Genetic Control of Sea-Urchin Development: A Chromatographic Study of Protein Synthesis in the *Arbacia punctulata* embryo. *J. Exp. Zool.* 164:1–22 (1966).

Hauschka, S. D., and I. R. Konigsberg, The Influence of Collagen on the Development of Muscle Clones. *Proc. Natl. Acad. Sci. U.S.* 55:119–126 (1966).

Ingram, V. M., *The Hemoglobins in Genetics and Evolution*. Columbia University Press, New York (1963).

Konigsberg, I. R., Clonal Analysis of Myogenesis. *Science* 140:1273–1284 (1963).

Markert, C. L., The Molecular Basis for Isozymes. *Ann. N.Y. Acad. Sci.* 151:14–40 (1968).

Markert, C. L., and H. Ursprung, *Developmental Genetics.* Prentice Hall, Englewood Cliffs, 1971.

Marks, P. A., and R. A. Rifkind, Protein Synthesis: Its Control in Erythropoiesis. *Science* 175:955–961 (1972).

Paul, J. *et al.*, The Globin Gene: Structure and Expression. *Cold Spring Harbor Symp. Quant. Biol.* 38:885–890 (1973).

Rutter, W. J., N. K. Wessells, and C. Grobstein, Control of Specific Synthesis in the Developing Pancreas. *Nat. Cancer Inst. Monogr.* No. 13, p. 51 (1964).

Schimke, R. T., Protein Turnover and the Regulation of Enzyme Levels in Rat Liver. *Nat. Cancer Inst. Monograph* 27:301–314 (1967).

Schimke, R. T., E. W. Sweeney, and C. M Berlin, The Roles of Synthesis and Degradation in the Control of Rat Liver Tryptophan Pyrrolase. *J. Biol. Chem.* 240:322–331.

Ursprung, H., and K. D. Smith, Differential Gene Activity at the Biochemical Level. *Brookhaven Symp. Biol.* 18:1–13 (1965).

Wessells, N. K., DNA Synthesis, Mitosis and Differentiation of Pancreatic Acinar Cells *in vitro. J. Cell Biol.* 20:415 (1964).

Wessells, N. D., *Tissue Interactions and Development.* W. A. Benjamin, Menlo Park, California (1977).

Wilt, F. W., The Beginnings of Erythropoiesis in the Yolk Sac of the Chick Embryo. *Ann. N.Y. Acad. Sci.* 241:99–112 (1974).

Yaffe, D., and H. Dym, Synthesis and Assembly of Myofibrils in Embryonic Muscle. *Curr. Topics in Develop. Biol.* 5:235–280 (1972).

CHAPTER 12

DIFFERENTIATION: HIGHER
ORDERS OF STRUCTURE

IN THE LAST TWO CHAPTERS, we have examined some of the changes that occur in nucleic acids and proteins in differentiating cell systems. When one looks at differentiating cells, one often sees major changes in cellular characteristics in addition to the observed changes in specific nucleic acids and proteins. For example, differentiating cells sometimes exhibit increases in the amount of endoplasmic reticulum, ribosomes, microtubules, and microfilaments. In this chapter we will look at how some of these subcellular structures are formed. We will stress that many of these cellular components self-assemble from pre-formed subunits when the chemical conditions in the cell are conducive to such assembly. In order to get a feeling for what is going on, we will not restrict our study to differentiating cells. Instead, we will focus in on the assembly of the components per se to better understand the specific factors that lead to differentiation of higher orders of structure in developing systems.

Protein Assembly

In the last two chapters, we stressed the processes of transcription of DNA to form messenger RNA and translation of this message into proteins. We did not focus in on the next level, that is, the foldings of the amino acid chains and the assembly of protein chains into functional entities that often consist of several protein subunits. Here, we'll briefly examine this assembly process to better understand how proteins become important structural and functional units that support the life of a cell.

Polypeptide chains are synthesized on polysomes, or groups of ribosomes. These chains of amino acids seldom remain as straight chains. Instead, the chains often twist into a so-called secondary structure (Fig-

folded chain

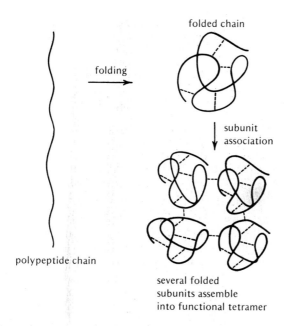

folding

subunit
association

polypeptide chain

several folded
subunits assemble
into functional tetramer

Figure 12–1. Example of formation of folded protein from polypeptide chain, followed by assembly of four of these subunits into a tetramer that is the functional unit of this generalized protein.

ure 12–1). The twisted structure is the functionally active form, as will be described later. What are the forces responsible for such "protein morphogenesis?" Hydrogen bonds, hydrophobic associations, electrostatic interactions, and covalent bonds are formed between certain amino acids in the polypeptide chains. These interactions mainly account for the folding of amino acid chains. It can be shown in the laboratory that folded proteins will unfold if conditions such as ionic strength, temperature, or pH are changed. These proteins will fold up again into their normal state if the original conditions are once again returned. It is in the folded state that most proteins display their functional properties, such as catalyzing reactions in the case of enzymes; maintaining cell and organelle structure in the case of ribosomal proteins, membrane proteins, and microtubule protein; regulating cell activities in the case of hormones and genetic repressors; binding and transporting molecules in the case of hemoglobin and hormone receptors; and binding of foreign materials in the case of antibodies, to name just a few. Recall that a single amino acid substitution resulting from a single gene mutation in the hemoglobin gene can cause sickle cell hemoglobin formation by changing the three-dimensional folding of normal hemoglobin.

It thus appears that the amino acid sequence in a protein chain is directly responsible for the way in which the chain folds, as a result of specific interactions that occur between given amino acids in the chain. Thus the assembly of a functional folded protein from a non-functional linear amino acid chain is determined by the specific groups present in the chain, and conditions of ionic strength and pH in which the chain resides.

We should also stress that many functional proteins do not consist of one single folded unit, but instead are composed of multiple units (Figure 12–1). The same types of interactions that account for the folding of single protein chains appear also to account for the assembly of protein subunits into larger complexes. By altering the conditions of ionic strength and pH or by exposing the protein to agents such as urea or detergents that break some of the bonds mentioned above, the protein aggregate will disociate into its component subunits. If the denaturing agents are removed and the conditions are returned to those present originally, the subunits will often fold and aggregate into the functionally active complexes. An example of such a complex is hemoglobin, which consists of four subunits. Hemoglobin does not actively bind oxygen unless all four folded protein chains are assembled into the tetramer with the iron-containing heme groups in their proper positions in each subunit. The assembly of proteins into folded configurations and aggregates of these folded chains is dependent upon the sequence of amino acids in the single protein chains. This sequence folds into the most stable configuration as a result of hydrogen bonds, hydrophobic associations, electrostatic interactions, and covalent bonds. The folded subunits in many proteins interact with each other as a result of the same types of interactions mentioned above. Conditions in the cell such as pH and ionic strength play an important role in controlling the assembly of specific cellular proteins. Now that we have a general idea about the nature of how functionally active proteins are formed, let us look at some specific examples of the assembly of cellular components that are above the protein level of structure.

Microtubules

We have mentioned in many parts of this text that cytoskeletal elements such as microtubules and microfilaments appear to be important components in developing systems. How do microtubules and microfilaments come about? Let us begin with microtubules. These components make up cilia, flagella, and mitotic spindles, and have been implicated in causing some of the changes in cell shape and organelle movement that occur

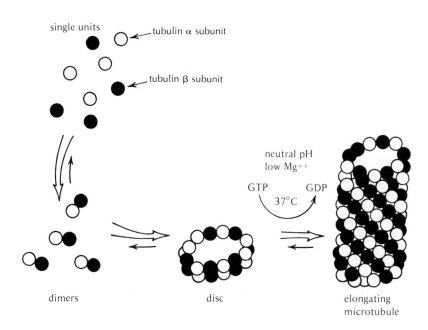

Figure 12–2.
Microtubule assembly.

during morphogenesis and in the everyday life of a cell. In Chapter 9 we noted that cytoplasmic microtubules appear to be associated with elongating cells. These cytoskeletal elements line up in the long axis of the cell. Cell elongation is prevented if the microtubule disrupting agent colchicine is present. Microtubules have been observed in elongating cells around the amphibian blastopore, in the neural plate, in cells moving into the primitive groove in bird embryos, and in outgrowing nerve cells. Colchicine prevents nerve cell outgrowth and inhibits cell elongation in the chick primitive streak and in the neural plate. Microtubules may directly force the cells to elongate or they may allow cytoplasm to flow in the direction of elongation by serving as tracts for cytoplasmic flow. These structures, therefore, are of major interest to us as developmental biologists. How are microtubules assembled?

Microtubules are cylinders of about 250 Å in diameter and are composed of globular subunits of the protein tubulin (Figure 12–2). Each subunit in turn consists of two proteins, α tubulin and β tubulin, that can be distinguished from each other by electrophoresis. Each of these protein monomers has a molecular weight of about 54,000 daltons. The amino acid sequence of the α and β components differ.

The assembly of microtubules can be studied *in vitro* to analyze the nature of the factors involved in the assembly process. The conditions of neutral pH, low Ca^{++}, low Mg^{++}, presence of GTP, and temperature of 37° C favor the assembly of tubulin into discs which assemble into elongating microtubules (Figure 12–2). Thus, when conditions are right, tubulin molecules assemble into microtubules. The exact nature of microtubule assembly *in vivo* is not well understood. It is likely, however, that the same factors that are required for microtubule assembly *in vitro* are involved in microtubule assembly *in vivo*. Other factors that may play a role in the arrangement of polymerizing microtubules in cells include the presence of initiation sites where microtubule assembly begins. Such sites include basal granules in cilia and centrioles in spindles. Centrioles may also serve as microtubule initiation sites in nondividing cells. Key factors that may control cell elongation, as described previously, could be conditions that set up microtubule initiation sites that serve to orient microtubule assembly in a specific axis of the cell.

Microfilaments

In Chapter 9 and in other portions of the text we mentioned that cytoplasmic microfilaments appear to be involved in cell narrowing, an important factor in morphogenesis and motility. Morphogenesis of tubular glands, for example, may be aided by bundles of microfilaments that could contract and exert force on the sides of cells, to make them narrow at one end. If a sheet of cells narrows at one side, the sheet will bend. This is one of many ways in which microfilaments may be involved in morphogenesis. Microfilaments may also play a role in controlling cytoplasmic division, nerve outgrowth, cytoplasmic streaming, invagination during gastrulation, smooth muscle contraction, cardiac muscle contraction, and cortical contraction in eggs. These processes are inhibited by the drug cytochalasin B that disrupts microfilaments. This drug has been used to suggest a role for microfilaments in many processes sensitive to the drug. Unlike colchicine which binds specifically to microtubule subunits and is not known to affect other cell processes, cytochalasin B affects many cellular processes, such as respiratory metabolism, membrane permeability, and protein and glycoprotein synthesis in addition to disrupting microfilaments. Cytochalasin B, therefore, is useful in providing preliminary evidence for a role of microfilaments in cellular processes, but can not be used to definitively establish such a role because of the multiple effects that this drug has on cells. How are microfilaments assembled in cells?

Microfilaments commonly are found in cells as 30–80 Å diameter rods. Other thicker 100–120 Å diameter filaments have also been observed attached to plasma membrane specializations called desmosomes. Very thick (up to 160 Å diameter) filaments have been observed in

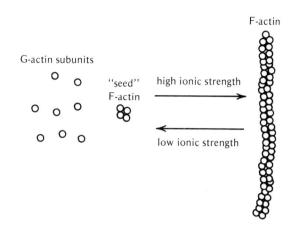

Figure 12–3. Self-assembly of F-actin microfilament.

nerve axons. The 30–80 Å filaments are similar to F-actin contained in thin filaments of muscle. Actin filaments (F-actin) consist of globular subunits (G-actin) (Figure 12–3). These subunits polymerize into F-actin filaments *in vitro* with increasing ionic strength. Small "seed" fragments of F-actin stimulate polymerization of G-actin into elongating F-actin filaments. Here again we see that microfilament assembly is dependent upon the presence of G-actin subunits and specific conditions of ionic strength. When the pre-formed subunits are present and when conditions are right, microfilaments assemble and become involved in the cellular processes and morphogenetic events that we have described throughout this text and have summarized above.

F-actin

G-actin

Based upon studies using a fluorescent antibody that specifically binds to given proteins associated with the cellular contractile apparatus, it has been shown that the protein α-actinin may serve as sites for the initiation of actin polymerization and attachment. Also the protein tropomyosin may assist in actin polymerization and stabilization. Thus, as in the case of microtubule assembly, microfilaments assemble from preformed subunits under specific ionic conditions. Initiation sites (possibly α-actinin) may play a role in governing the position of polymerizing microfilaments in cells.

Flagella

The bacterial flagellum serves as another model system that illustrates the nature of self-assembly of structure from pre-formed subunits. Flagella are composed of a globular protein called flagellin. Thousands of

flagellin

Figure 12–4. Self-
assembly of the bacterial
flagellum.

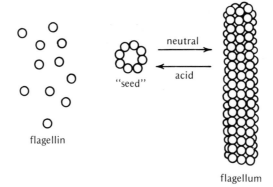

these flagellin subunits assemble to produce the flagellum (Figure 12–4). Flagella dissociate into the subunits by lowering the pH. Bringing the pH back to neutrality results in the repolymerization of the subunits into the flagellum (Figure 12–4). Here again we see that when subunits are present and conditions are right, polymerization occurs. Assembly of the flagellum, as with microtubules and microfilaments, is speeded up by the presence of a small cluster of aggregated subunits. This is another example of how specific proteins contain within themselves all that is needed to assemble into higher orders of structure as long as relatively simple pH or ionic requirements are met.

Ribosomes

Throughout this text we have looked at differentiating cell systems. We saw that with differentiation we often find an increase in the number of cytoplasmic ribosomes in cells. They appear to be required for the synthesis of the many new proteins that appear during the differentiation process. It is therefore of interest to briefly examine the assembly of ribosomes and see if the formation of these key cellular organelles also occurs in a spontaneous way, similar to that described for the other components discussed in the chapter so far.

The ribosome is a magnificent complex of nucleic acid and protein that functions in the synthesis of protein. Let us focus on the well-studied bacterial ribosome that consists of ribosomal RNA combined with many (20–40) separate proteins. The two subunits dissociate at low

Mg^{++} concentrations. Upon raising the Mg^{++} concentration, the subunits reassemble into the active ribosome that is able to synthesize protein *in vitro*. The two subunits can not synthesize proteins unless they are assembled into the intact ribosome.

Now let us look at one of the subunits, the 30S component, to see how it is assembled. The 30S subunit consists of one ribosomal RNA molecule and about 21 different proteins. If the specific proteins and the ribosomal RNA molecules are combined, 30S particles assemble, and when combined with the other subunits, the 50S components, functional ribosomes are produced. Thus, assembly of ribosomes from RNA and proteins can take place *in vitro* under relatively simple conditions. It is likely that specific ribosomal RNA and specific ribosomal proteins contain in themselves all that is needed for ribosome assembly when relatively simple environmental conditions are met. It has been shown that some of the specific ribosomal proteins are required for the assembly process, while other proteins appear to be associated with maintaining structural integrity or function of the ribosome. Most of the ribosomal proteins must be present if stable, functional ribosomes are to be assembled. It may be that an important control mechanism in ribosome assembly involves synthesis of one or more of the specific ribosomal proteins at specific times in the development and differentiation of cells.

Summary

In summary, many higher orders of structure in cells appear to be able to self-assemble. The information for such assembly or polymerization exists within the subunits of these structures. If relatively simple conditions are met, these subunits polymerize into functional cellular components such as specific enzymes, microtubules, microfilaments, flagella, and ribosomes. It should be stressed that self-polymerization from preformed subunits is not the only way in which cellular components are assembled. In many cellular organelles, specific components are added to already existing structures. For example, membrane systems in mitochondria and chloroplasts seem to involve addition of specific components to pre-existing units. Differentiation of some of these membranes appear to occur as a result of addition of new enzymes or proteins to the existing lipid bilayer. Assembly of many basic cellular components, does, however, appear to be controlled by the amino acid sequence of the subunit proteins. Thus, basic types of chemical interactions such as hydrogen bonding, hydrophobic interactions, electrostatic interactions, and covalent bonding, between amino acids within the subunits and between the subunits, lead to the spontaneous formation of stable units. It is likely that such self-assembly plays a major role at least in the beginning stages of the morphogenesis and differentiation of many cellular organelles.

Readings

Anfinsen, C. B., Principles that Govern the Folding of Protein Chains, *Science* 181:223–230 (1973).

Caspar, D. D., Design and Assembly of Organized Biological Structures. In *Molecular Archetecture in Cell Physiology*, K. Hayashi and A. G. Szent-Gyorgi, eds., Prentice Hall, Englewood Cliffs, New Jersey, pp. 191–207 (1966).

Grant, P., *Biology of Developing Systems.*, Holt, Rinehart & Winston, New York (1978).

Johnson, K. A. and G. G. Borisy, The Equilibrium Assembly of Microtubules *in vitro*. In *Molecules and Cell Movements*, S. Inoue and R. E. Stephens, eds., Raven Press, New York, pp. 119–139 (1975).

Kushner, D. J., Self Assembly of Biological Structure. *Bacteriol. Rev.* 33:302–345 (1969).

Lazarides, E., Actin, α-Actinin and Tropomyosin Interaction in the Structural Organization of Actin Filaments in Non-Muscle Cells. *J. Cell Biol.*, 68:202 (1976).

Nomura, M., Assembly of Bacterial Ribosomes. *Science* 179:864–873 (1973).

Olmsted, J. B., and Borisy, G. G., Microtubules. *Ann. Rev. Genet.* 8:411–470 (1973).

Oosawa, F., M. Kasai, S. Hatano, and S. Asakura, Polymerization of Actin and Flagellin. In *Principles of Biomolecular Organization*, G. E. W. Wolstenhome and M. O'Connor, eds., Churchill, London, pp. 273–303 (1966).

Osborn, M., and K. Weber, Cytoplasmic Microtubules in Tissue Culture Cells Appear to Grow from an Organizing Structure Towards the Plasma Membrane. *Proc. Nat. Acad. Sci. U.S.* 173:867–871 (1976).

Raff, R. A., and H. R. Makler, The Non-Symbiotic Origin of Mitochondria. *Science* 177:575–582 (1972).

Reinert, J., and H. Ursprung, eds., *Origin and Continuity of Cell Organelles*. Springer-Verlag, New York (1971).

Tilney, L. G., and J. Goddard, Nucleating Sites for the Assembly of Cytoplasmic Microtubules in the Ectodermal Cells of Blastulae of *Arbacia punctulata*. *J. Cell Biol.* 46:564–575 (1970).

Wessells, N. K., B. S. Spooner, J. F. Ash, M. D. Bradley, M. A. Luduena, E. L. Taylor, J. T. Wrenn, and K. M. Yamada, Microfilaments in Cellular and Developmental Processes, *Science* 171:135–143.

CHAPTER 13

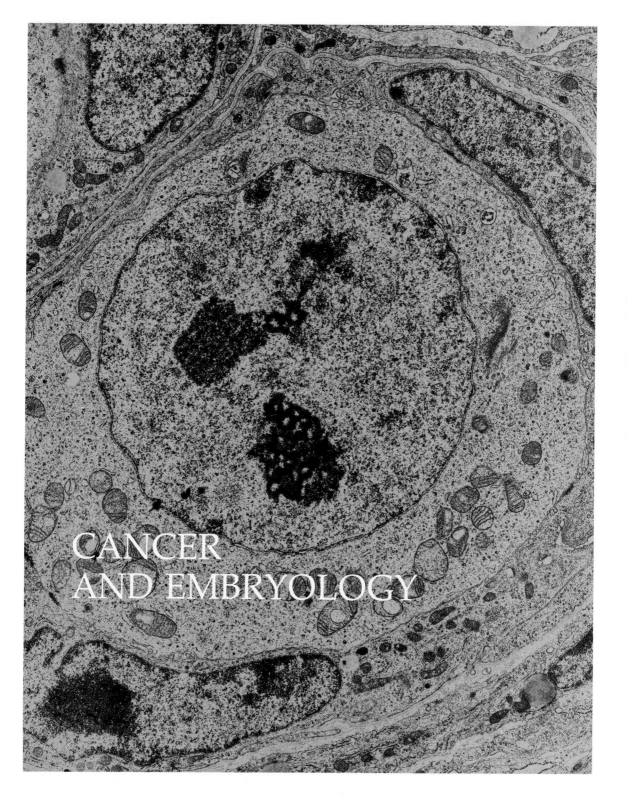

CANCER
AND EMBRYOLOGY

WHY ARE WE DISCUSSING cancer in a book on embryology? As will be seen, cancer cells and certain embryonic cells are similar in many ways. A study of cancer and embryology will help us focus upon many concepts that we have developed in this text. This chapter will tie together many of these concepts and show how an understanding of certain characteristics of embryonic cells is leading to important insights into a major disease, cancer.

The story of cancer is multi-faceted. A complete discussion of the topic includes causes, types of tumors, diagnosis, treatment, and cellular aspects. It is in this last category that the topic of cancer cells versus embryonic cells appears. Thus, we will begin our discussion with causes and then unravel the entire story, ending with the cellular aspects. In this way, we will develop an understanding of the cancer problem as a whole, rather than just dealing only with the cellular aspects. Discussion of the clinical aspects of cancer will, in a book such as this one, be brief.

Causes of Cancer

Cancer is a disease in which cells grow in an unregulated manner. They may eventually kill the individual by growing into blood vessels, causing hemorrhage, or interfering with some vital function associated with a specific organ of the body. Most of us realize that many things appear able to cause cancer. These causative factors include radiation, viruses, and chemicals. Let us say at the outset that many of the agents that cause cancer may act on the chromosomes of the cells. Certain genes that code for proteins able to transform cells into a cancerous condition may become activated by the cancer-causing factors listed above. These factors may also act by damaging the DNA so that normal genes are altered enough to transform the cell into a cancerous state. In the case of some viruses, cancer may be caused by direct insertion of the viral genome into the host cell genome. As can be seen by the "iffy" nature of this discussion, the mechanisms by which specific agents cause cancer are uncertain at best. There is little doubt, however, about which agents will definitely cause cancer and it is this question that we will turn to next.

Radiation

X-rays and ultraviolet radiation clearly cause cells to become transformed into tumor cells. This can be demonstrated in culture. When cells, for example, are exposed to X-ray irradiation in culture, they become transformed into tumor cells that grow in an unregulated manner. That is, unlike normal cells, these transformed cells do not stop growing or stop moving when they are in contact with other cells. Instead, they continue to divide and move. So, tumor cells lack contact inhibition of growth and movement. These properties can be easily observed in culture and are examples of the criteria used to determine if cells have been transformed into the cancerous state. These cells can also form tumors when inoculated into animals. We will come back to these and many other cellular aspects of cancer later on.

contact inhibition

Evidence that radiation directly causes cancer in man is substantial. For example, the survivors of the atom bomb blasts in Japan have a high incidence of leukemia, a cancer of the blood-forming tissues. The frequency of leukemia is proportional to radiation dose exposure during the blasts. Also, a high incidence of leukemia is observed in people who have received large doses of radiation for treatment of conditions other than leukemia. Radiologists using some of the early X-ray equipment, which allowed large amounts of radiation leakage, also exhibit high rates of incidence of leukemia. Ultraviolet radiation from the sun has been implicated in causing human skin cancer in a variety of studies. It is likely that radiation damage of the chromosomes is a direct cause of cancer.

Virus

Although virus particles have been found to be associated with some human tumors, it is difficult to obtain proof that virus is a major cause of human cancer because of the limited availability of human tissue for

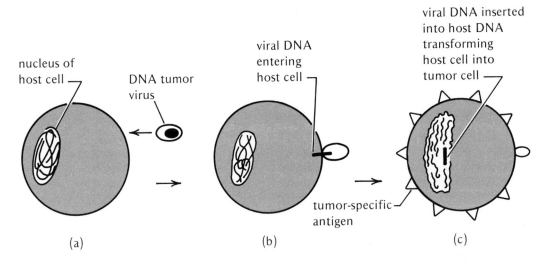

Figure 13–1. Viral transformation.

experimental studies. There is little doubt that virus can cause cancer in animals, however, and it is likely that viral genes can also be a causative factor in the development of human cancer. When we talk about virus we must consider two major types: DNA virus and RNA virus. The DNA virus consists of a core of DNA surrounded by a protein coat. In the RNA virus, the core is RNA and the coat is also protein. Lipid is associated with the coats of some virus particles. Let's briefly consider the DNA tumor viruses first, and then turn to RNA tumor viruses.

DNA tumor virus   Two extensively studied DNA tumor viruses are polyoma virus and simian virus 40 (SV 40). These viruses typically attach to the cell surface and inject their DNA core into the cell. Once inside the cell, the viral DNA can either initiate production of many additional viral particles, leading to cell death, or the viral DNA can incorporate itself into the host cell's genome (Figure 13–1). When this latter event occurs, the cell may become transformed into a cancer cell. Viral genes that have been incorporated into the host genome may induce cancer by coding for new proteins that alter the behavior of the cells, or the insertion of the viral genome itself may damage the host DNA so that processes such as regulation of growth are impaired.

RNA virus   RNA viruses have also been implicated as causative agents of cancer. A well studied cancer-causing RNA virus is Rous sarcoma virus (RSV), which induces tumors in chickens when extracted from chicken tumors and inoculated into tumor-free birds. This can be demonstrated in culture by infecting normal chick cells with RSV and observing their transformation into the cancerous state. The transformation is seen in terms of the lack of contact inhibition of growth and movement of the cells (Figure 13–2) and the ability of these cells to form tumors when inoculated into chicks.

The means by which RNA tumor viruses transform cells into cancer cells has been the subject of intensive recent investigations. Briefly, it appears that the RNA virus attaches to the cell, and the virus then injects its RNA into the cell. Viral DNA is synthesized on the viral RNA template by an enzyme called reverse transcriptase (RNA-dependent DNA polymerase). This enzyme forms DNA on an RNA template, contrary to the once accepted dogma that DNA can be synthesized only on a DNA template. This viral DNA made from the RNA template now can insert into the host DNA, just as the DNA from DNA virus does. This viral DNA then can induce cancer by coding for proteins that alter cell behavior such as new cell surface proteins, or by altering control of cell division.

reverse transcriptase

A recent theory (oncogene theory) proposes that all cancer is the result of viral genes that have been incorporated into the host species' genome long ago. These genes are proposed to be passed on from generation to generation. Chemicals, radiation, and other carcinogens act by "turning on" or activating these genes. Thus, according to this theory, all of us have genes that code for proteins that can transform our cells into cancer cells. Whether or not we get cancer will depend upon whether or not these genes become activated, and whether or not our bodies can fight off those cells that do become transformed. This theory does offer a sort of unified explanation of all cancer. It explains how many seemingly different agents can cause cancer (by activating the cancer genes) and how there can be a genetic "pre-disposition" of certain individuals to develop certain tumors. Such a genetic tendency can simply be explained in terms of the numbers of these cancer-producing genes termed "oncogenes" in different families, and in terms of the ease of activating these genes.

oncogene theory

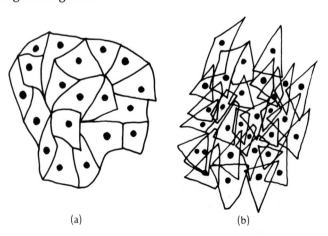

Figure 13–2. Cultures of normal and transformed cells. a. Normal cells show contact inhibition. b. Polyoma virus transformed cells overlap each other, showing decreased contact inhibition.

(a)   (b)

Figure 13–3. Industrial chemicals believed to cause (or suspected of causing) cancer in humans*

Chemical	Target Organ(s)	Route of Exposure
4-Aminobiphenyl	bladder	inhalation, oral
Arsenic compounds	skin, ? lung	oral, inhalation
Asbestos (crocidolite, amosite, chrysolite, and anthophyllite)	lung and chest cavity gastrointestinal tract	inhalation, oral
Auramine	bladder	oral, inhalation skin
Benzene	bone marrow	inhalation, skin
Benzidine	bladder	inhalation, oral skin
Bis(chloromethyl)ether	lung	inhalation
Cadmium oxide and sulphate	prostate, ? lung	inhalation, oral
Chromium (chromate-producing industries)	lung	inhalation
Hematite (mining)	lung	inhalation
2-Naphthylamine	bladder	inhalation, oral
Nickel (nickel refining)	nasal cavity, lung	inhalation
Soot and tars	lung skin (scrotum)	skin contact
Vinyl chloride	liver, brain, lung	inhalation, skin

*This table does not include many other chemicals that have been shown to produce cancers in animals, but whose effects in humans are not definitely known. From *Cancer and the Worker*, N.Y. Acad. Sci., 1977, p. 12.

Is there any evidence to support the oncogene theory? The evidence so far is suggestive but not unequivocal. What is some of the evidence in support of the oncogene theory? Many agents that cause cancer, such as methylcholanthrene or radiation, will not only transform normal cells into cancer cells but will also induce the appearance of type C RNA virus particles (type C is simply a classification of virus structural type) in some of these cells, and will induce the appearance of virus-specific protein antigen in these cells. These results suggest that cancer-causing agents appear able to promote virus gene expression in cells that apparently contained no visible virus. These sorts of experiments, however, do not prove that the viral genome is the cause of cancer. Appearance of viral particles in treated cells may only be a secondary effect of the X-ray or chemical treatment.

It is not clear if all cells contain cancer-causing viral genes in their genomes. Some studies with identical twins suggest that cells in all individuals do not necessarily contain oncogenes. Using identical twins, one of whom had leukemia, and a technique in which type C DNA synthesized from type C RNA virus (using reverse transcriptase) was hybridized with the DNA from the identical twins, it was shown that only DNA from the identical twin who had leukemia would pair with or hybridize with the viral DNA. This suggests that the leukemia patient had viral DNA genes while the healthy twin did not. Such a study suggests that viral genes were not present in the fertilized egg, but were picked up or formed by only one twin at some later time.

It can be said that certain viruses cause cancer, although it should be stressed that the mere identification of virus particles in tumor cells is no proof that the virus caused the tumor. It remains to be seen if virus is an important cause of cancer in man. Certain viruses called herpes viruses are associated with many human diseases such as cold sores, mononucleosis, and certain genital infections. Similar viruses have also been identified in human tumors such as Burkitt's lymphoma. Virus isolated from these human tumors can cause tumors in animals. It still remains to be determined, however, if the herpes virus was the actual cause of the human tumor. It is likely that in future years evidence will indicate that viruses are a cause of human cancer. Such experiments are difficult, however, because we can not inject human beings with virus to determine if the virus causes cancer. Rather good evidence, however, is being obtained from carefully performed animal experiments.

Chemicals

Cancer-causing substances are called carcinogens. Many chemical carcinogens have been identified and we will review some of them here. These chemicals may cause cancer by altering the genetic information as discussed previously. Proof that a given chemical causes cancer is obtained by exposing cells in culture or living animals such as mice to these chemicals. If a chemical, at a carefully determined concentration,

carcinogens

causes cancer in mouse cells, it is likely that the substance will cause cancer in human cells because mammalian cells are very similar. The concentrations of chemicals used in the living animal experiments are based upon weight of the animal versus weight of a human being, and lifespan of the animal versus that of a human being. In other words, concentrations of substances given to animals are similar to those that could be expected to build up in the human body over a long period of time. Thus, when we hear that animals are sometimes given large doses of chemicals in these tests or that animal tests are just not relevant to human cancer, we should remember that most chemicals that cause cancer in animals have been found to be a cause of cancer in humans. Animal tests are valid. Mammalian cells are very similar from species to species.

Animal tests, however, are not the only sources of evidence for whether or not a chemical can cause cancer. A great deal of data is available from human studies examining the incidence of cancer in specific occupational groups, smokers, and other groupings of people. These studies have provided extensive evidence that cigarette tar, asbestos, benzene, nickel, vinyl chloride, 2-naphthylamine, hematite, chromium, cadmium oxide, cadmium sulphate, bis(chloromethyl) ether, benzidine, auramine, arsenic compounds, 4-aminobiphenyl, soot, and tars can cause cancer in humans (Figure 13–3). Many other substances have also been implicated in causing human cancer, such as chloroform and nitrosamines. We should mention that the American Cancer Society has estimated that fifty percent of cancer deaths could be eliminated if cigarette smoking ceased. Such a dramatic statement is a guess, but an educated guess and just points to the massive numbers of cancer cases that may be caused by things that we eat, breathe, or touch. In fact, some cancer specialists suggest that as much as eighty-five percent of all cancer may be induced by environmental factors!

Before leaving the topic of chemical agents that cause cancer, let's say a word or two about the difficulty of testing potential carcinogens. Not all cancer-causing agents cause cancer immediately. Some only cause cancer after cells have been exposed to other substances. Some agents work in a two-step process. One substance sets up a pre-cancerous condition, while another substance is needed to actually cause the formation of a visible tumor. Some studies show that even months or years after application of the first agent on the skin of a mouse, the second agent will induce a tumor on that spot. If the first agent is not applied, the second does not promote tumor development. These sorts of findings illustrate how difficult it is to determine if an agent is a carcinogen. It may be that many potential carcinogens go undetected because the right second agent was not used in any tests.

The difficulty in determining whether a substance is hazardous is also illustrated by the sodium nitrite story. Sodium nitrite, a widely used preservative in certain meat products, is not a known carcinogen. Nitrosamine, a product of the reaction between sodium nitrite and amine, however, is a potent carcinogen. For many years there has been a controversy regarding the formation of nitrosamine in the human body after ingestion of sodium nitrite. Some studies indicate that nitrosamine is formed. It is difficult, however, to determine if the amount of nitrosamine formed in this way is hazardous. Recent studies do indicate that nitrosamine is actually formed when bacon containing sodium nitrite is cooked. As a result of these new findings, federal regulatory agencies are presently working on stringent guidelines for the use of sodium nitrite in bacon and meat products that are heated at high temperatures. We can conclude that until there is solid data available, suspected carcinogens are often not eliminated from the environment. Decisions to regulate hazardous substances are difficult and are often based upon economic considerations along with the health-related factors. There is little doubt, however, that efforts to control carcinogens in the environment are clearly increasing, so we can end this section on a somewhat optimistic note.

Tumor Types

Tumors are abnormal growths. Some tumors are slow growing, do not spread, and do not endanger the organism. Warts and moles are examples of such tumors. These slow growing, nonspreading growths are termed benign tumors. Benign tumors can, however, change into tumors that can invade body tissues, often ending in death. Tumors that spread and could cause death are termed malignant tumors. Malignant tumors spread by two major means. The tumor cells may crawl to nearby tissues. Such spreading movement is called invasion. Tumors also spread by invading blood vessels, and the malignant cells are carried by the blood to distant sites in the body. Such spread by means of the bloodstream is called metastasis. We will consider the mechanisms of tumor spread in the section on cellular aspects.

There are many terms used to classify tumors. Carcinomas are tumors derived from epithelial tissue. Sarcomas are derived from

benign tumors

malignant tumors

invasion

metastasis

carcinomas

sarcomas

leukemia

melanomas

teratomas

connective tissue. Leukemia is a type of tumor of the blood-forming tissues in which there is an increased production of certain white blood cells. Melanomas are derived from pigment cells. Teratomas are tumors originating from germ cells, and often develop into embryo-like growths consisting of many types of differentiated tissues. It might be mentioned here that teratomas are very like embryos in some respects. Some investigators believe that these tumors represent abnormally activated eggs or primordial germ cells that begin to develop like normal embryos before they go awry. It may be that some of these tumors could really form normal embryos if they grew in the right place in the body. Suffice it to say here that teratomas are so embryo-like that they offer magnificent material for studies involving the relationship of embryonic cells and cancer cells, the nature of the benign versus the malignant state, and the nature of differentiation. We'll soon see how these remarkable tumors are helping to shed much light in these key areas.

Diagnosis and Treatment

Before turning to the cellular aspects of cancer, to make this story complete, let's say a few words about diagnosis and treatment of cancer.

Many tumors, if treated before they spread, are curable. Even after spreading, treatment can cure many tumors. It is much more difficult, however, to treat a tumor that has spread because instead of dealing with only one tumor, one then must deal with numerous secondary tumors at sites often distant from the original or primary tumor. Thus, the problem of successful treatment of many cancers boils down to one of diagnosing tumors early in their development.

Certain cancers are now almost always curable, even though they were major killers not too long ago. An example of a success story in the cancer field is that of cervical cancer in human females. This form of cancer was a major killer and now is often completely curable because of early diagnosis made possible by development of the PAP test. This simple test involves taking a smear of cervical cells and examining them with a microscope. Cancer cells, precancerous cells, and normal cells are easily observed. This test allows treatment of this form of cancer, often before a real tumor develops, so the cancer can be cured with freezing or surgery before any spread has occurred.

Other tumors, however, are often difficult to detect early. For example, it has been estimated that some lung tumors and breast tumors may grow for a period of ten years before they are detectable. Present methods of detecting these tumors involve techniques such as the use of diagnostic X-rays. These techniques often are unable to detect a tumor much smaller than about one centimeter in diameter. Some tumors take about ten years to develop to the one centimeter stage that contains about one billion cells. At this stage, tumor spread has sometimes already occurred. Thus, some types of cancer that currently kill a lot of people, such as lung and breast cancer, may not be detectable until they have been growing for ten years. It is likely that in the near future detection of these tumors will improve with improving technology. Such improved diagnosis should facilitate an increased cure rate as seen in the cervical cancer story.

Cancer is treated by many well established and some new experimental methods. The well established methods include surgery, radiation therapy, and chemotherapy. The newer experimental methods include immunotherapy and bone marrow transplants.

Surgery is the treatment of choice if the tumor can be removed without excessive danger to the individual. Surgery is a curative treatment of many tumors, especially those that have not spread to distant parts of the body.

Localized radiation is the treatment of choice in some tumors. For example, cancer of the voice box is often cured with radiation treatment with little impairment of speech. Although very effective in treating the tumor, surgery in this case often drastically impairs speech function. Localized radiation therapy can cure cancer in carefully controlled use. As mentioned earlier, radiation can, in other instances and under other conditions, cause cancer.

Chemotherapy is the method of choice in treating many nonlocalized cancers, such as leukemia and tumors that have spread. Chemotherapy utilizes chemicals that interfere with the processes of living cells. Some chemicals mimic metabolites needed by the cells in the synthesis of DNA or RNA. These substances are used by the cell, producing defective nucleic acid leading to cell death. Some drugs inhibit protein synthesis and others inhibit respiration. In treating cancer, drug dose must be carefully regulated because these agents are also harmful to normal cells. But because tumor cells usually are in a constant state of growth, these drugs tend to become incorporated into these cells more easily than into normal cells. Individuals undergoing chemotherapy, however, often display side effects such as hair loss and nausea as a result of the toxicity of these drugs. Many cancers such as certain leukemias have been treated very effectively with chemotherapy. In fact, while a decade or two ago leukemia was considered not curable, now many people are free of the disease for five, ten, or fifteen or more years as a result of modern chemotherapy.

tumor-specific antigens

Immunotherapy is an experimental technique that takes advantage of the body's natural defenses against tumor cells. Cells often display new surface antigens called tumor-specific antigens. These antigens can be recognized by the body's immune system just as the body can recognize and destroy invading bacteria. The white blood cells of the body recognize the invading bacteria as foreign and destroy them. White blood cells also appear able to recognize tumor cells and are able to destroy them. For some not well understood reasons, in some individuals the body's immune system does not function properly. A poorly functioning immune system could increase the likelihood that an individual will develop cancer or of developing persistent forms of infectious diseases. Immunotherapy is based upon the idea that cancer may be controlled by stimulating the immune system, in much the same way as the Salk vaccine protects against polio. Various laboratories around the world are experimenting with immunotherapy as a treatment for cancer. Patients used in these studies are usually those who have little chance of recovery using conventional treatments. These patients are usually inoculated with a substance that acts as a generalized stimulus of the immune system, or with a combination of such a substance with living or dead tumor cells or tumor cell surface antigen material. The substance usually used as a generalized stimulus of the immune system is BCG (Bacillus of Calmette and Guerin), a weakened strain of tuberculosis bacteria that has been used to treat tuberculosis. This material appears to activate the immune system. Vaccines are also made with living or dead tumor cells, or parts of tumor cells. Immunotherapy seems to be potentially very promising for treatment of cancer. Difficulties with the method, however, are that many tumors have different tumor-specific antigens and that it is very difficult to obtain large quantities of pure human tumor cells for the vaccines. It is likely that a different tumor vaccine would be needed for the treatment of each specific tumor.

bone marrow transplants

The last form of treatment that will be discussed is the experimental method of bone marrow transplants. This method is used experimentally to treat certain terminal cases of leukemia. Leukemia is a disease in which certain white blood cells are produced in massive numbers by the bone marrow. If the bone marrow producing these leukemia cells is destroyed, leukemia should be cured. The catch is that white blood cells are also essential to fight disease. Bone marrow transplants involve destruction of the patient's bone marrow with strong whole body radiation treatment. Leukemia cell forming tissue is destroyed. Then healthy bone marrow removed with a syringe from an identical twin or a person with an identical tissue type is inoculated into the bones of the patient. This marrow colonizes the bones and begins producing normal white blood cells. This method is very drastic because the patient is subjected to high radiation doses that can have severe effects. Also, such transplants can only be made between identical twins or between other persons that share identical tissue types, with the presence of similar antigens on the surfaces of the body cells. If the tissue types are not identical, one may observe the "graft versus host" reaction in which the new white blood cells produced by the marrow transplant recognize the rest of the body as foreign and begin to destroy it. This rejection reaction is similar to the rejection of a heart or kidney transplant. In this case, however, the transplanted cells destroy the host's body rather than the host's white blood cells destroying the transplanted kidney or heart.

Many cancers are curable by conventional techniques. New experimental techniques such as immunotherapy and bone marrow transplants and better diagnostic methods offer new hopes for the future. We will turn to the cellular aspects of cancer now that we have a broad understanding of the many aspects of the disease.

Cellular Aspects: Cancer Cells versus Embryonic Cells

Tumor Spread and Embryonic Cell Migration

We have already mentioned that one of the major problems in dealing with malignant tumors is that cells separate from these tumors and spread to other parts of the body. In some ways this spread resembles certain embryonic cell migrations. Recall that primordial germ cells move from other regions in the embryo to the gonad area. Neural crest cells home to many areas of the body. Optic nerve cells migrate to specific regions of the optic tectum.

It is possible that the same mechanisms that appear to control embryonic cell migration may also control tumor spread. We discussed the experiments of Barbera, Marchase, and Roth, who showed that specific adhesive recognition may control homing of optic nerves from the retina to specific regions of the optic tectum (Chapters 7 and 9). These experiments involved rotating dorsal or ventral retina cells with dorsal or ventral chunks of optic tectum. In a similar way, Nicolson and Winkelhake showed that cells from primary tumors may form secondary

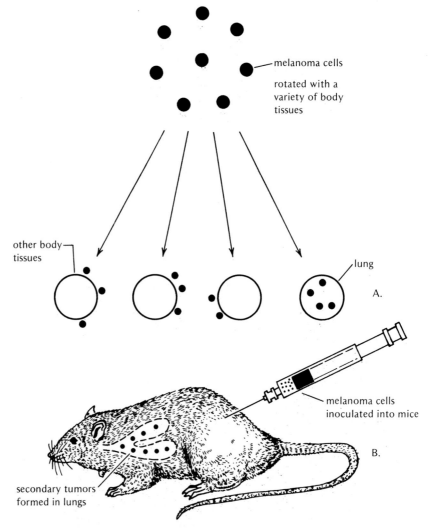

Figure 13–4. Melanoma tumor cells home to the lung and also stick best to lung tissue. A. Melanoma cells stick best to lung tissue when rotated with a variety of body tissues. B. Melanoma cells from secondary tumors most often in the lung. Based on experiments by Nicolson and Winkelhake, *Nature* 255:230–232 (1975.)

tumors at specific sites in the body because the tumor cells stick best to these specific sites. These experiments were done by rotating cells from tumors, such as malignant melanomas, with body tissue cells. It was found that the melanoma cells stuck better to lung cells than to other body cells (Figure 13–4). The lung is the site at which most secondary tumors develop from the line of spreading melanoma used in the experiments (Figure 13–5). Thus, adhesive recognition may control the sites of secondary tumor formation just as it may control homing of embryonic cells to specific regions of the embryo.

Another factor involved in tumor spread is the initial separation of cells from the primary tumor. Coman demonstrated, using microneedles

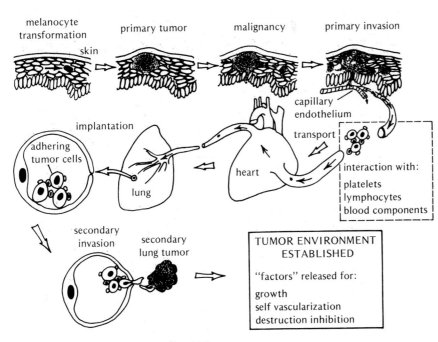

Figure 13–5. Metastatic spread of skin melanoma to lung. From Nicolson et al., *Cell and Tissue Interactions*, pp. 225–241, Raven Press, 1977.

Opp, Fig. 13.5

to separate cell pairs, that cells of tumors separate more easily than cells in the normal tissues from which the tumors are derived (Figure 13–6). Thus, cell adhesion, or more correctly tissue cohesion, of tumors is reduced. Experiments by Oppenheimer's and Roseman's groups suggest that reduced adhesiveness of some tumor cells may result from a decreased ability of the cells to synthesize, store, or transport L-glutamine, an amino acid that is required in the formation of complex carbohydrates needed for cell adhesion in the tumor lines studied (mouse ascites tumors) (Figures 13–7, 13–8).

Another factor that may be of importance in tumor spread is the finding that many tumor cells exhibit increased production of proteases, enzymes that catalyze the hydrolysis of proteins. Increased quantities of protease at the cell surface may assist the tumors in spreading and invading other regions of the body. Most tumor cells also exhibit mobile cell

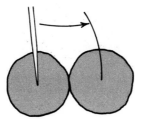

Figure 13–6. Cell pairs separated with microneedles. Tumor cell pairs required less force to be separated than normal cell pairs of the same tissue type. Based upon experiments by Coman (1944).

Figure 13–7. Mouse ascites (grown in body cavity) tumor cells rotated in different media. (this page) Media without L-glutamine. (opposite page) Media with L-glutamine. Mouse ascites tumor cells stick together (aggregate) when rotated in media containing L-glutamine. This suggests that these tumor cells have difficulty in synthesizing, storing or transporting L-glutamine. Most normal cells do not require added L-glutamine to adhere under these conditions. From Oppenheimer et. al., *Proc. Natl. Acad. Sci. U.S.* 63:1395–1402 (1969).

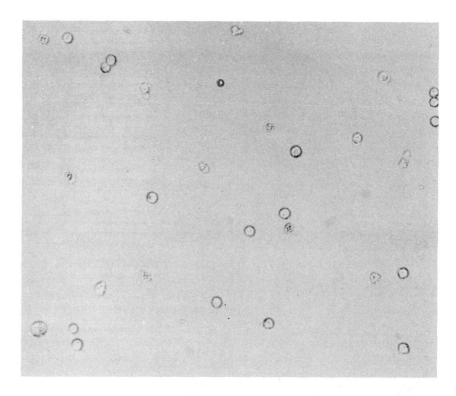

surface receptor sites. This can be demonstrated by using fluorescent tagged lectins that bind to carbohydrate-containing cell surface receptor sites. Using the tagged molecules, the receptors can be moved to one pole of the cell, just as was found by Roberson, Neri, and Oppenheimer for migratory embryonic cells (Chapter 3). Thus, both tumor cells and certain embryonic cells possess cell surfaces that do not restrict movement of specific receptors. These mobile cell surfaces may help tumor and embryonic cells to migrate.

Many normal cells contain cytoskeletons made up of organized bundles of microfilaments and microtubules. This can be observed using fluorescent antibodies that bind to the actin of microfilaments and the tubulin of microtubules. Most tumor cells, however, display a diffuse fluorescence pattern when stained with these antibodies. These results suggest that although the subunits of microfilaments and microtubules are present in tumor cells, these subunits are not organized into distinct microfilaments and microtubules. Microfilaments and microtubules may prevent cell surface receptor site mobility in normal cells by anchoring some of the cell surface sites from within the cell. In tumor cells, however, lack of an organized cytoskeleton may play a role in increasing cell

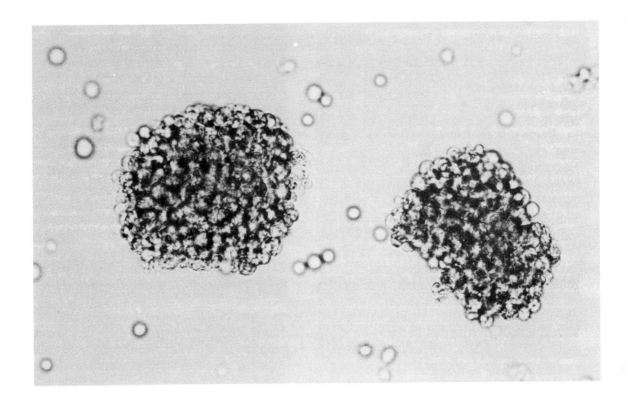

surface receptor site mobility and may permit increased cell migratory ability and continued growth.

We have already mentioned that tumor cells show less contact inhibition of growth and movement than many normal cells. This factor may be important in tumor spread. Cell-cell contact in normal tissues stops cell movement and growth. In tumors, spread may occur partly as a

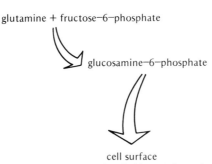

glutamine + fructose–6–phosphate

glucosamine–6–phosphate

cell surface
glycoproteins, glycolipids or polysaccharides
required for cell adhesion

Figure 13–8. Synthesis of cell surface molecules required for adhesion of mouse ascites tumor cells. The reaction is based upon many experiments with mouse ascites tumors. These pathways may be of general importance in controlling cell-cell adhesion.

result of the inability of contact to inhibit cell growth and movement. Contact inhibition may be controlled by cellular cyclic AMP. Virus-transformed tumor cells often contain less cyclic AMP than nontransformed cells. Some experiments show that contact inhibition of growth and movement of tumor cells can be restored by treating the tumor cells with derivatives of cyclic AMP. Some work suggests that cyclic AMP may allow cells to become contact inhibited by influencing the cytoskeleton. The relationship between the cell surface, cyclic AMP, the cytoskeleton, and cell growth and migration is a stimulating topic that is probably of major relevance to the cancer problem. One possible hypothesis regarding the nature of cell transformation might be that genetic changes cause cell surface changes, which in turn lower the levels of cellular cyclic AMP (probably resulting from lowered levels of the membrane bound enzyme, adenyl cyclase). Low cyclic AMP levels may then be responsible for many of the other cellular changes associated with the malignant state. It is likely that major breakthroughs in our understanding of the process of tumor cell transformation will come from studies in this area in the near future.

We should realize that many theories have been proposed to explain the mechanisms responsible for the altered behavior of cancer cells, such as unregulated growth and spread. In this chapter only a few of these have been described. Let us conclude this section by mentioning one last theory; that proposed by Holley. He suggests that a major factor that causes unregulated growth in tumor cells is increased uptake of nutrients from the medium or surrounding regions of the tumor. He suggests that if cells have a constant supply of the sugars, amino acids, vitamins, and other substances needed for growth, they will continue to grow. This hypothesis is supported by findings that do indeed indicate that many types of tumor cells take up significantly higher levels of nutrients than their normal counterparts. The permeability of the plasma membrane for many nutrients, therefore, appears increased in tumor cells. Whether this altered permeability is a major cause or just represents a secondary effect of the transformed state remains to be determined.

Embryo-like Tumors or Tumor-like Embryos?

We have already mentioned that teratomas are tumors derived from germ cells. These tumors often are indistinguishable from early embryos and may in fact develop rather normally like embryos if they are maintained in the proper place in the body. Teratomas and other tumors often also resemble embryos in terms of the presence of specific cell surface antigens (embryonic antigens). This suggests that some of the same genes are active in tumors that are active in embryos. Normal adult cells seldom display these antigens.

Mouse teratomas can be isolated from the testes of some strains of young mice and can be transplanted to other mice or maintained in culture. When teratoma cells are inoculated into the body cavities of mice, they develop into numerous clusters called "embryoid bodies". These free-floating clusters resemble normal 6-day old mouse embryos and differentiate into many tissues if grown in the eye cavity of mice. If maintained in the body cavity, however, embryoid bodies continue to proliferate rapidly. A variety of experiments indicate that these teratomas are derived from germ cells. Mintz and Illmansee isolated teratoma cells and injected these cells into early mouse embryo blastocysts from another strain of mouse. These blastocysts containing the teratoma cells were then implanted into the uteri of other female mice. The embryos developed into normal tumor-free adult mice, with some of their tissues derived from the normal mouse embryo cells and others derived from the teratoma cells (Figure 13–9). The teratoma cells had markers specific for their genotype and the normal blastocyst cells also had specific markers. The markers included strain-specific enzyme patterns, hemoglobin,

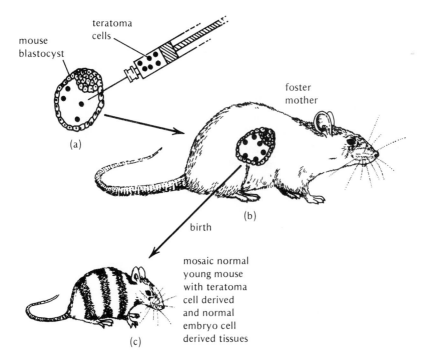

Figure 13–9.
Development of normal mouse from teratoma cells and normal mouse embryo cells. (a) Teratoma cells (black) are inoculated into normal mouse blastocyst embryo (white). (b) The embryo (including teratoma cells) is grown in the uterus of a female mouse. (c) The mouse developing from the combination consists of normal tissues, some derived from teratoma cells (black), some from normal embryo cells (white). Based on experiments by Mintz and Illmansee.

Figure 13–10. Some cellular changes associated with cancer cells.

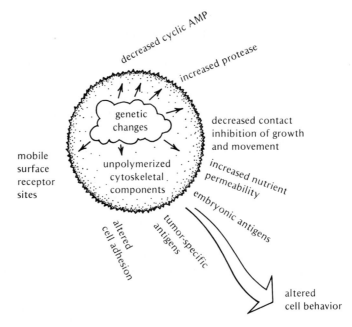

immunoglobulins, and pigment. Thus, normal mice developed with some of their cells originating from a type of tumor! The important thing to remember about this experiment is that teratoma tumor cells can differentiate into normal adult tissues. Here we see the ultimate relationship between normal embryo cells and certain tumor cells. Both can divide and remain relatively undifferentiated under some conditions, while under others they can differentiate into normal adult tissues.

In fact, normal six day old mouse embryo cells will develop into teratomas if implanted into the testes of adult mice! Older embryo cells, however, do not form tumors under these conditions. Here again we see that young embryo cells and certain tumor cells are very similar and can grow in an unregulated way or can differentiate into normal adult tissue depending upon the environment in which the cells reside.

Teratomas are being used in many studies designed to investigate the nature of the malignant versus the benign state and the nature of the conditions required for cell growth and cell differentiation. Teratoma cells provide excellent material for these studies because they can be obtained in enormous numbers by growing them in the body cavity of mice or in tissue culture. These cells can remain very malignant if maintained in the free-floating form in the body cavity, or can differentiate into normal tissues if implanted into embryo blastocysts or under the skin. For these reasons it is likely that teratomas will provide us with many answers to the relationship between the malignant and benign state, and between cell growth and cell differentiation.

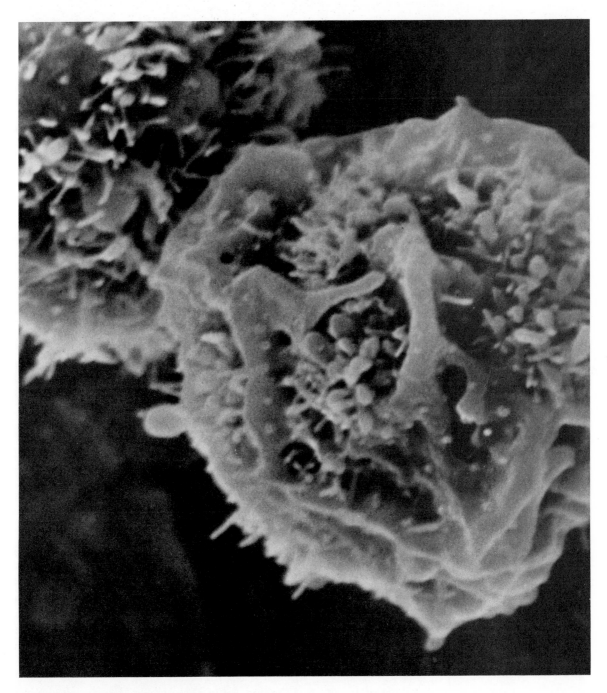

Figure 13–11 (above and overleaf). Mouse ascites tumor cells. Note the active cell surfaces.

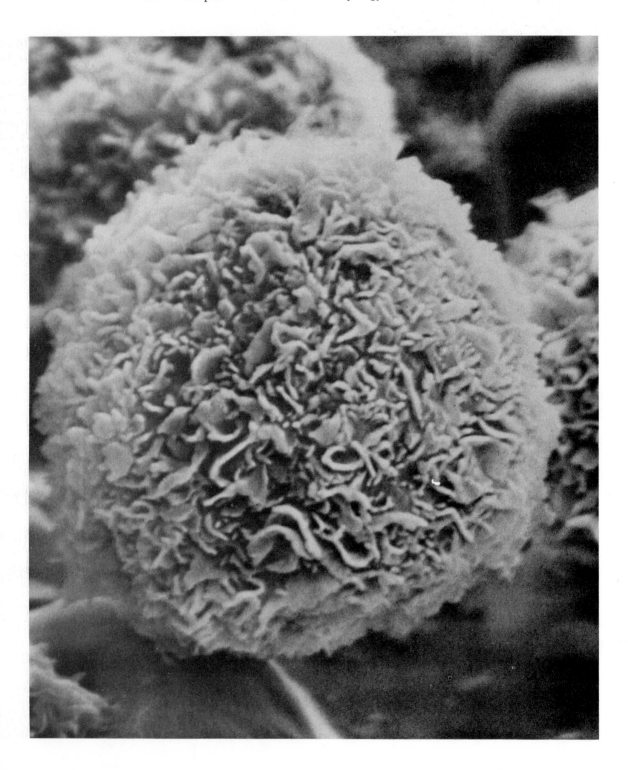

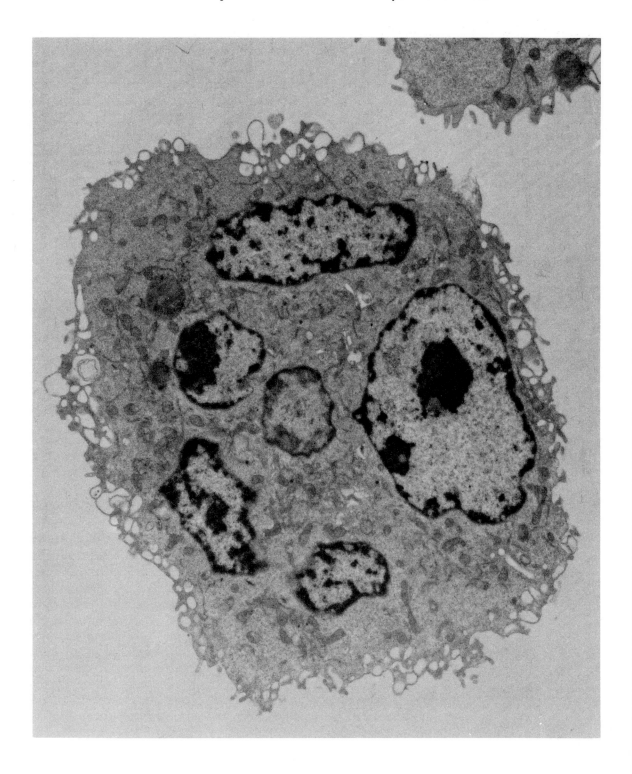

Figure 13–12. Some changes found after transformation to cancerous state. From Nicolson, G., *Biochem. Biophys. Acta,* 458:1–72 (1976). The reader is referred to this article for further discussion about the topic.

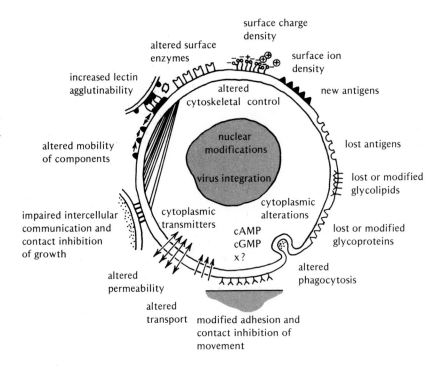

Summary

Many agents can cause cancer. These agents probably act on the genes. Genetic changes result in a variety of altered cellular properties that cause altered cell behavior (Figures 13–10, 13–11, 13–12). It is clear that embryonic cells resemble tumor cells in many respects. Let us keep in mind, however, that some tumors do not appear to possess certain embryonic cell characteristics and it is unlikely that activation of embryonic genes is the only reason for all cancer. Such a conclusion is supported by the appearance of tumor-specific antigens (in addition to embryonic antigens) on the surfaces of tumor cells. These antigens are usually not found on embryonic cells. The similarities between many types of tumor and embryonic cells, however, are also quite significant as illustrated by teratoma experiments, cell surface receptor site mobilities, and other characteristics described in this chapter. It is clear that the study of embryology and the study of cancer are inextricably intertwined. Information from one area often leads to a better understanding of the other.

Readings

Baltimore, D., Tumor Viruses. *Cold Spring Harbor Symp. Quant. Biol.* 39:1187–1200 (1975).

Baltimore, D. Viruses, Polymerases and Cancer. *Science* 192:632 (1976).

Coman, D. R., Decreased Mutual Adhesiveness, A Property of Cells from Squamous Cell Carcinomas. *Cancer Res.* 4:625–629 (1944).

German, J. ed., *Chromosomes and Cancer*. John Wiley and Sons, New York (1974).

Heidelberger, C., Chemical Carcinogenesis. *Annu. Rev. Biochem.* 44:79 (1975).

Holley, R. W., A Unifying Hypothesis Concerning the Nature of Malignant Growth. *Proc. Natl. Acad. Sci. U.S.* 69:2840 (1972).

Koprowski, H., Ed., *Neoplastic Transformation: Mechanisms and Consequences.* Pergamon, Elmsford, N.Y. (1977).

Markert, C. L., Neoplasia: A Disease of Cell Differentiation. *Cancer Res.* 28:1908–19 (1968).

Nicolson, G. L. Transmembrane Control of the Receptors on Normal and Tumor Cells. II. Surface Changes Associated with Transformation and Malignancy. *Biochim. Biophys. Acta* 458:1–72 (1976).

Nicolson, G. L., Cell and Tissue Interactions leading to Malignant Tumor Spread (Metastasis). *Amer. Zool.* 18:71–80 (1978).

Oppenheimer, S. B., M. Edidin, C. W. Orr, and S. Roseman, An L-Glutainime Requirement for Intercellular Adhesion. *Proc. Nat. Acad. Sci. U.S.* 63:1395–1402 (1969).

Robbin, R., I. N. Chow, and P. H. Black, Proteolytic Enzymes, Cell Surface Changes and Viral Transformation. *Adv. Cancer Res.* 22:203–260.

Sheppard, J. R., Restoration of Inhibited Growth to Transformed Cells by Dibutyryl Adenosine 3', 5'-Cyclic Monophosphate, *Proc. Nat. Acad. Sci. U.S.* 68:1316–1320 (1971).

Stevens, L. C., Spontaneous Parthenogenesis and Teratocarcinogenesis in Mice. *33rd. Symposium, The Society of Developmental Biology*, Academic Press, New York (1975).

Weber, K., E. Lazarides, R. E. Goldman, A. Vogel and R. Pollack, Localization and Distribution of Actin Fibers in Normal, Transformed and Revertant Cells. *Cold Spring Harbor Symp. Quant. Biol.* 39:363–369 (1975).

GLOSSARY

This glossary consists mainly of terms used in the text, and is not intended to be a complete listing of anatomical or embryological terms but instead is designed to aid the student in reading this book. More complete discussions of the terms in the context of developmental biology are given in the text. This glossary should not preclude the use of an unabridged dictionary, because definitions are given here in the context of embryology.

acinus: (pl. **acini**) Minute saclike structure.

acrosome: The anterior tip of a mature sperm cell that functions to aid the sperm in penetrating through the outer egg coats and in establishing connection with the egg cytoplasm.

actin: Protein subunit of microfilaments.

actinomycin D: An RNA synthesis inhibitor.

adhesion-promoting factors: Molecules that cause cells to stick together.

adrenal gland: Endocrine gland on or near the kidney.

adult hemoglobin: Hemoglobin found in adults.

albumen: White of an egg such as that of birds; a type of protein.

allantois: Extraembryonic saclike extension of the hindgut of amniotes, serving excretion and respiration.

amnion: Inner membrane surrounding the embryo.

amniotes: Vertebrates possessing an amnion during development.

amniotic cavity: Space between the amnion and the embryo proper.

amphibia: Class of vertebrates intermediate in many characteristics between the fishes and reptiles, which live part of the time in water and part on land.

amphioxus: Cephalochordates. Small, fishlike creatures that are often called lancelets or lancet fish.

anamniotes: Vertebrates that lack an amnion during development.

anaphase: A stage in mitosis or meiosis in which the chromatids of each chromosome separate and move to opposite poles.

androgens: Male sex hormones.

angiosperm: Member of a class of plants having the seeds in a closed seed vessel.

animal hemisphere: Region of the egg where the nucleus resides.

anneal (reanneal): Reassociation of nucleic acid strands upon slow cooling.

anther: Pollen-bearing part of a stamen.

antibody: Protein that immunizes the body against a specific foreign substance.

antifertilizin: Sperm cell surface molecule that binds to egg fertilizin.

antigen: Substance that stimulates the production of and reacts with antibodies.

anucleolate: Without nucleoli.

anura: Amphibia without tails, such as frogs and toads.

anus: Rear opening of digestive tube.

aorta: A main trunk of the arterial system.

apical ectodermal ridge: Thickened epidermis at the tip of limb buds.

archenteron: Primitive or embryonic digestive tube.

archenteron roof: Prospective notochord.

area opaca: Peripheral zone of chick blastoderm which is attached to yolk underneath.

area pellucida: Relatively transparent central region of chick blastoderm underlaid by the subgerminal space.

atrium: Heart chamber. Sometimes also applied to a chamber in other organs.

autonomic: Self controlling and independent of outside influences.

auxin: A plant growth hormone, indolacetic acid.

aves: The class of birds.

axon: The outgrowth of a nerve cell, that conducts impulses away from the cell body.

B-cell (bursa lymphocyte): Lymphocyte active in humoral immunity.

benign tumor: Slow growing, nonspreading growth that does not harm the host.

bilateral cleavage: Cleavage pattern that results in the formation of an embryo with a right and left side that are mirror images of each other.

bile: Digestive fluid secreted by the liver.

bindin: Protein isolated from sperm acrosomal granules that may function in helping the sperm to recognize and adhere to the egg cell surface.

bipolar neuron: Nerve cell with an axon at one end and a dendrite at the other.

blastema: Primitive aggregation of cells from which an organ develops.

blastocoel: Cavity present in the blastula-stage embryo.

blastocyst: Blastula-like stage of the mammal embryo.

blastocyst cavity: Blastocoel-like cavity in mammalian blastula.

blastoderm: Primitive cellular plate of early embryos.

blastodisc: A plate of cytoplasm at the animal pole of the egg.

blastomere: One of the cells that results from cleavage of the zygote.

blastopore: Opening into the archenteron from outside the embryo.

blastula: Embryonic stage in which the embryo is a hollow ball with a cavity. In some embryos the blastula stage resembles a cap of cells rather than a ball.

block to polyspermy: Prevention of more than one sperm from entering the egg.

bone marrow transplant: Replacement of an individual's diseased bone marrow with bone marrow from a compatible donor, after destroying the host individual's marrow with radiation and chemicals.

Bowman's capsule: Expanded and invaginated portion of kidney tubule that contacts the glomerulus.

BUdR: 5-bromodeoxyuridine, a thymidine analog.

bursa of Fabricius: Gland involved in the immune system of birds.

calyx: Outermost series of leaflike parts of a flower.

capsella: Genus of weedy flowering plant, known as shepherd's purse.

carbohydrate binding proteins: Lectins; proteins that can combine with carbohydrates.

carcinogen: A cancer-causing agent.

carcinoma: Tumor derived from epithelial tissue.

carpel: A simple pistil or seed vessel.

cartilage: Gristle-like skeletal tissue.

cDNA: Complementary DNA; DNA made using RNA as a template.

cell-cell recognition: Factors that allow interaction between specific cells and not other cells.

cell motility: Cell movement.

cellular immunity: Direct attack by lymphocytes on a foreign cell.

centriole: Minute cytoplasmic granules surrounded by a zone of gelated cytoplasm. Microtubule-containing organelle located at the spindle poles in dividing cells or embedded at the cell periphery, and forming the basal portion of a cilium or flagellum.

centrolecithal eggs: Eggs with yolk concentrated in the interior, such as insect eggs.

cerebellum: Brain region derived from the roof of the metencephalon.

cerebral hemispheres: Brain regions derived from the roof of the telencephalon.

chalaza: The tissue supporting the yolk in a bird's egg.

chemotaxis: Movement towards increasing concentration of a chemical.

chondroblast: Embryonic cartilage-forming cell.

chondrocyte: A cartilage cell.

chordata: Phylum of animals with a permanent or transient notochord.

choroid coat: An outer coat of the eyeball that develops from mesenchyme.

chorion: A surface coat exterior to the plasma membrane in the eggs of fishes and tunicates. The term is also used to describe an extraembryonic membrane made of somatopleure that surrounds some embryos.

chromatid: One of the two daughter strands of a duplicated chromosome.

chromosome: A deeply staining rod or thread in the cell nucleus that contains the genes.

ciliary body: Structure involved in controlling the shape of the lens of the eye.

cilium: A whiplike locomotor organelle produced by a centriole.

cleavage: Division of the fertilized egg.

cleft: A split (in embryonic tissue).

cleidoic eggs: Eggs such as those of birds that are insulated from the environment by albumen, membranes, and shell.

clitoris: Organ composed of erectile tissue, the homologue in females of the penis.

cloaca: Combined urogenital and rectal receptacle.

clonal selection hypothesis: Theory of production of antibodies, that holds that cells that produce a given antibody all derive from a single cell; contact with an antigen causes that specific cell to divide, producing daughter cells that recognize that antigen.

coelom: Body cavity.

colchicine: Drug that disrupts microtubules.

collagen: A fibrous protein present in connective tissue.

Competition-hybridization experiment: Experiment in which cold (non-radioactive) RNA is used to compete with radioactive RNA for binding sites on a DNA molecule. Such experiments are used to suggest base sequence similarities and differences among nucleic acid molecules.

contact guidance: Cell growth in response to mechanical components or to oriented components of the substratum.

contact inhibition: The inhibition of growth or movement of cells upon contact with other cells.

contractility: The ability to draw together into a more compact form.

conus arteriosus: Truncus arteriosus. Anterior-most portion of the heart.

coracoid process: Bony structure extending from the shoulder blade to the breastbone.

cornea: Transparent front covering of the eye.

corolla: The circle of flower leaves, usually colored, forming the inner floral envelope.

corona radiata: Follicle cell layer surrounding an ovulated mammalian oocyte.

corpus luteum: Endocrine capsule formed from an ovulated follicle.

cortex: Outer portion of an organ or part.

cortical granule: Structure in the surface cytoplasm, just below the plasma membrane of eggs, that functions in the fertilization reaction and in formation of a fertilization membrane.

CoT: Concentration × Time with value in moles per liter × seconds.

cotyledons: Embryo leaves.

crossing over: The exchange of segments of genetic material between the chromatids of homologous chromosomes in the prophase of meiosis.

crystallins: Proteins in the lens of the eye.

cycloheximide: A protein synthesis inhibitor.

cytochalasin: Drug that disrupts microfilaments but also has other effects on cellular processes.

cytotrophoblast: Inner or cellular layer of the trophoblast.

dendrite: A receiving branching process of a nerve cell.

dermatome: An embryonic skin segment. Also used to specifically denote the outer region of the somite that gives rise to the dorsal dermis of the skin.

dermis: Inner or lower layer of skin.

diabetes: Disease resulting from lack of insulin production.

diencephalon: Posterior portion of forebrain.

digit: Finger or toe.

diploid: The normal number of chromosomes in all body cells except the gametes.

diplotene: The stage in the prophase of meiosis, following pachytene, in which the chromosomes are distinctly double and begin to separate from one another.

distal: Farthest from the center or median line.

DNA tumor virus: Tumor-causing virus with a DNA core.

dorsal: Relating to the top or back.

dorsal lip of the blastopore: Prospective notochord; the first region to enter embryo during gastrulation.

Down's syndrome: Trisomy 21 or Mongolism; individuals with three instead of two chromosomes in chromosome pair number 21.

ductus deferens: The sperm duct.

duodenum: The first division of the small intestine.

ectoderm: Outer germ layer of the embryo.

egg: The mature haploid female gamete.

embryology: The study of the origin and development of the embryo.

embryonic field: An embryonic area that has the ability to regulate itself, and possesses the four properties given on page 186.

embryonic hemoglobin: Hemoglobin found in embryos that differs in structure from adult hemoglobin.

embryonic induction: Stimulation of differentiation in one tissue as a result of interaction with another tissue.

endocardium: Inner lining of heart cavities.

endoderm: Innermost germ layer of the embryo.

endometrium: Lining of the uterus.

epiblast: Upper embryonic region of the blastoderm.

epidermis: The outer epithelial portion of the skin.

epididymis: First convoluted portion of the excretory duct of the testis, passing from above downward along the posterior border of the testis.

epimere: Somite. Dorsal region of mesoderm on each side of the neural tube consisting of myotome, dermatome, and sclerotome.

equal cleavage: Division that divides zygote into cells of similar size.

erythropoietin: Glycoprotein hormone that stimulates synthesis of globin messenger RNA.

esophagus: Portion of gut between pharynx and stomach.

estrogen: Hormone that influences estrus or produces changes in the sexual characteristics of female mammals.

estrus: Period of sexual excitement in the female.

eucaryotic cell: Cell of higher organisms (those above the bacteria and blue-green algae).

extraembryonic coelom: Cavity outside the embryo proper that is continuous with the body cavity (coelom) and is surrounded by extraembryonic membranes.

eye cups: Cup-like structures formed by invagination of the optic vesicles.

F-actin: actin filament.

fallopian tube: Oviduct.

fate map: Projection of what specific areas on earlier embryo will become at later stages.

femur: Thigh bone or upper leg bone.

fertilization: Union of egg and spermatozoan.

fertilizin: Acid mucopolysaccharide in egg jelly that binds sperm.

fetus: Embryo of an animal before birth; embryo in the uterus at later stages in development.

fibroblast: Primitive mesenchymal cell giving rise to connective tissue.

fibula: External and smaller of the two bones of the lower leg.

filopodia: Long cellular projections.

flagellin: A globular protein that makes up flagella.

flagellum: Long, threadlike structure that protrudes from the cell body.

fluid-mosaic model of the cell surface: Theory that describes the cell surface as composed of a lipid bilayer with mobile proteins that are attached to and imbedded in the lipid.

follicle: Vesicular body in the ovary containing the developing oocyte.

follicle cells: Cells that surround oocytes and function to nourish and protect the developing oocytes.

follicle-stimulating hormone: A hormone produced by the anterior pituitary gland that functions in the control of reproductive events. It plays a major role in stimulating the maturation of the oocyte and its follicle.

freeze-fracture technique: Cell microscopy technique that involves splitting of the membrane down the middle and peeling away of one of the lipid bilayers. The surface of the cleaved membrane is coated with heavy metal, forming a replica of the contours of the fractured surface. The replica is examined with an electron microscope.

Fucus: A brown seaweed.

FUdR: 5-fluorodeoxyuridine, a DNA synthesis inhibitor.

gall bladder: Sac-like vessel associated with the liver, which stores bile.

gamete: An egg cell or sperm cell.

ganglion: Aggregation of nerve cells along the course of a nerve.

gastrula: Embryonic stage, following the blastula, consisting of a layered sac.

gastrulation: Transformation of a blastula into a layered embryo, the gastrula.

gene: DNA segment, the hereditary unit in chromosomes, with a base sequence that codes for a polypeptide chain.

gene cloning: Production of many copies of a given gene.

germinal epithelium: Surface of genital ridge; outer part of primitive gonad.

globin: Polypeptide chains of hemoglobin.

glomerulus: Tuft of capillary loops at the beginning of each kidney tubule.

glomus: Network of fine blood vessels associated with the ciliated funnels of the pronephric tubules.

glucagon: Pancreatic hormone involved in regulating blood sugar.

glycogen: A complex carbohydrate (polysaccharide) that serves as a store of energy-rich carbohydrate.

glycosaminoglycans: Sugar polymers consisting of uronic acids and amino acids.

glycosyl transferase: Enzyme that catalyzes the transfer of single sugar residues from nucleotide sugars to the growing ends of carbohydrate chains.

glycosyl transferase-carbohydrate acceptor model: Cell adhesion model that states that cell-cell adhesion may be mediated through glycosyl transferases on one cell binding to carbohydrate chains on an adjacent cell.

gonad: A reproductive gland; the ovary or testis.

gonadotropic: Refers to a hormone that influences the gonads.

graafian follicle: Mature mammalian ovarian follicle.

gray crescent: A grayish surface cytoplasmic region of the amphibian embryo, exposed soon after fertilization, that gives rise to the dorsal lip of the blastopore. Prospective notochord.

gymnosperm: Member of a class of plants whose ovules and seeds are not enclosed in a case, as certain evergreens.

haploid: Half the chromosome number of normal body cells. Mature sperm and egg cells have haploid chromosome number. At fertilization the normal number of chromosomes (diploid) is restored in the zygote.

heme: Nonprotein, iron-containing chemical group that binds oxygen.

hemoglobin: Red blood cell protein (globin) combined with a non-protein heme group that binds oxygen.

Henson's node: Anterior thickened end of the primitive streak.

histone: A major type of chromosomal protein.

holoblastic cleavage: Total cleavage; divisions pass through the entire zygote.

homologous chromosome pairs: Chromosomes of identical size and shape of maternal and paternal origin.

humerus: Upper arm bone.

humoral immunity: Immunity through synthesis of free soluble antibody that combines with foreign antigens.

hyaline layer: Clear layer that is present, for example, at the surface of sea urchin embryos, derived from cortical granule material.

hybridization experiments: In nucleic acid technology, those experiments in which comparisons of nucleic acid base sequences can be made by measuring the binding of one nucleic acid strand with another.

hypoblast: Lower embryonic region of blastoderm.

hypomere: The most ventral subdivision of the mesoderm consisting of somatic and splanchnic mesoderm.

hypothalamus: Ventral portion of the thalamus of the diencephalon of the brain.

ileum: Lowest division of the small intestine.

immunoglobulin: An antibody.

incomplete cleavage: Cleavage in which divisions do not pass through the entire zygote.

indifferent stage of sexual development: Stage at which the gonad has not yet differentiated into a testis or ovary.

induction: The process whereby the interaction of an inducing tissue with a responding tissue causes differentiation of the responding tissue.

inner cell mass: Inner part of the mammalian embryo that forms the embryo proper.

instructive information: Genetic information provided by an inducing tissue to a responding tissue.

insulin: Pancreatic protein hormone functioning in carbohydrate metabolism.

integral membrane protein: Protein that is very strongly associated with or deeply imbedded in the lipid bilayer of the cell membrane.

intermediate mesoderm: Mesomere. The region of mesoderm between the epimere and hypomere.

intestine: The digestive tube passing from stomach to anus.

invagination: The process of folding or buckling in so that an outer surface becomes an inner surface, as in gastrulation.

invasion: The spreading of tumor tissue by cellular crawling to nearby tissues.

iris: Pigmented disk-like diaphragm of the eye that controls the size of the pupil.

islets of Langerhans: Insulin-producing cells of the pancreas.

isolecithal egg (oligolecithal egg): An egg that possesses a small amount of evenly distributed yolk.

isozymes: Enzymes that exist in multiple molecular forms.

jejunum: Middle division of the small intestine.

karyotype: Chromosome size, shape, and number.

keratin: A hard, relatively insoluble protein, present largely in cutaneous (skin) structures.

kinetin: A plant hormone.

Klinefelter's syndrome: A condition resulting from an XXY karyotype.

Lactate dehydrogenase (LDH): Enzyme that catalyzes the interconversion of pyruvate and lactate.

lampbrush chromosomes: Extended diplotene chromosomes that appear during oogenesis in some species. These chromosomes possess loops that develop perpendicular to the long axis, and which are actively synthesizing RNA.

lateral: On the side.

lateral lips of the blastopore: Side lips of the blastopore.

lectin: Carbohydrate-binding protein.

lens-forming ectoderm: Prospective lens.

leptotene: Stage in the prophase of meiosis immediately preceding homologous chromosome pairing (synapsis) in which the chromosomes appear as fine threadlike structures.

leukemia: Tumor of blood forming tissues in which white blood cell number greatly increases.

leukocyte: A white blood corpuscle.

Leydig cells: Interstitial cells of the testis.

liquor folliculi: Liquid in ovarian follicle.

liver: Largest glandular organ in vertebrates, which secretes bile and is active in metabolism.

luteinizing hormone (LH): A hormone produced by the anterior pituitary gland that functions in the control of reproductive events. It plays an important role in causing ovulation to occur; in the male it stimulates interstitial cells in the testis to synthesize male sex hormones.

lymph node: A glandlike body involved in the immune system.

lymphocyte: A white blood cell, formed in lymph node tissue, involved in the immune response.

lysins: Enzymes present in acrosomal granules that aid sperm in penetrating the outer egg coats.

macromeres: Large sized cleavage blastomeres.

macrophage: A large blood cell that engulfs and destroys other cells.

malignant tumor: Spreading tumor that can cause the death of the host.

Malpighian body: Bowman's capsule plus its glomerulus; the functional renal unit.

mammalia: Highest class of living organisms, including all the vertebrates that suckle their young.

marsupialia: Order of mammals featuring an abdominal pouch for nurture of the young.

medulla: Inner portion of an organ or part. In the brain, the region that forms from the myelencephalon.

meiosis: Divisions in the germ cell line that eventually result in the formation of haploid gametes.

melanin: Dark pigment in skin.

melanoma: Tumor derived from pigment cells.

melanophore: Cell containing melanin.

menstruation: Periodic discharge of blood, necrotic tissue and glandular secretions from the uterus; a part of the reproductive cycle in female primates.

meristem: Undifferentiated plant tissue that gives rise to stem and root tissue.

meroblastic cleavage: Cleavage that involves only a small cytoplasmic area at the animal pole of the egg.

mesencephalon: Midbrain.

mesenchymal factor: Factor that stimulates cell division and differentiation of pancreas epithelium.

mesenchyme: Embryonic connective tissue.

mesendoderm: Combination of prospective mesoderm and prospective endoderm.

mesentery: Double layer of peritoneum enclosing an organ.

mesocardium: Mesentery (two-layered membrane) supporting the heart.

mesoderm: Middle germ layer of the embryo.

mesomeres: Medium sized cleavage blastomeres. Also, the intermediate mesoderm, the region of mesoderm between the epimere and hypomere.

mesonephric duct: Duct connecting mesonephric tubules and cloaca.

mesonephric tubules: Kidney tubules of adult fish and amphibians.

mesonephros: Second stage in the development of the amniote kidney; functional kidney of adult fish and amphibians.

messenger RNA: RNA molecules containing the genetic information needed to code for the synthesis of a polypeptide chain.

metamorphosis: Marked change in form and structure in an animal in its development from embryo to adult.

metanephros: Last stage in development of the amniote kidney.

metaphase: The stage of mitosis or meiosis during which the chromosomes lie in the central plane of the spindle.

metastasis: Tumor spread by entering the bloodstream.

metencephalon: Anterior portion of the hindbrain.

microfilaments: Rod-like elements composed of the protein actin.

micromeres: Small sized cleavage blastomeres.

micropinocytosis: Intake of very small fluid droplets by cells.

microtubules: Cylindrical units composed of globular subunits of the protein tubulin.

microvilli: Fine projections or extensions of the surface of cells.

mitochondrion: Cytoplasmic organelle envolved in energy metabolism.

Mongolism: Trisomy 21 or Down's syndrome; a condition appearing in humans with three instead of two chromosomes in chromosome pair number 21.

morphogenesis: Development of form and structure.

morphology: The study of the form or structure of organisms.

morula: Solid ball of blastomeres (about the 16-cell stage) that precedes the blastula stage.

Mullerian duct: Oviduct.

mutant: An organism or gene that has undergone mutation (chromosomal change that can be hereditary).

myelencephalon: Posterior portion of the hindbrain.

myoblast: Cell that gives rise to muscle.

myocardium: Muscle layer of the heart.

myofibril: Contractile muscle fiber.

myosin: A fibrous protein composed of two identical subunits wound around each other.

myotome: Prospective or actual muscle segment. Also used to specifically denote the inner part of the somite that forms back muscles.

myotube: Muscle tubes containing many muscle nuclei in a common cytoplasm.

nephrocoel: Internal cavity of the nephrotome.

nephrostome: Funnel-shaped opening by which a kidney tubule communicates with the nephrocoel.

nephrotome: Segmented portion of intermediate mesodem giving rise to the kidney tubule.

nerve growth factor: Protein that stimulates nerve outgrowth from ganglia.

nests of oogonia: Clusters of oogonia surrounded by follicle cells.

neural crest: Tissue that forms from the region of ectoderm (neural folds) between prospective epidermis and prospective neural tube.

neural folds: Elevated ridges of the neural plate.

neural groove: Trough formed by the bending up of the neural plate.

neural plate: Embryonic region that becomes the nervous system.

neural retina: Inner sensory layer of the eye cup.

neural tube: Tube, derived from the neural plate, that forms the nervous system.

neuroblast: Embryonic nerve cell.

neuron: Cellular unit of the nervous system.

neurulation: The stage or process of neural tube formation.

non-histone protein: Low molecular weight (7,000–15,000 daltons) proteins associated with DNA. Most are phosphorylated and acidic.

non-notochordal mesoderm: Mesoderm (middle germ layer) that gives rise to a variety of derivatives other than the notochord.

notochord: Fibrocellular rod constituting the primitive skeletal axis.

nucleolus: Dense spherical granule in the cell nucleus composed of ribosomal RNA and protein.

nucleosomes: Bead-like particles making up eucaryotic chromosomes.

nucleus: Large body within the cell containing the chromosomes that include the hereditary information. The nucleus also usually contains one or more nucleoli.

nurse cells: Cells in organisms such as insects that function to nourish the developing oocyte.

olfactory: Relating to the sense of smell.

oligolecithal (isolecithal) egg: An egg that possess a small amount of evenly distributed yolk.

ontogeny: Development of the individual.

oncogene theory: Theory that all cells of animals contain cancer genes (oncogenes) that can transform cells into cancer cells if the genes are activated.

oocyte: Stage in the maturation of female gamete.

oogenesis: Development of egg cells.

oogonia: Primordial egg cells that give rise to oocytes and eggs.

opistonephros: Adult kidney in anamniotes.

optic chiasma: Where the nerve fibers from both eyes cross.

optic tecta: Visual interpretation centers in the midbrain.

optic vesicle: Outpocketing of the diencephalon of the brain that forms the retina of the eye.

organizer: Spemann and Mangold's term for the dorsal lip of the blastopore.

organogenesis: The formation of organs.

ostium tubae: Funnel shaped openings of the oviducts.

otic: Relating to the ear.

oviduct: Tube that transports the egg and embryo.

ovule: In plants, the body in the ovary that upon fertilization gives rise to the seed.

ovum: Egg.

pachytene: The stage in meiotic prophase, immediately following pairing of homologous chromosomes (synapsis), in which the chromosome threads become shortened and thickened and crossing over occurs.

pancreas: Abdominal digestive and endocrine gland.

parathyroid: Endocrine gland adjacent to the thyroid.

parthenogenesis: Development without fertilization.

penis: Male organ of copulation.

pericardial: Surrounding the heart.

perichondrium: The fibrous membrane that covers cartilage.

peripheral membrane protein: Protein that is loosely associated with the cell membrane.

peritoneum: Lining of the body cavity and covering of the organs.

permissive information: Information that turns on genes in a responding tissue.

petal: A division of the corolla, the set of modified leaves, usually colored, of a flower.

pharynx: Anterior portion of the foregut.

phloem: Plant tissue that conducts sap.

physiological polyspermy: A method of fertilization in some organisms whereby many sperm enter the egg, and all but one disintegrate after entry.

pigmented retina (pigmented layer of the eye): Outer layer of eye cup.

pinocytosis: Intake of fluid droplets by cells.

pistil: Seed-bearing organ of flowering plants.

pituitary: Endocrine gland at the base of the forebrain that secretes several growth hormones.

placenta: Organ of physiological communication between mother and fetus in mammals.

placode: Thickened platelike area of epithelium that gives rise to some structures.

plasma cell: A mature antibody-secreting B lymphocyte.

polar body: Tiny nonfunctioning cell with little cytoplasm produced during meiotic divisions in egg cells.

polar lobe: A cytoplasmic extrusion that is formed in some spirally cleaving embryos.

pollen: Male gamete of seed plant.

polyploid: More than the normal number of chromosomes.

polysome: Group of ribosomes active in protein synthesis.

polyspermy: Fertilization of an egg by more than one sperm.

pouches: Sac-like structures that form in embryos.

primary mesenchyme cells: Cells, derived from the micromeres, that form the skeletal elements in sea urchins and related embryos.

primary oocytes: Cells that form as a result of growth and DNA duplication of oogonia. These cells precede the first meiotic division.

primary sex cords: Inner part of primitive gonad.

primary spermatocytes: Cells that form as a result of growth and DNA duplication of spermatogonia.

primary tumor: Tumor in a host that develops from transformation of normal cells into tumor cells.

primates: Highest order of mammals, including humans, monkeys, and lemurs.

primitive groove: Indentation in the midline of the primitive streak.

primitive streak: Thickening in surface of some embryos at the beginning of gastrulation.

primordial germ cells: Primitive cells that give rise to the gametes.

proamnion: Crescent shaped area around head of early bird embryos.

procaryotic cell: Primitive bacteria or blue-green algae cell.

proctodeum: Terminal portion of rectum formed in the embryo by an ectodermal invagination.

progesterone: A hormone of the corpus luteum that helps induce changes in the uterine lining, preparing it for development of the embryo.

pronephric duct: The duct connecting the pronephric tubules and the cloaca.

pronephric tubules: The functional units of the kidney of aquatic larvae.

pronephros: First stage in development of amniote kidney: the functional kidney of fish and amphibian embryos.

pronucleus: The nucleus of the egg or sperm prior to fertilization.

prophase: Early phase of nuclear division characterizaed by condensation of the chromosomes and movement of the chromosomes towards the equator of the spindle.

prosencephalon: Forebrain.

prospective (presumptive) region: A region of an early embryo that will become a specific structure in the more advanced embryo.

prostate: Gland that surrounds the beginning of the urethra in the male, that in mammals produces a secretion that activates sperm in the seminal fluid.

protease: Enzyme that catalyzes the hydrolysis of protein.

protein kinase: An enzyme that catalyzes the phosphorylation of soluble yolk (phosvitin) into an insoluble form that can be stored in the egg.

proteoglycans: Glycosaminoglycans linked to proteins.

protoplast: Wall-free plant cell.

proximal: Nearest the trunk or point of origin.

pupil: Opening in the front of the eye that allows light to enter.

puromycin: A protein synthesis inhibitor.

radial cleavage: Cleavage that divides the embryo into an upper and lower tier of cells. The upper cells lie exactly over the lower cells. The cells are uniformly distributed around the polar axis of the egg.

radius: Forearm bone.

rectum: Terminal portion of large intestine connecting with the anus.

regeneration blastema: Undifferentiated cells together with their epidermal covering that accumulate at the surface of a wound.

regional inducing specificity: The ability of a specific region of a structure to induce a specific differentiation in a responding tissue.

renal: Pertaining to the kidney.

reptilia: Class of vertebrates including the alligators, crocodiles, lizards, turtles, and snakes.

rete cord cells: Cells that take part in testis formation in males.

rete testis: Netowrk of tubules in the testis, formed from rete cord cells.

retina: Light-sensitive layer of the eye.

reverse transcriptase: An enzyme that catalyzes the the synthesis of DNA using RNA as a template.

rhizoid: Rootlike plant outgrowth functioning in support and nourishment.

rhombencephalon: Hindbrain.

ribosomes: Particles on which protein synthesis occurs. The active protein synthesizing system is the polysome, which consists of a group of ribosomes.

RNA tumor virus: A tumor-causing virus with an RNA core.

rods and cones: Photoreceptor cells in the retina of the eye.

root apical meristem: Embryonic tissue at the tip of plant roots that is capable of giving rise to adult tissue.

sarcoma: Tumor derived from connective tissue.

sclera: Tough outer coat of eyeball that develops from mesenchyme.

sclerotome: The somite region that gives rise to the vertebral column.

scrotum: Sac containing the testes.

secondary mesenchyme cells: Cells that form the major portion of mesoderm in the sea urchin embryo.

secondary oocyte: Product of the first meiotic division of the primary oocyte, containing half the number of chromosomes of the primary oocyte.

secondary sex cords: Clusters of cells containing the oogonia; nests of oogonia.

secondary spermatocyte: Product of the first meiotic division of the primary spermatocyte, containing half the number of chromosomes of the primary spermatocyte.

secondary tumor: Tumor that develops from spread of a primary tumor.

selective adhesiveness: Specific adhesion between certain cells but not others.

selective gene amplification: Selective copying of certain genes so that many (sometimes thousands) of DNA copies of specific genes are formed.

self-regulation: The ability of a structure to direct or control its development.

seminiferous tubules: Sperm-forming tubules of the testis.

sepal: An individual leaf of the calyx.

Sertoli cells: Cells in the testis that support and nourish developing sperm.

shoot apical meristem: Embryonic tissue at the tip of plant roots that is capable of giving rise to adult tissue.

sinus venosus: Heart chamber that receives venous blood.

somatic mesoderm: Hypomere mesoderm that is in close contact with ectoderm.

somatopleure: The combination of somatic mesoderm with ectoderm.

somite: Epimere. Dorsal region of mesoderm consisting of myotome, dermatome, and sclerotome.

sorting out: The redistribution of cells in a mixed clump so that like cells group together.

species-specific differentiation of DNA: In this text, refers to specific base sequences of DNA that are similar in a species but vary among different species.

spermatid: Product of the second meiotic division of the secondary spermatocyte. Spermatids have the haploid chromosome number and differentiate into mature sperm cells.

spermatocyte: Stage in the maturation of male gametes preceding the spermatid stage.

spermatogenesis: Development of sperm cells.

spermatogonia: Primordial sperm cells that give rise to spermatocytes and sperm.

spermatozoan: The fully mature haploid sperm cell that differentiates from the spermatid; the male gamete.

spermiogenesis: Transformation of spermatids into spermatozoa.

splanchnic mesoderm: Hypomere mesoderm that is in close contact with endoderm.

splanchnopleure: The combination of splanchnic mesoderm with endoderm.

spiral cleavage: Cleavage pattern resulting from oblique spindle orientations in which daughter cells in upper tiers lie over the junctions of the lower cells.

stage-specific competence: The competence of a given embryonic stage to respond to an inducer.

stage-specific differentiation: The presence of some specific characteristics (such as RNAs and proteins) in some embryonic stages but not in others.

stamen: Pollen-bearing organ of a flower.

stigma: The part of the pistil of a flower that receives the pollen.

stomach: Portion of gut between the esophagus and the small intestine.

stomodeum: Ectodermal invagination that forms the mouth cavity.

style: The prolongation of a carpel or ovary of a flower that bears the stigma.

sublethal cytolysis: Theory that embryonic induction can be caused by damage to cells without embryonic death.

superficial cleavage: Incomplete cleavage in centrolecithal eggs, where the cleavage does not extend all the way through the egg.

synapsis: Pairing of homologous chromosomes of maternal and paternal origin in the prophase of meiosis.

T-cell (thymus lymphocyte): Lymphocyte that plays a major role in cellular immunity.

telencephalon: Anterior portion of the forebrain.

teleosts: Modern bony fishes.

telolecithal: An egg with yolk that is concentrated at one pole. A *moderately* telolecithal egg has yolk that is substantially more concentrated in the vegetal hemisphere; a *strongly* telolecithal egg has all the yolk in the vegetal hemisphere, and all the non-yolky cytoplasm as a cap atop the yolk.

telophase: The last stage in meiosis and mitosis during which the chromosomes become reorganized into two new nuclei.

teratoma: Tumor derived from reproductive cells, that often develops into embryo-like growths.

tetrad: A complex of four chromatids, two from each of two homologous chromosomes. Tetrads are present during the prophase of meiosis.

thalamus: Side walls of the diencephalon.

thymus: Ductless gland in the neck involved in the immune system.

thyroid: A gland or cartilage in the neck.

tibia: The inner and thicker of the two bones of the lower leg.

tissue-specific differentiation: Refers to specific characteristics (such as specific RNAs and proteins) taken on by some tissues and not others.

total (holoblastic) cleavage: Divisions that pass through the entire zygote.

trachea: Air tube extending from the larynx to the bronchi.

transcription: The synthesis of RNA from DNA gene template.

transfer RNA: RNA molecule that transports amino acids to their proper places on the messenger RNA molecule during the translation process.

translation: Synthesis of proteins using genetic information coded in messenger RNA.

trisomy 21: Down's syndrome or Mongolism. Possession of three instead of two chromosomes in chromosome pair number 21.

trophoblast: Outer layer of the mammal embryo.

truncus arteriosis: Conus arteriosis. Anterior most portion of heart.

trypsin: Proteolytic enzyme.

tryptophan pyrrolase: Enzyme that catalyzes the opening of the indole ring of tryptophan.

tumor: Abnormal growth.

tumor-specific antigen: Antigen found primarily on or in tumor cells.

tunica: An enveloping layer.

ulna: The inner and larger of the two bones in the forearm.

unequal cleavage: Cleavage that results in new cells of unequal size.

ureter: Tube conducting urine from the kidney to the bladder.

ureteric bud: Beginning of the formation of the ureter.

urethra: Canal leading from the bladder, discharging urine externally.

urodela: Amphibia with long tails, such as salamanders and newts.

uterus: Womb. Hollow muscular organ in the mammalian mother in which the embryo develops into the fetus.

vagina: Genital canal in the female extending from the uterus to the vulva.

vas deferens: Sperm duct.

vegetal hemisphere: The half of the egg, opposite the animal hemisphere, in which yolk may accumulate. The egg nucleus resides in the animal hemisphere.

ventral: Relating to the bottom or belly.

ventral lip of the blastobore: Lower or belly region adjoining the blastopore; last region to enter the blastopore during gastrulation.

ventricle: A cavity in the brain or the heart.

virus: Infective agent usually composed of a nucleic acid core and a protein coat.

vitelline membrane: A surface coat exterior to the plasma membrane of some eggs.

vulva: The entrance to vagina; the external genital organ of the female.

Wolffian duct: Mesonephric duct.

wrist: Part of arm between hand and forearm.

Xenopus: African clawed toad (a frog).

xylem: In higher plants, the portion of a vascular bundle made up of woody tissue, functioning in water transport.

yolk: Food reserve in the egg.

yolk plug: The center of the circular blastopore in amphibian embryos, consisting of yolky endoderm cells.

yolk sac: Baglike extraembryonic membrane extending from the midgut of bird, fish, and reptile embryos.

zona pellucida: A surface coat exterior to the plasma membrane of mammalian eggs.

zone of polarizing activity (ZPA): An area of limb bud mesoderm that exists near the posterior junction of the limb bud with the body, and that appears to play a key role in determining limb orientation.

zygote: Fertilized egg.

zygotene: The stage in the prophase of meiosis in which the chromosomes arrange themselves in homologous pairs.

zymogen: Storage or inactive form of an enzyme.

INDEX